Advances in Photoelectrochemical Water Splitting
Theory, Experiment and Systems Analysis

Energy and Environment Series

Editor-in-chief:
Heinz Frei, *Lawrence Berkeley National Laboratory, USA*

Series editors:
Nigel Brandon OBE FREng, *Imperial College London, UK*
Roberto Rinaldi, *Imperial College London, UK*
Vivian Wing-Wah Yam, *University of Hong Kong, Hong Kong*

Titles in the series:
1: Thermochemical Conversion of Biomass to Liquid Fuels and Chemicals
2: Innovations in Fuel Cell Technologies
3: Energy Crops
4: Chemical and Biochemical Catalysis for Next Generation Biofuels
5: Molecular Solar Fuels
6: Catalysts for Alcohol-Fuelled Direct Oxidation Fuel Cells
7: Solid Oxide Fuel Cells: From Materials to System Modeling
8: Solar Energy Conversion: Dynamics of Interfacial Electron and Excitation Transfer
9: Photoelectrochemical Water Splitting: Materials, Processes and Architectures
10: Biological Conversion of Biomass for Fuels and Chemicals: Explorations from Natural Utilization Systems
11: Advanced Concepts in Photovoltaics
12: Materials Challenges: Inorganic Photovoltaic Solar Energy
13: Catalytic Hydrogenation for Biomass Valorization
14: Photocatalysis: Fundamentals and Perspectives
15: Photocatalysis: Applications
16: Unconventional Thin Film Photovoltaics
17: Thermoelectric Materials and Devices
18: X-Ray Free Electron Lasers: Applications in Materials, Chemistry and Biology
19: Lignin Valorization: Emerging Approaches
20: Advances in Photoelectrochemical Water Splitting: Theory, Experiment and Systems Analysis

How to obtain future titles on publication:
A standing order plan is available for this series. A standing order will bring delivery of each new volume immediately on publication.

For further information please contact:
Book Sales Department, Royal Society of Chemistry, Thomas Graham House, Science Park, Milton Road, Cambridge, CB4 0WF, UK
Telephone: +44 (0)1223 420066, Fax: +44 (0)1223 420247
Email: booksales@rsc.org
Visit our website at www.rsc.org/books

Advances in Photoelectrochemical Water Splitting
Theory, Experiment and Systems Analysis

Edited by

S. David Tilley
University of Zurich, Switzerland
Email: david.tilley@chem.uzh.ch

Stephan Lany
National Renewable Energy Laboratory, CO, USA
Email: stephan.lany@nrel.gov

and

Roel van de Krol
Helmholtz Zentrum Berlin, Germany
Email: roel.vandekrol@helmholtz-berlin.de

ROYAL SOCIETY OF CHEMISTRY

Energy and Environment Series No. 20

Print ISBN: 978-1-78262-925-2
PDF ISBN: 978-1-78262-986-3
EPUB ISBN: 978-1-78801-446-5
ISSN: 2044-0774

A catalogue record for this book is available from the British Library

The Royal Society of Chemistry is a charity, registered in England and Wales, Number 207890, and a company incorporated in England by Royal Charter (Registered No. RC000524), registered office: Burlington House, Piccadilly, London W1J 0BA, UK, Telephone: +44 (0) 207 4378 6556.

For further information see our web site at www.rsc.org

Printed in the United Kingdom by CPI Group (UK) Ltd, Croydon, CR0 4YY, UK

Preface

Research activities in photoelectrochemical (PEC) water splitting have risen dramatically in recent years. Although the lion's share of research efforts has centered on the synthesis and characterisation of PEC materials with an aim of improving the solar conversion efficiency, new directions within the field have emerged, ranging from theoretical studies for new materials discovery all the way to systems analysis on the gigawatt scale. Our objective, as editors, was to bring together the leading experts in the field to provide an authoritative and forward-looking survey of the different approaches to PEC water splitting, with particular emphasis on key issues and unsolved problems. This book summarises the present-day challenges that need to be solved before PEC water splitting can make the transition from the laboratory to large-scale application.

Theory plays a very important role in the further development of high performance PEC systems, through *e.g.* prediction of band alignments with the electrolyte, assessment of the stability of different materials at different pH and under bias, and the discovery of novel materials with attractive band gaps, stability, carrier transport and charge transfer. Although great strides have been made with empirical investigation of various semiconductors as well as combinatorial studies, theoretical approaches have the potential to greatly accelerate progress in this field. The first part of the book highlights some of these theoretical efforts.

Recent experimental developments comprise the second part of the book. Fundamental understanding of water splitting electrodes under operation are critical to the long-term goal of stable and efficient systems, and two chapters address this topic. The following chapters focus on the low cost and scalable synthesis of water splitting materials, as well as bipolar membranes, which allow increased flexibility of the electrolyte for each half reaction, greatly influencing the overall stability of the PEC cell.

Energy and Environment Series No. 20
Advances in Photoelectrochemical Water Splitting: Theory, Experiment and Systems Analysis
Edited by S. David Tilley, Stephan Lany and Roel van de Krol

Published by the Royal Society of Chemistry, www.rsc.org

The final part of the book focuses on systems analysis, which provides a roadmap of where we hope the fundamental research will lead us. These analyses provide important information for prioritising different avenues of present-day basic research, including PEC cell architecture and practical matters for scaling up these systems.

With a wide view of the field—ranging from theory all the way to systems analysis—we feel that the reader will gain a crucial insight into the remaining challenges for the PEC field, and find guidance for present-day research aims.

I would like to thank the authors of this book, as well as my co-editors Dr Stephan Lany and Prof. Roel van de Krol for their time and effort in this project. It is my hope that this book will become a key text that will inspire current researchers as well as newcomers to the field towards the advancement of practical systems to the market place.

S. David Tilley
Zurich

Contents

Energy and Environment Series No. 20
Advances in Photoelectrochemical Water Splitting: Theory, Experiment and Systems Analysis
Edited by S. David Tilley, Stephan Lany and Roel van de Krol

Published by the Royal Society of Chemistry, www.rsc.org

CHAPTER 1

The Challenge of Water Splitting in View of Photosynthetic Reality and of Research Trends

H. TRIBUTSCH[†]

Bio-Mimetics in Energy Systems Program, Carinthia University for Applied Sciences, Europastrasse 4, 9524 Villach, Austria
Email: helmut.tributsch@alice.it

1.1 Introduction

For half a century and two generations, physical chemistry and electrochemical science have been confronted with the challenge of artificial water splitting using solar light. The dream of a sustainable energy technology based on a cheap photocatalyst providing hydrogen from water splitting has stimulated a lot of research initiatives and has seen the rise of several new research disciplines: semiconductor photoelectrochemistry, interfacial science, photocatalysis, nano-material science, dye solar cell development and research on quantum size phenomena. It became clear that the real challenge is not liberating hydrogen from water. For this much-investigated process, cheap catalytic electrodes such as Ni–Mo alloys or $NiMoN_x$ catalysts,

[†]Retired from: *Free University Berlin, Fachbereich Biologie, Chemie, Pharmazie, Takustr. 3, 14195 Berlin, Germany.*

Energy and Environment Series No. 20
Advances in Photoelectrochemical Water Splitting: Theory, Experiment and Systems Analysis
Edited by S. David Tilley, Stephan Lany and Roel van de Krol

Published by the Royal Society of Chemistry, www.rsc.org

on carbon supports, were identified, which can successfully replace the best performing Pt, since they keep the overpotential loss below 100 mV.[1–3]

The big challenge turned out to be the evolution of oxygen, which involves the transfer of four electrons from water. A lot of progress in understanding basic mechanisms has been achieved over the years and many interesting new materials and catalysts were discovered. While scientists rightly feel that significant advances have been achieved and fully justified repeated funding initiatives towards solar energy utilisation, there are, however, some critical problems: few practically thinking analysts feel that an economically feasible photocatalytic water splitting technology is close to application. As outlined in Chapter 10, it is the much lower cost of hydrogen from fossil fuel, which is an important handicap, while the engineering aspects of solar water electrolysis are quite advanced.

Major technical obstacles towards cost reduction are the long-term instability of wet solar energy converting interfaces, the instability of catalysts, too high overpotentials for oxygen evolution, the formation of aggressive radical intermediates, theoretical problems in describing, understanding and tailoring multi-electron transfer, practical geometries for water splitting technology, the difficulty in handling high proton turnover rates and the cost of applied (noble) metal catalysts.

Facing a progressive deterioration of the environment our industrial civilisation requires convincing scientific and technical progress towards sustainability in energy technology. Here, the solar water splitting strategy remains an unequalled option, which, through photosynthesis, has also shaped and safeguarded the climate on earth.

To better understand the difficulties, and in order to learn what evolution has solved more elegantly and practically with photosynthesis and its fuel circuits than modern research, this analysis aims at a comparative evaluation. The challenges of energy conversion handled in nature should be compared with both past and present scientific and technical efforts aimed at solving the same problems. The aim is to identify key areas, into which extension of our knowledge is desirable or even unavoidable. An analysis of why such challenges have not yet been addressed before will be included. Figure 1.1 compares a simplified scheme for the solar water splitting strategy in nature (plants plus animals, left) with a scheme for a programmed technical solar hydrogen economy (centre). To the right a more general presentation of nature's energy-material strategy is given, which also emphasises its elegance in dealing with sustainable materials.

Nature has implemented a series of physical chemical technologies which cannot, at present, be reproduced technically: electronic processes are used for photon energy harvesting and fuel cell operation, because this is unavoidable, but proton gradients and proton currents are preferred for energy storage and production of mechanical and chemical energy (ATP). Nature is also, evidently, much more efficient in dealing with catalysis and, by fixing carbon dioxide, also provides materials for living activities. How the entire biological energy-material strategy could be approached bio-mimetically for

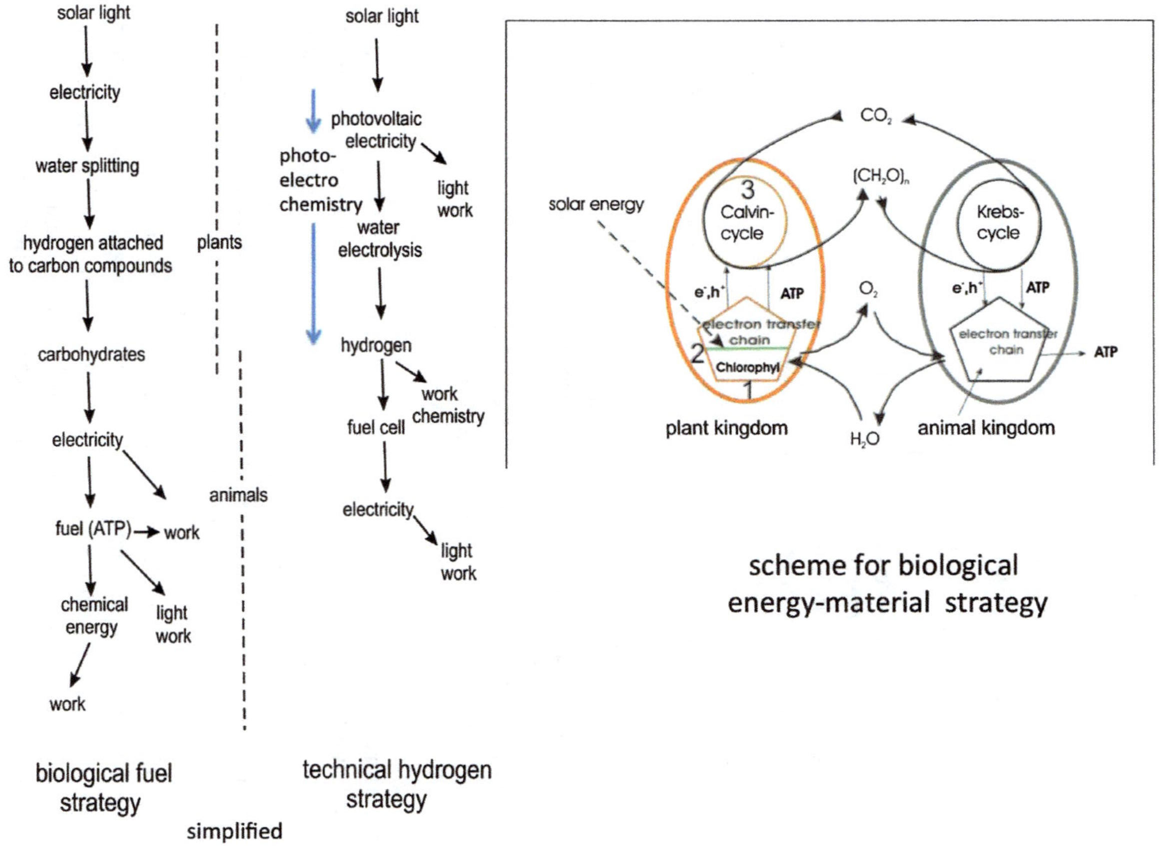

Figure 1.1 The scheme to the right explains the energy-material strategy applied in living nature. To the left two schemes are shown which compare essential energy steps in biology and in projected artificial hydrogen technology (centre).

human civilisation has been discussed elsewhere.[4] For the simplified artificial solar hydrogen energy technology (Figure 1.1, centre) the key challenge is the cost of sustainable hydrogen from water splitting. It is at present approximately five times more expensive than hydrogen from natural gas. Making the process of photo-induced water splitting more efficient and cost-effective is here the key challenge.

1.2 The Evolution of Natural Photosynthetic Water Splitting: The Most Remarkable Facts

Oxygen-evolving photosynthetic-bacteria started to exist on earth around 3.2–3.5 billion years ago.[5] Before their arrival, anoxygenic photosynthesis already existed. Here, molecules with greater negative oxidation potentials than water were used as electron donors. For major phyla of photosynthetic bacteria like Cyanobacteria it is accepted that at the very beginning of water splitting photosynthesis worked more or less as at present. The evolutionary discoveries for sustainable energy conversion were so well functioning that they were conserved and no simpler or more elegant water splitting mechanism has been discovered since. What were these pioneering inventions which evolution introduced? Let us first, for comparison, look at a modern artificial water splitting system:[6] a photovoltaic cell supplying electricity *via* two wires for water electrolysis into oxygen and hydrogen. Photo-generated electrons and holes are separated by a thermodynamic potential gradient, permanently imprinted into a silicon or related junction, fabricated at high temperature. In photo-electrochemical cells the junction is present in the semiconductor interface bordering the liquid. From the generated electricity, electrons are extracted from water at a catalytically optimised electrode interface.

What nature solved differently is well known: materials are not fabricated at high but at ambient temperature *via* dynamic self-organisation processes. No permanent thermodynamic potential gradient is used for the separation of photo-generated charges. Charge separation occurs kinetically, *via* specially tailored molecular mechanisms, and occurs at 3 ps in purple bacteria, strangely speeding up to a shorter time of 0.9 ps at 1 K. When the excited species is transferring an electron the singlet state is, among other temporary changes, converted into a triplet state *via* a simultaneous inversion of a nuclear spin (Figure 1.2, left). A back reaction of the electron transferred is strongly suppressed. When a far from equilibrium thermodynamic kinetic model for the electron transfer was calculated[7] considering autocatalytic feedback, a negative effective activation energy, a very fast rate and a negligible reverse reaction resulted. The activated complex behaves as a transient 'dissipative structure' (term used by Prigogine's group to describe self-organised order built up at the expense of energy dissipation and entropy generation). It obtains the energy not from entropy fluctuations, but from the reaction itself. The feedback is assumed to occur *via* the protein

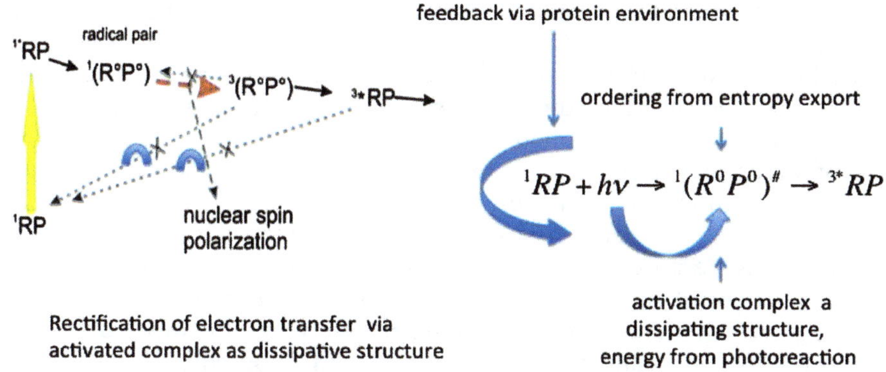

Figure 1.2 Nature in photosynthesis does not use an imprinted potential barrier for charge separation during water splitting but generates perfectly rectified electron transfer *via* a singlet–triplet transition in the excited radical pair facilitated by simultaneous nuclear spin polarisation (left). Electron transfers of this type, which considers polarisation and feedback of the environment leading to an activated, dissipative complex, is not accessible *via* Marcus theory (right).

environment (mechanism explained in Figure 1.2, right). Solar energy conversion here follows the principle of kinetic charge separation: not a thermodynamic potential gradient, but rectifying molecular-electronic mechanisms are responsible for charge separation.

Also dye solar cells, based on nano-material photo-electrochemistry, work due to kinetic charge separation. They only function well with a poorly reversible redox system such as I^-/I_3^-. The power output P of such kinetic solar cells, as well as the primary solar energy conversion process in photosynthesis, follow different laws, as compared with classical solar cells with inbuilt potential gradient. The mechanism has been derived from the principle of least action.[8] It turns out that the power, P, of such kinetic solar cells is proportional to the rate constant w and the chemical affinity A of the system (eqn (1.1)). The latter (A) describes the distance from equilibrium and increases as the distance from equilibrium increases. Similarly, the rate constant (w) also increases as the back reaction is suppressed. This confirms that the kinetic solar cell only works well, when electron transfer is highly rectified and operation is pushed far from equilibrium.

$$P \approx wA \tag{1.1}$$

This will also be true for primary solar energy conversion in photosynthesis and is not the case for classical water splitting photoelectrochemical cells. There is a big advantage from such a biological strategy. By designing molecularly rectified electron transfer it is also possible to operate solar cells with soft materials produced at ambient temperature. They do not have to be stable against internal ion movement, which would gradually degrade any imprinted thermodynamic potential gradient for charge separation.

Another remarkable difference between the biological and the technical approach is the nature of the electrochemical oxidation process of water. We are, in photosynthesis, definitively dealing with a multi-electron process of water oxidation. For the oxidation of water to molecular oxygen ideally a minimum of four times 1.23 eV of energy would have to be supplied. However, when a first electron is extracted approximately 2.8 eV would already be required for generating an OH radical. Only an efficient catalyst can avoid such an activation barrier by properly binding and handling intermediates. The best technical electrodes for water oxidation, such as RuO_2, typically require, at 12 mA cm^{-2} of current density an overpotential of 250 to 350 mV. In contrast, an overpotential of only 60 mV is estimated for photosynthetic water splitting at a comparable current density for water oxidation[9] (Figure 1.3) and should attract our special attention (see below). What could nature have accomplished to bypass limitations, which science and technology have experienced on the basis of electrochemical research and what is exactly the experimental evidence for

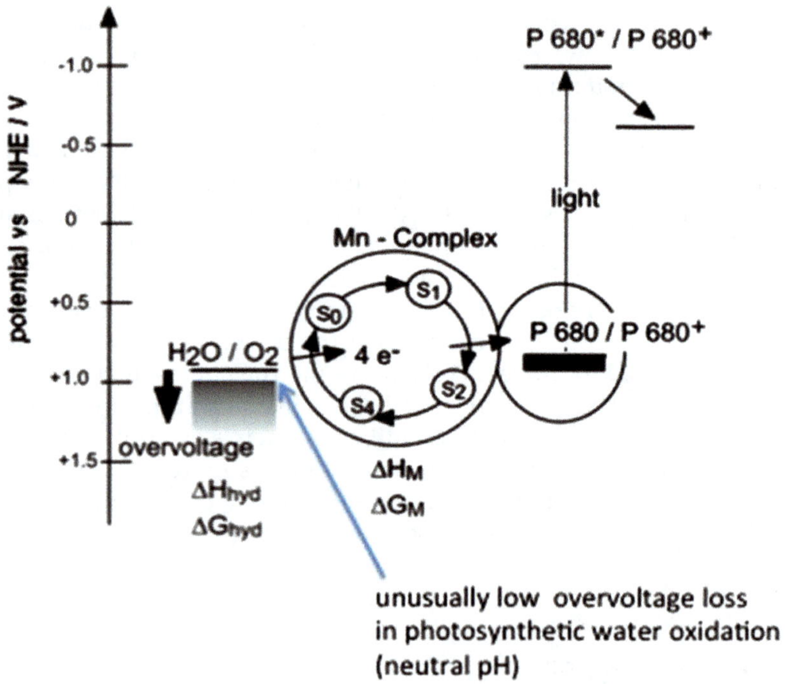

Figure 1.3 Potential scheme explaining the energetic conditions for oxygen evolution in photosynthesis (at neutral pH). Photoexcitation in the reaction centre leads to a stepwise oxidation of the Mn_4O_5Ca-complex, which then induces oxygen evolution at an estimated current density of 12 mA cm^{-2} with only 60 mV overpotential losses.

such extraordinary electrochemical behaviour in photosynthesis? According to the author this is the most central question within the water splitting initiative. It should give an answer to both mysteries, that of high oxygen evolution efficiency and that of the use of the abundant transition metal manganese for catalysis.

1.2.1 The Missing Overpotential in Photosynthesis: What Is the Evidence?

The photooxidation process of water in photosynthesis is initiated by electron transfer from chlorophyll P680 to Plastquinone Q_A, *via* several intermediates, which generates the cationic radical $P680^+$. This abstracts all together step-by-step, *via* a tyrosine Y_Z, four electrons from the manganese containing catalyst, whereby O_2 is evolved during the last step.[10,11]

As pointed out in detail by Watanabe *et al.*[9] there is a paradox linked with the oxygen evolution capacity of the oxygen evolution system: The redox potential of the $Chla/Chla^+$ couple has been extensively measured. Six independent measurements, for example, confined a redox potential value in the following range (eqn (1.2)):[9]

$$E_0(\text{Chla/Chla}^+) = +0.807 \pm 0.045 \text{ V } vs. \text{ NHE} \tag{1.2}$$

Dimerisation of chlorophyll is known to shift the potential even more negative so that complex formation cannot easily be used as an argument for a more positive shift of the redox potential. However, the reversible potential of the H_2O/O_2 couple and thus for oxygen evolution is (eqn (1.3)):

$$E_0 = +1.230 \text{ V} - 0.059 \text{ pH} \tag{1.3}$$

which, for a pH = 6 or pH = 5 solution in the thylakoid internal liquid, would yield a value of (eqn (1.4)):

$$E_0 \text{ (pH: 6)} = +0.876 \text{ V} \quad \text{and} \quad E_0 \text{ (pH: 5)} = +0.935 \text{ V} \tag{1.4}$$

respectively. The values obtained are surprisingly close to that of the $Chla/Chla^+$ couple and do not account for electrochemical energy losses. Geometrical estimations within the photosynthetic system have shown that under typical conditions of photosynthetic solar energy conversion, the manganese centres carry effective oxidation current densities of 12 $mA\,cm^{-2}$.[9] Under such conditions even the best artificial catalysts for oxygen evolution (*e.g.* RuO_2) require an overpotential of 250–350 mV.[9,12] But the overpotential loss in photosynthetic oxygen evolution is estimated to only 60 mV.

Up to 300 mV of oxidation potential are missing, when photosynthetic oxygen evolution is compared to classical electrochemical experience, not speaking of the use of the abundant transition metal manganese for catalysis.

To understand how nature could accomplish electrochemical processes, which require a significantly smaller overpotential than typical technical electrochemical processes, some general considerations on energy limitations for water splitting are helpful. In laboratory experiments on water splitting with platinum black and RuO_2 electrodes it was shown that more than 90% of photovoltaic energy can be converted into chemical energy of hydrogen.[13]

These experiments demonstrate that by combining efficient photovoltaic devices with efficient water electrolysis using noble metal catalysts overall efficiencies for solar energy conversion in the range between 10 and 30% are technically possible. In this mentioned experiment[13] (Figure 1.4), which combined a 20% efficient laboratory tandem solar cell with electrolysis *via* noble metal electrodes at 90% efficiency, an 18% total efficiency for hydrogen generation was confirmed in the laboratory. This appears to be a very high efficiency and it is a quite positive prospect for technical solar electrolysis of water, when comparing it to the 3% energy conversion efficiency into biomass for productive C4 plants during their growth period, and 0.5% biomass efficiency for a sugar cane field with three harvests over the year. But nature in photosynthesis follows not only the aim of energy conversion, it also turns over materials, provides energy for living activities themselves and deals with many issues of survival. Our technology has the advantage of being able to design water splitting mechanisms for energy conversion only, and could therefore become much more efficient than a living organism. However, we would have to

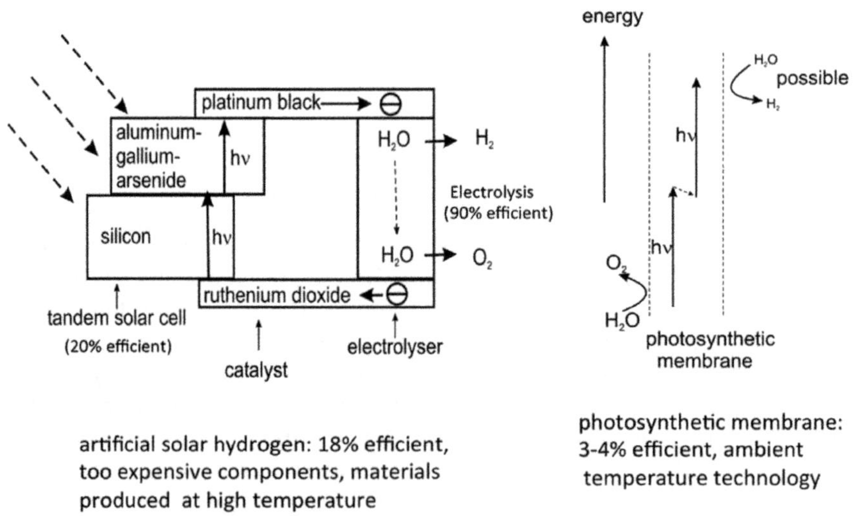

Figure 1.4 Schematic illustration of photovoltaically driven electrolysis cell (left), which yielded 18% efficiency for hydrogen evolution using a 20% efficient tandem photovoltaic cell.[13] To the right the qualitatively equivalent process in a photosynthetic membrane is sketched.

learn to do that without expensive photovoltaic technology and without noble metal electrodes. We need therefore to explore principles, which allow implementation of cheap photocatalytic water splitting. Here biomimetic strategies could help us. For this reason, we should learn to understand the principles, involved in the natural energy conversion process. Water electrolysis itself can approach 100% efficiency, when the energy is adequately provided. This is possible because the energy conversion efficiency for water splitting is determined by the ratio of enthalpy turnover (enthalpy change) for water splitting ΔH_{hyd} and for catalyst (manganese complex) function ΔH_M (see Section 1.3.1 below). The question now arises, how such a ratio of enthalpies could be manipulated to yield an electrochemical efficiency higher than empirically known from technical electrochemistry and the properties of the best technical catalyst for water splitting, RuO_2 electrodes.

1.3 How Can Photosynthetic Water Oxidation Be More Efficient Than Technical?

The oxygen evolution reaction in photosynthesis (photosystem II + water oxidation system) is, because of its relevance for life and as a model for regenerative energy conversion, one of the most intensely studied processes in science.[14–20] But it is also one of the most intriguing ones, to which numerous theoretical studies have been dedicated.[21,22] After the approximate nature of the manganese complex became known, many artificial manganese complexes have been synthesised to mimic oxygen evolution.[21] However, the efforts were essentially in vain. Theoretical calculations even suggested so high energy barriers for oxygen evolution that equilibrium structures of manganese clusters were considered not to be able to oxidise water.[22] Distorted non-equilibrium clusters were consequently invoked.[22] In spite of these apparent complications the oxygen evolution reaction from water is, in photosynthetic literature, typically considered to be preceded by a stepwise oxidation of the manganese complex. This essentially occurs *via* manganese states. The three to four residues within the surrounding amino acids are expected to essentially stabilise the cluster and to adjust charges. Various concepts on the function of Mn_4O_5Ca clusters and their interaction with water have been developed.[23] Basically, the interaction of the nanostructured Mn_4O_5Ca cluster with water should be understood as an ordinary electrochemical oxidation reaction of water. It should, however, not be overlooked that manganese compounds have never shown catalytic activity in any relevant scientific-technological process. It was for this reason, and because classical (Marcus) electron transfer theory is not allowing to consider multi-electron transfer without intermediates, that self-organised multi-electron transfer catalysis has been suggested as a possible mechanism of oxidative water splitting.[24,25] Why is it necessary and why must irreversible thermodynamics be invoked?

1.3.1 Thermodynamics of Photo-induced Water Splitting

The thermodynamic limit of efficiency η_{hyd} for driving the hydrolysis reaction has been investigated in some detail in the past with respect to technical electrolysis.[26] The efficiency was found to be determined by eqn (1.5):

$$\eta_{hyd} = \frac{\Delta H_{hyd}}{\Delta G_{hyd} + \text{losses}} = \frac{\Delta G_{hyd} + T\Delta S_{hyd}}{\Delta G_{hyd} + \text{losses}} \tag{1.5}$$

with ΔH_{hyd} and ΔG_{hyd} being the enthalpy and free energy change, respectively, for the reaction $H_2O \rightarrow H_2 + \frac{1}{2} O_2$ (the losses could arise from resistance losses or overpotentials). Eqn (1.5) means that the temperature dependent entropy change occurring during water splitting has to be accounted for in the overall reaction.

Now it has to be considered that the manganese complex will mediate the photogenerated energy (the photogenerated holes). Some of the introduced energy (from the reaction center) will also be dissipated during the build-up and termination of the water splitting process which requires the free energy $\Delta G_{hyd} + \text{losses} = \Delta G_M + \text{losses}$ (ΔG_M determines the free energy which has to be provided by the manganese complex for water splitting to which energy losses have to be added that occur during hydrolysis and which also have to be supplied by the manganese complex) (Figure 1.3). For comparison: if water would be photo-oxidised by a semiconductor, ΔH would correspond to the forbidden energy gap and ΔG to the (much smaller) light intensity dependent photo-induced splitting of the Fermi-level. The energy lost due to relaxation of charge carriers from the energy bands will contribute to $T\Delta S$. A catalytic semiconductor photoelectrode, the energy band gap of which is well matched to the energetic requirements of electrolysis will involve a much smaller $T\Delta S$ loss than a photoelectrode with excess band gap.

The efficiency for providing the necessary energy for hydrolysis *via* the manganese complex will thus be (eqn (1.6)):

$$\eta_M = \frac{\Delta G_M + \text{losses}}{\Delta H_M} \tag{1.6}$$

with ΔH_M being the enthalpy with $\Delta H_M = \Delta G_M + T\Delta S_M$, and $\Delta G_M + \text{losses}$ the sum of free energy provided plus additional losses during hydrolysis, which will eventually occur. The total efficiency (η_{tot}) will thus become (eqn (1.7)):

$$\eta_{tot} = \eta_0 \eta_M \eta_{hyd} \sim \eta_M \eta_{hyd} \sim \frac{\Delta G_M + \text{losses}}{\Delta H_M} - \frac{\Delta H_{hyd}}{\Delta G_{hyd} + \text{losses}} \tag{1.7}$$

where η_0 considers the remaining losses in the complex photosynthetic water splitting system, and η_{hyd} only considers the anodic part of the hydrolysis reaction, which occurs in the manganese complex (eqn (1.8)):

$$2H_2O \rightarrow O_2 + 4H^+ + 4e^- \tag{1.8}$$

Since water splitting will occur when the free energy generated in the manganese complex matches the free energy requirement for hydrolysis, that is when, in eqn (1.7), $\Delta G_{hyd} + \text{losses} = \Delta G_M + \text{losses}$ the total thermodynamic efficiency for photosynthetic oxygen evolution will be (eqn (1.9)):

$$\eta \sim \frac{\Delta H_{hyd}}{\Delta H_M} \tag{1.9}$$

or, since ΔH_{hyd} and ΔH_M are given thermodynamic quantities, and since the necessary free energy has to be provided by the manganese complex ($\Delta G_M \approx \Delta G_{hyd}$):

$$\eta \sim \frac{\Delta H_{hyd}}{\Delta G_{hyd} + T\Delta S_M} \tag{1.10}$$

Eqn (1.10) is an interesting result, since it clearly shows that the entropy loss or turnover ($T\Delta S_M$) in the manganese complex is a most critical factor for the efficiency of water splitting. This seems trivial at the first glimpse, because the dynamics of the manganese complex will define its catalytic activity. The situation will become more interesting when the following question is asked: how would evolution of the chemistry and dynamics of the manganese complex have responded to such a boundary condition?

1.3.2 How Did Evolution Optimise Photosynthetic Water Oxidation?

It should now be investigated how nature, during evolution, could have improved oxygen liberation from water. At first, of course, a decrease of the entropy loss ($T\Delta S_M$) in eqn (1.10) would have improved the efficiency (Figure 1.5). The contribution of entropy production to the overpotential in electrochemical processes has been discussed in the literature.[27] A stepwise improvement of the manganese complex to reduce the entropy losses would have been necessary. A better matching of electronic states for positive charges in the manganese complex with respect to states in the Chla/Chla$^+$ redox system would be an example. But would the minimisation of this quantity have finally stopped the energetic improvement of the manganese complex? It is proposed that this was not the case. The manganese complex could have 'learned' to invert the sign of entropy, that is to generate negative entropy (negentropy), chemical electronic order (within the manganese complex) (Figure 1.5), which is possible through self-organisation. In this way, the process would have shifted from reversible to irreversible thermodynamics. Practically this could have worked in the following way: the mineral hollandite ($Mn_4CaO_9 \cdot 3H_2O$), believed to have been adapted for early biological water splitting activity, would have been incorporated into a dynamically self-organised molecular energy conversion process with

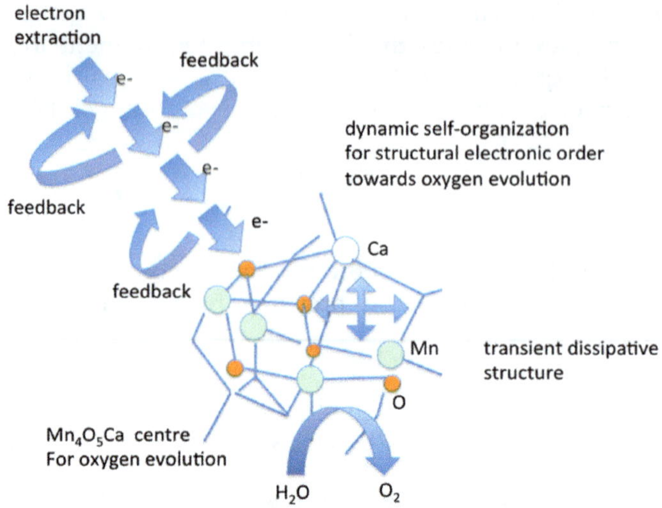

Figure 1.5 To reduce the overpotential loss for water splitting in photosynthesis the manganese Mn_4O_5Ca complex was allowed to develop, *via* feedback processes, an activation complex in the form of a dissipative (dynamically ordered) structure. With this complex a multi-electron transfer, without intermediates, is possible.

feedback processes allowing local entropy reduction at the expense of increased entropy production elsewhere.

The positive sign of the entropy term in eqn (1.10) would have gradually changed to a negative one (eqn (1.11)):

$$\eta \sim \frac{\Delta H_{\text{hyd}}}{\Delta G_{\text{hyd}} - T\Delta S_{\text{Neg}}} \tag{1.11}$$

thus, permitting a further increase of efficiency η (in eqn (1.11) the sign of ΔS has been inverted to show the possibility of further decreasing the denominator. ΔS_{Neg} has therefore to be introduced as a positive quantity). This may have happened by allowing feedback processes *via* the protein environment of the manganese complex. A negative activation energy facilitated by an activation complex functioning as a dissipative structure allowed to build-up favourable temporary dynamic order within the manganese complex for efficient oxygen evolution. Mathematically this can be shown in a similar way as demonstrated for the activated complex in primary photosynthetic charge separation.[7] However multi-electron transfer would be the proceeding reaction.[24,25] Nature would, after systematically decreasing entropy losses in the manganese complex, have simply discovered that increasing dynamic order and generating negentropy within an activation complex, functioning as a dissipative structure, would have allowed to continue optimisation. It would have been optimisation towards maximum entropy turnover, within the given restraints, for the entire photosynthesis system.

What were the preconditions for such further improvements? First the process had to shift to a situation sufficiently far from thermodynamic equilibrium, a condition which was not difficult to fulfil in the combination with the second condition: autocatalytic feedback loops *via* the protein environment would have been needed to allow dynamic build-up of order during the last step of the manganese complex cycle, during which oxygen is evolved. To accomplish this some energy would have to be channelled through this feedback loop to maintain the structural order during catalysis.

The option for evolution of water oxidation efficiency, evident from eqn (1.10), to follow up the decrease of entropy losses by changing the sign of the entropy term (eqn (1.11)) implies a very 'elegant' strategy: some energy is converted into structural order (or information, negative entropy) which may significantly improve catalytic activity (such property of information handling is also known from other catalytic centres in biology[28]). In this way, a catalytically non-suspicious manganese catalyst may have been upgraded to become highly catalytic. However, a shift from a reversible thermodynamic situation (eqn (1.10)) to irreversible thermodynamics would also be a transition to new efficiency relations, so that eqn (1.11) with entropy inverted to negentropy can only be understood as a formula indicating a trend of development.

The explanation proposed here for this paradox of missing overpotential would be, that evolution indeed succeeded in developing dynamic self-organisation by supplying the manganese complex with dynamic steric order or information (negentropy) and thus significantly increasing efficiency as evident from transition from eqn (1.10) to eqn (1.11). The entire process will consequently not be subject to reversible thermodynamics any more, but to non-linear irreversible thermodynamics. Local transient entropy reduction, or entropy export, of course, occurs at the expense of overall entropy production within the photosynthetic system.

1.3.3 How Could Such a Self-organisation Mechanism Be Experimentally Dealt With?

The key claim from the presented analysis is that during photosynthetic oxygen evolution the manganese complex is not generating entropy but negentropy. It is manifesting itself in molecular electronic order within an activation complex, which is generated as a self-organised, 'dissipative structure'. Approximately 0.2–0.3 V of potential difference, which correspond to 0.2–0.3 eV per electron, which are not available as oxidation potential, should therefore be made available as negentropy for the oxygen evolution process. This would amount to up to $-105\ \mathrm{J\,mol^{-1}\,K^{-1}}$, that is 105 entropy units of negentropy. To give a point of reference, approximately 300 entropy units of decreased entropy have been detected through thermodynamic studies to occur during the light induced Bacteriorhodopsin cycle.[15,29] Such a significant entropy decrease involves a quite substantial increase of order, and may be compared to an entropy increase, that accompanies the unfolding of a protein like lysozyme (though in the opposite direction). This means that a search should be undertaken for a transient entropy decrease occurring during photosynthetic oxygen evolution or with model systems using calorimetric techniques. The magnitude of negentropy change required would be reasonable and significantly smaller than that observed during the bacteriorhodopsin photoreaction. Also in this case the negentropy gain has been associated with a synergetic mechanism of energy transduction.[30]

The arguments developed in this contribution directly support the proposal that the kinetic efficiency of photosynthetic oxygen evolution is based on transient, dynamic self-organisation. In order to show that the activation complex becomes a 'dissipative structure' (a local decrease of entropy) it is sufficient to assume that a small portion of the available energy is used *via* a feedback loop involving the protein environment of the manganese complex. The mathematical procedure will be more or less the same as for the calculation of the self-organised activation complex in primary photosynthesis (Figure 1.2).[7]

As indicated in Figure 1.5, the proposed mechanism for photosynthetic oxygen evolution which achieves a transient build-up of negentropy (structural order or information) will involve autocatalysis. After accumulation of the required charges in the manganese complex, the transfer of the first electron from water will stimulate the transfer of the second and so on. Some energy will of course be needed for the feedback loops *via* the protein environment. It will be ultimately derived from photon energy. Calculations show that sufficient non-linear feedback will lead to self-organisation of electrons during electron transfer and catalysis.[24,25] The first electron slaves the other electrons which means that one mathematical equation will describe the interdependent behaviour of all transferred electrons. No radical intermediates will be formed that way and a real multi-electron transfer, as described by eqn (1.8) will be possible. Such a multi-electron process is excluded by classical (Marcus) theory, which only allows subsequent individual electron transfer steps and does not consider polarisation and feedback coupling *via* the environment. Order is built up during such a self-organisation process, equivalent to negentropy. This negentropy, corresponding to a negative activation complex in form of a dissipative structure,[7] will increase the efficiency of oxygen evolution as indicated in eqn (1.11) and may be the ultimate reason why manganese clusters are catalytic in photosynthesis but not behaving exceptionally well in traditional electrochemical experiments (where autocatalysis and self-organisation does not occur). Manganese has apparently been selected by nature to provide optimal coordination chemistry, whereby four manganese centres within a protein matrix are better for building up and releasing order (negentropy) than two manganese centres. The manganese complex may, in fact, have also evolved for the purpose of withholding oxygen evolution during the charging process, until, when the autocatalytic process (*via* the protein environment) provides negentropy and order, the four-electron transfer from water occurs. The proposed autocatalytic nature of the oxygen evolution process would also have the consequence that this mechanism would be highly non-linear. In fact, clear oscillative oxygen evolution transients have been reported for photosynthetic systems.[31]

An important consequence from these considerations would also be that understanding the presently evidenced structure of the manganese complex would not yet reveal its full function. Its dynamics and its non-linear reaction behaviour has to be evaluated and quantified before the reactivity of this unparalleled catalyst can be understood. This situation also explains why the large number of artificial manganese cluster compounds did not give evidence for high catalytic activity (see also below).

Mathematically the expected strategy in nature has been calculated and it was found that extraction of the first electron from water indeed leads to a 'slaving' of the following ones. An energy efficient multi-electron transfer becomes possible. One can talk about self-organised electron transfer (Figure 1.5).[24,25]

Such a notion of self-organised electron transfer for multi-electron catalysis appears to open a new frontier, but faces a quite fundamental theoretical problem, when considering quantum mechanical calculation. When electrons move in a molecular system, there will be an immediate effect on electron density around. There will be a polarisation of the environment. However, with the now well-established Marcus theory for electron transfer[32,33] consideration of such a polarisation is not possible. It consequently does not allow consideration of feedback mechanisms either. This would in addition require a distinction between 'before' and 'after', which is not evident for a quantum process (into the time dependent Schrödinger equation classical time was introduced *via* perturbation theory).

1.4 Progress with Artificial Photo-electrochemical Water Splitting

Here is an evaluation of the trends and state of artificial water splitting. The first realistic interest in light induced water splitting arose with the advent of photoelectrochemistry and the study of oxides with large energy gaps such as ZnO, TiO_2 and $SrTiO_3$. TiO_2 turned out to be reasonably stable and became the prototype of a photocatalyst releasing oxygen from water.[34] Many material studies followed, including nitrides and oxynitrides and oxides doped for band gap tailoring, in the hope to learn about the art of catalytic water splitting.[35] However TiO_2 and related oxide materials simply extracted a first electron from water at a potential of +2.8 V forming a radical, instead of extracting all four electrons near the thermodynamic potential of 1.23 V. Such a mechanism of radical formation automatically eliminates the ability of a system for high energetic efficiency, since 1.5 eV of energy are already dissipated during the first electron transfer step. TiO_2 nevertheless received a tremendous amount of attention, which finally led to its successful secondary application in self-cleaning interfaces.[36,37] Because of the need to use visible light for solar energy conversion many water splitting studies were focussed on semiconductors with lower energy gaps such as Si, GaAs, GaP, CdS, CdTe, InP, $CuInSe_2$. Much experience has been accumulated.[38–40] But many open questions remain, for example as to a practical geometry of water splitting systems. A coiled up, light transparent tube with an internal photovoltaic/catalytic membrane was proposed, which is separating hydrogen generation from oxygen generation. With water in the tube the gases could be transported to a fuel cell and heat to a heat exchanger.[40]

While hydrogen evolution and oxygen evolution was observed and optimised with numerous materials, a persistent problem turned out to be electrode degradation in contact with aqueous solutions. The reasons were quite understandable. The holes generated in the valence bands of these semiconductors were simultaneously eliminating chemical bonds from the electrode material. Whenever holes were allowed to accumulate in the interface the integrity of the crystalline electrode structure was consequently bound to collapse. An additional problem turned out to be reaction products such as elemental hydrogen that reacted, for example, with surface bound P, in InP, to yield toxic PH_3 gas. Only selected materials were proposed to be practical for hydrogen evolution. Quite a lot of research on water splitting catalysis still focuses on photo-induced hydrogen evolution, considering the oxidation of chemicals as sacrificial processes. As a more recent review with 267 references shows,[41] a main emphasis is focussed on TiO_2 and other oxides such as ZrO_2, $KTaO_3$, $SrTiO_3$ or $BiVO_4$, which show conduction bands suitable for hydrogen evolution, when supported by added catalysts or dopants. An example is TiO_2 covered by gold particles.[42,43] The involvement of TiO_2 excitons with Au plasmons is invoked for explaining efficient hydrogen evolution. This may be scientifically interesting, but also shows some exaggeration in the orientation of modern water splitting research as compared to photosynthetic reality. When nature aims at hydrogen evolution, which occurs under special conditions, hydrogenases are used as catalysts, which apply abundant transition metals. An impressive example of a useful hydrogen evolution catalyst is an 'iron only' catalyst operating in certain bacteria (*Clostridium pasteuranium*, *Desufovibrio* spp., *Scenedesmus obliquus*). Turnover frequencies for hydrogen evolution of 10 000 s^{-1} have been measured.[44] The overpotential is very low, comparable to and even lower than for platinum. How is nature able to apply abundant transition metals for efficient energy catalysis? The apparent trick is that the bonding around the catalytic centre, a few Fe–S clusters embedded within a protein, is stabilised through coordination *via* strong CO and CN bonds. Catalytic activity is facilitated while the catalytic centre is chemically preserved and stabilised. We need to understand more.

Catalysis of light induced oxygen evolution, however, remained a significant challenge. One research strategy was inspired by bio-mimetic considerations: electron transfer catalysis in photosynthetic structures is always handled *via* centres of abundant transition metals. Their involvement makes sense because they can engage in coordination chemical mechanisms. Water does not necessarily react with a photo-generated hole in an interface. But if interfacial transition metals are oxidised by holes water species from the electrolyte will engage in coordination reactions. Hole reactions *via* d-states involve the attachment of water species. This requires transition metal compounds with valence bands derived from d-states of transition metals. Identified materials with energy bands suitable for harvesting visible solar light included RuS_2, FeS_2, PtS_2, ZrS_2, MoS_2, $MoSe_2$, WS_2, and WSe_2.[45,46]

The photoelectrode RuS_2 actually turned out to oxidise water to molecular oxygen with a quantum efficiency of up to 60% while remaining stable.[47] Also, illuminated FeS_2 reacted with water. However, the OH^- species attached to the generated interfacial Fe^{3+} ion is passed on to the $(S_2)^{2-}$ sulfur pairs yielding $S_2O_3{}^{2-}$ species, which are soluble. As Figure 1.6 explains, this is the consequence of the energy state density distribution in the valence band. While the transition metal d-states of ruthenium and iron in RuS_2 and FeS_2 are forming the upper edge of the semiconductor valence band, the sulfur 3p states below are quite prominent in FeS_2. This provides the possibility to shift attached water oxidation species to sulphur. A soluble product, sulfate is thus finally formed instead of molecular oxygen. Also, the freshly split Van der Waal surface of MoS_2 liberates molecular oxygen. But when increasingly step sites are formed, $SO_4{}^{2-}$ is generated instead of O_2. Such experience demonstrates that light induced coordination chemical reactions may lead to efficient oxygen liberation from water. No radical intermediates are formed and electrons are successively extracted from water *via* the coordination chemical complex formed. However, the

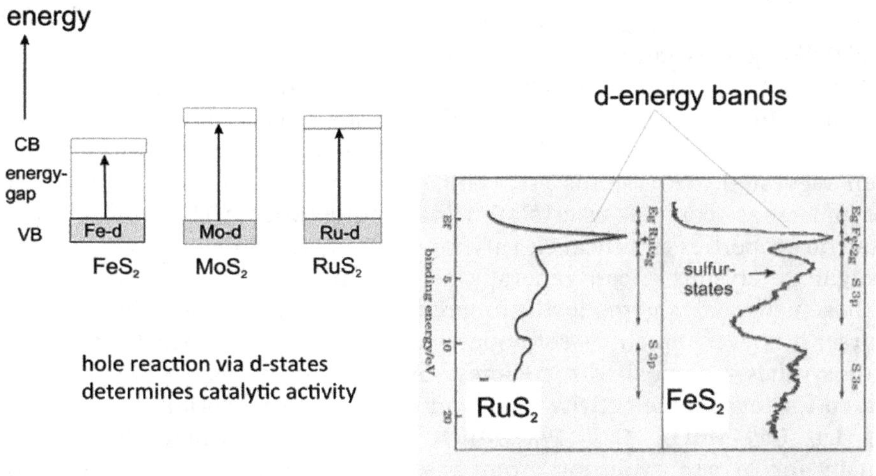

Figure 1.6 The valence band edges in the energy scheme of RuS_2, MoS_2 and FeS_2 show strong transition metal d-character and facilitate coordination-type interaction with water species. When there is an overlap with S-p states, as in FeS_2, corrosive side reactions leading to $SO_4{}^{2-}$ are observed.

intermediate water species attached to the metal may switch to the sulfur or other catalyst components, when the corresponding states are energetically or spatially available for reaction. The conclusion to be drawn is that a quite high degree of electronic and steric order is required for a successful and efficient water oxidation. Ruthenium is, of course, a too expensive noble metal for water oxidation catalysis. But both, as photoelectrode in RuS_2 and as dark catalyst as RuO_2, it provides the presently most efficient catalyst for oxygen evolution. RuO_2 undergoes under high pressure a phase transition from rutile, the most common phase, to marcasite, pyrite and fluorite structure. They yield metallic properties, with the exception of the fluorite structure, which shows semiconducting behaviour with a band gap of 0.47 eV.[48] Theoretical calculations show that in the metallic rutile conductor a sharp Ru-d peak is placed just below the Fermi level and clearly above a weak O-p distribution.[49] This corresponds to the specific density of state distribution measured for the valence band of RuS_2 (Figure 1.6).

Positive electronic charges generated by photoexcitation have consequently to be handled *via* pure transition metal d-states when electrons are to be extracted from water without side products. In addition, ruthenium offers oxidation states III, IV, and V, which are needed for accommodating four electrons for the liberation of one molecule of oxygen from water. A similar range of accessible oxidation states is only offered by Mn, which nature selected for the oxygen evolution cluster in photosynthesis, since it is the 12th most abundant element in the earth crust. Also, Mn in the photosynthetic water oxidation complex is mostly active *via* its oxidation states III and IV in the storage of positive charge, with a probable participation of Mn V during the final oxygen evolution step. Even though many details are known about manganese redox chemistry, and major efforts were developed towards understanding multi-electron transfer, a consistent concept of catalytic oxygen evolution is not yet available.[50,51]

Recently, more and more attention is attributed to crystallised models for the Mn complex in photosynthesis such as $CaMnO_3$, $CaMn_2O_5$, $Ca_{1-x}Pr_xMnO_3$ and $Ca_{2-x}Pr_xMnO_4$.[52–54] This is not surprising, since it had been suggested that systems preceding cyanobacteria originally assimilated the manganese complex as crystalline hollandite ($Mn_4CaO_9 \cdot 3H_2O$).[55] While catalytic properties significantly vary for different stoichiometries of calcium manganate crystals it can generally be said that oxygen evolution activity of these perovskites is modest compared to RuO_2, and photoactive RuS_2. Of course, it is even much more modest compared to the $CaMn_4O_y$ centre of photosynthesis. Deposited manganese oxide could be electrochemically activated for reasonable activity (1 mA cm^{-2} at an overpotential of 450 mV).[56] For $La_{0.5}Ca_{0.5}MnO_3$, $Ca_{0.75}Pr_{0.25}MnO_3$ or $Ca_{1.75}Pr_{0.25}MnO_4$ the current density for oxygen evolution from water observed at an overpotential of approximately 460 mV was 50 μA cm^{-2}.[53] This should be compared to a current density of 12 mA cm^{-2} observed with RuO_2 at an overpotential of 250–300 mV, or in photosynthetic oxygen evolution estimated to occur at an overpotential of only 60 mV. The manganese oxide model catalysts also always show a significant degree of corrosive deterioration. Theoretical calculations indicate that in the energetic region, where electrons are accepted from water, calcium manganite O 2p states show significant overlapping with Mn 3d t_{2g} states.[57] This would explain the involvement of oxygen states from the manganese oxide in electron transfer reactions with the consequence of lower rates and electrochemical corrosion activities.

Another important research initiative concerns, of course, molecular studies of metal–organic manganese clusters in relation to the function of the oxygen evolution complex in photosynthesis. A review from 2004 already discusses 171 manganese clusters including 54 Mn_4 clusters, on the basis of 410 references.[58] Achieving a working analogue for the active manganese cluster in photosynthesis is considered one of the greatest challenges in bioinorganic chemistry. Until today this aim could not yet be approached in spite of sophisticated experimental tools and much progress in theoretical understanding of Mn–oxo complexes. The positions of Mn-3d states are apparently not optimised for reaction and they simply do not show the expected favourable water oxidation properties. The poor catalytic activity of artificial Mn–cluster complexes has already longer time ago been theoretically addressed.[22] As explained *via* eqn (1.11) positive charges accumulating in them have low interest in reacting with water until they become involved in self-organisation for a build-up of temporary order (information) within an activation complex functioning as dissipative structure for energy efficient water oxidation. Such a (low entropy) activation complex may expose the required Mn-d states (not overlapping with oxygen states) for electron extraction from water. A theoretical model for self-organisation,[7] when applied to Mn complexes, even suggests a negative activation energy, explaining the 'missing' overpotential for photosynthetic oxygen evolution.

1.5 Bio-mimetic Approaches Require Progress in Non-equilibrium, Irreversible Thermodynamics

This review has sketched the path of energy in photosynthetic water splitting and compared it to the rise in knowledge and practical experience in artificial photoelectrochemical water splitting. Half a century in this field has seen the development of understanding semiconductor junctions, of photo-induced interfacial reactions, of catalyst function, of nanomaterial properties and of mechanisms of bioinorganic chemistry. During the same period, many new materials became available and were studied with increasingly sophisticated experimental techniques. Significant progress in artificial water splitting has been achieved and an on-going progress is seen, as this book is documenting. But is it always a progress in the required direction?

It is presently not at all sure, whether technological water splitting, as we are investigating and handling it today, will ever reach the qualitative level we see in nature. This is disturbing, considering the much higher overall efficiency which could be reached *via* an artificial technology, which could be better focussed on energy output. Photosynthetic oxygen evolution has reached a level, which has facilitated evolution of higher life and shaped our climate. If nature would not have been able to implement photosynthetic water splitting of that quality, humans would simply not exist. The challenge for industrial society to imitate that process has therefore to be taken seriously. It cannot be excluded, that efforts in artificial water splitting based on mechanisms shaped by reversible thermodynamics will reach a technological standard, which enables a breakthrough in harvesting solar fuel. This is, in fact, the projected aim of on-going projects. But the author is sceptical because nature and present research are based on different limiting conditions and are operating in different ranges of thermodynamics (Figure 1.7). When relying on present research trends, a breakthrough in artificial water splitting would have to occur with a less sophisticated and different kind of technology. Nature has efficiently exploited the possibilities of non-linear irreversible thermodynamics for water splitting and has been relying on such technology for more than 3 billion years. During that period, no simpler technology close to reversible thermodynamics has been discovered by nature. There are obvious reasons: dynamic self-organisation gives access to abundant materials, to temporary stability and to dissipative

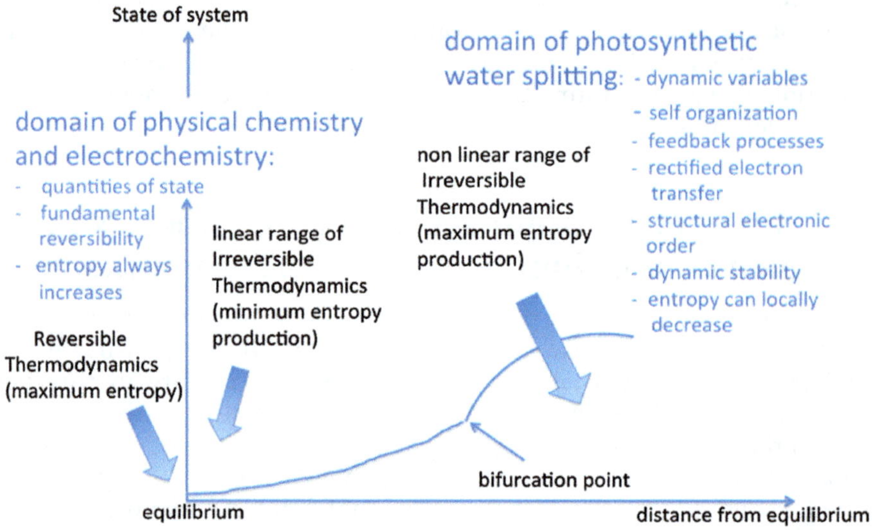

Figure 1.7 A scheme for the state of the system, with an axis showing the distance from equilibrium is used to show where present electrochemistry and physical chemistry is addressing problems of water splitting (equilibrium thermodynamics and linear range of irreversible thermodynamics), and where biological evolution has found appropriate solutions (non-linear range of irreversible thermodynamics).

structures (negentropy, information) for much improved catalysis. The problem is: present research is still struggling to even understand the foundations of irreversible thermodynamics. In spite of significant progress in science generally it has not even been possible to derive basic entropy laws for non-linear irreversible thermodynamics. It should not be forgotten that even the empirical second law of thermodynamics could not yet be derived from fundamental principles, based on time-invertible fundamental mechanism and time-neutral elementary particles.

At the beginning of this chapter it was explained that essential energy strategies, which nature has successfully applied for energy harvesting (rectified electron transfer, kinetic solar energy conversion, self-organised catalysis *via* activation complexes functioning as dissipating structures, self-organised materials) are presently not attractive or accessible for dynamic research and development activities. In spite of some early efforts dating back 25 years[7,8,23,25] there are not sufficient research activities in direction of self-organised catalysis. Why are, for example, so many manganese cluster compounds studied[58] to find out that they are catalytically not really attractive, but practically no efforts developed to get them to form functioning dissipative structures, as proposed for photosynthesis?[7] What is the reason for such a situation and how can one overcome it?

1.5.1 A Paradigm Change Towards a Fundamental Time Arrow Is Needed

According to the author the answer is relatively simple, but radical: all fundamental laws that have been identified to govern physics and chemistry are time invertible. They are designed to work both in positive and negative time direction. The properties of elementary particles are time neutral. Time is explained statistically as an entropy arrow by mathematically deriving an entropy increase for particles subject to time-invertible laws. However, this was only achieved by drastic simplification of the originally very complex mathematical situation. With the introduction of this simplification, information was thrown out of the system. Today one knows that information involves energy turnover, which means that energy was thrown out of the system *via* the simplifications introduced. The calculated statistical entropic time arrow is the result of mathematical manipulation. Anyway, the same mathematical formalism should also allow a system to develop into the past. This has never been observed (Loschmidt paradox). The applied considerations are consequently not consistent.

Self-organisation and the build-up of order for activation complexes as dissipative structures (ordered structures build-up *via* energy dissipation) requires feedback processes to occur. Feedback processes need a ‘before’ and an ‘after’. They can in a transient catalytic complex, which is building up order, not rely on a statistical time mechanism. How should such a probabilistic, statistical time be read by a self-organised feedback reaction anyway? Feedback processes require a fundamental time orientation *via* an energy driven flow of action. In addition, quantities of state have to become dynamic variables and this should also be reflected in quantum processes during electron transfer. The fact that everything is moving into one direction is not considered as convincing evidence for a time arrow for actual science, which believes time to be 'an illusion' (comment by A. Einstein), and should not enter into crucial formulas. The author of this chapter, in two recent publications[59,60] addressed the problem and showed that a fundamental time arrow can already be derived from the important principle of least action. When the time integral over energy, action, is split up in infinitesimal sections, energy and time in it still have to minimise. A scalar energy (an energy, which has the ability to perform work, but no interest) and a time which is just an ordering parameter, cannot do that. They are just numbers. They have to be oriented in order to minimise, which also means that the principle of least action expresses nature's profound property of being fundamentally oriented. Free energy has, within the limitation of the system concerned, the 'property to decrease its presence per state'. It thus drives time, which is the flow of action generated by energy conversion. The process, which is believed to bring a snowflake to form, free energy approaching a minimum, should just be accepted as a fundamental law. It was shown that, on the basis of such a dynamic energy, which implements a time arrow, rational explanations can be derived for relevant quantum and cosmic phenomena.[59,60] On the basis of such a paradigm change leading to a fundamentally time-oriented universe, counter intuitive and bizarre natural phenomena can be understood rationally and differently.[61] A fundamentally irreversible nature can simply not easily be described by time reversible laws.

Those phenomena attributed to the statistical time arrow however actually exist. A compressed gas indeed becomes an expanded, statistically distributed, chaotic one. The distribution of gas particles becomes a more probable one with less information content. The entropy S of this gas, thereby, following the statistical time arrow towards increasing entropy, grows from S_1 to S_2 by ΔS. When multiplying the entropy with the temperature T of the system, however, one realises that this means, that the non-useful (entropic) energy of the system is simply increasing by the non-useful (entropic) energy $T\Delta S$. Since energy cannot be created from nothing and cannot be lost (1st law of thermodynamics) this means that the statistical time arrow is simply explaining that original free (available) energy, contained in the more ordered system, is being converted into non-available (entropic) energy. Instead of saying that a system assumes a more probable state of disorder (compared to a time-neutral initial state) it is therefore entirely equivalent saying that (original) free energy has the tendency or drive to decrease its presence per state while generating entropic, non-available energy ($T\Delta S$). Such a phenomenon is actually observed and can accordingly be interpreted without claiming time-neutral gas particles. This however is exactly the proposed definition of the 'fundamental' energy driven time arrow.[59] These simple considerations are, besides of the conclusions derived from the principle of least action, another support for an energy concept, in which energy is not dormant, but dynamic (within the restraints of a system). It can also generate statistical disorder (simulate a statistical time arrow) while converting energy and producing action and entropy.

Such a paradigm change from a statistical to a fundamental time arrow is justified and reasonable, especially since it makes research accessible for the domain of irreversible thermodynamics.

Important consequences for physical chemistry would be that quantities of state become dynamic, entropy generation becomes rate controlling, and non-linear irreversible phenomena consequently develop towards maximum entropy production within the restraints of the system concerned.[60] This would not only allow to adequately deal with self-organised energy phenomena, but would also explain why evolution conduced the manganese based oxygen evolution catalyst towards maximum performance *via* phenomena of irreversible thermodynamics. Photosynthetic oxygen evolution from water reflects a situation of maximum entropy turnover within the restraints of the system. It is achieved by allowing a static manganese catalyst with sluggish response for oxygen evolution to develop an activation complex in form of a dissipative structure for efficient multi-electron extraction from water. Here the protein environment must provide support *via* feedback activities[7] and some energy for the well-ordered, dissipative transient activation complex must be deduced from the energy providing primary photoreaction.

What would, in terms of research strategy, be the advantage of adopting the idea of a fundamentally oriented, irreversible world? The advantage would be both practical and psychological. First, all presently tried and tested concepts for physical chemistry and electrochemistry could remain valid, can be considered to be limiting cases for situations where time orientation can be neglected. However, the possibility to access the driving forces, the dynamic variables for irreversible processes, and to handle them as nature did in photosynthetic processes, will open productive new theoretical and experimental routes towards self-organised energy conversion mechanisms. A growing research community, which is now restricted by the dogma of time neutral elementary particles and a probabilistic time concept, would attempt to access opportunities from irreversible thermodynamics. At the long term this may help to reduce costs for hydrogen from photoelectrochemical water splitting *via* more elegant and more efficient mechanisms (compare Chapter 11).

A simple example may underline this necessity to introduce a fundamental time arrow into physical chemical considerations. The self-similar structure of a cauliflower is considered to be the product of dynamic self-organisation with energy flowing through the system while feedback processes function. The mechanism for its generation cannot be derived from basic time-invertible principles and a statistical time arrow. The formation of an ice crystal is considered to occur *via* passive self-organisation. The explanation and the experience is that it occurs because the free energy of the system is minimising within its restraints. However, such behaviour cannot be derived from basic principles either, since, according to present scientific standard, only the generation of statistical disorder is creating the time orientation. If, however, the decrease of free energy per state (which makes the ice crystal grow) is assumed to drive time (as also derivable from the principle of least action),[59,61] both phenomena are immediately accessible. In other words: the well-known mechanism, by which an ice crystal forms *via* a minimisation of free energy should be recognised as a fundamental law that opens irreversible thermodynamics for research and understanding. As a consequence, free energy, up to now handled as a quantity of state, which has no interest in performing work, would be accepted as a dynamic variable implementing the flow of time. It is the reduction of free energy, which accompanies the growth of the ice crystals and, through the action generated, determines the visible time flow. A fundamental time arrow results, which is driven by a minimisation of free energy per state, as well as a time, in form of a flow of action, which allows feedback processes for dynamic self-organisation. The second law of thermodynamics can, of course, also immediately be derived from such a paradigm change. Its derivation was not possible up to now, which should encourage steps towards such a fundamental time arrow.

Our present understanding and handling of physical chemical and electrochemical mechanisms, using concepts of reversible thermodynamics and quantities of state would still be valid, but as a limiting case. However, relevant paradoxes in quantum physics and cosmology would simply disappear, because they result from the assumption of a fundamentally time invertible world.[60,61]

Without such a courageous step in favour of irreversible thermodynamics there would be, within the research community, neither the tolerance nor the interest in addressing the complex challenges for water splitting as handled in nature. To give an example: only when quantum phenomena can be understood as dynamic processes in a fundamentally oriented world[58] it will be possible to consider, *via* quantum mechanical mechanisms, polarisation and feedback processes for electron transfer, kinetic rectification, and self-organisation.

If such a door towards new insight and exploration is not opened, water splitting efforts, based on concepts of thermodynamic equilibrium, may finally turn out to lead into a dead-end. It would be a road which nature herself, during three billion years of evolution, has not found passable. We therefore need a paradigm change in understanding and dealing with energy and time for successfully joining and addressing nature's far from equilibrium water splitting strategy. Bio-mimetic considerations request that.

Is there also an experimental frontier for research to get started with the handling of negentropy for catalysis? Let us look at the well-known Belousov–Zhabotinsky reaction, which, when certain parameters are adjusted, develops beautiful waves and rings until its chemical energy (malonic acid) is exhausted.[62] This self-organised reaction is used as a simple model system explaining build-up of order in early life. Three billion years ago life learned to harvest solar energy. Also, the Belousov–Zhabotinsky reaction can do that following the identified principle of maximum entropy production (within the restraints of the system).[60] It is sufficient to add the dye tris(bipyridine)ruthenium(II)chloride for replacement of dark induced by light induced electron transfer. Now light energy flows into the oscillating system and illumination patterns determine the spatial and temporal course of oscillating self-organisation. Early chlorophyll driven life forms may have indeed started like that. Now the light induced self-organised reaction should be coupled to nano-structured interfaces. This could best be achieved *via* the dye tris(bipyridine)ruthenium(II)chloride within the geometry of nano-structured dye solar cells. When the parameters are found within which photo-induced oscillating self-organisation continues in such an environment, attention should be focussed on the fate of photo-generated holes. They should be channelled into potential catalysts for water oxidation. In trying to maximise entropy turnover, and in presence of suitable chemical agents the system may attempt to self-organise the water oxidation reaction. The first steps in experimental learning may proceed.

References

1. J. M. Jaksić, N. V. Krstajić, B. N. Grgur and M. M. Jaksić, *Int. J. Hydrogen Energy*, 1998, **23**, 667–681.
2. M. M. Jaksić, *Int. J. Hydrogen Energy*, 2001, **26**, 559–578.
3. W.-F. Chen, K. Sasaki, C. Ma, A. I. Frenkel, N. Marinkovic, J. T. Muckerman, Y. Zhu and R. R. Adzic, *Angew. Chem. Int. Ed.*, 2012, **51**, 6131–6613.
4. H. Tributsch, Energy Bionics: The Bio-Analogue Strategy for a Sustainable Energy Future, in *Carbon-Neutral Fuels and Energy Carriers*, ed. N. Z. Muradov and T. N. Veziroglo, CRC Press, Taylor & Francis Group, 2012, pp. 415–464, ISBN 978-1-4398-1857-2; H. Tributsch, *Erde, wohin gehst Du? Solare Bionik-Strategie: Energie-Zukunft nach dem Vorbild der Natur*, Shaker Media, Aachen, 2008, p. 346, ISBN 978-3-86858-044-0.
5. R. E. Blankenship, Early Evolution of Photosynthesis, *Future Perspect. Plant Biol.*, 2010. DOI: 10.1104/pp.110.161687; http://www.plantphysiol.org/content/154/2/434.full.
6. *Photoelectrochemical Water Splitting, Springer Briefs in Energy*, ed. Z. Chen, E. Miller and H. N. Dinh, 2013, DOI: 10.1007/978-1-4614-8298-7_3.
7. L. Pohlmann and H. Tributsch, *J. Phys. Chem.*, 1993, **97**, 11318–11323.
8. H. Tributsch, *J. Solid State Electrochem.*, 2009, **13**, 1127–1140.

9. T. Watanabe, M. Kobayashi and T. Sagara, Electrochemical and photochemical Interpretation of oxygen evolution process, in *Current Research in Photosynthesis: Proceedings of the VIIIth International Conference on Photosynthesis Stockholm, Sweden, August 6–11, 1989*, ed. M. Baltscheffsky, 1990, Cluver Academic Publishers.
10. H. Kretschmer and H. T. Witt, *Biochem. Biophys. Acta*, 1993, **1144**, 331.
11. K. H. Rhee, E. P. Morris, D. Zheleva, B. Hankamer, W. Kühlbrandt and J. Barber, *Nature*, 1997, **389**, 522.
12. C. Iwakura, K. Hirao and H. Tamura, *Electrochim. Acta*, 1977, **22**, 335.
13. S. Licht, B. Wang, S. Mukerji, T. Soga, M. Umeno and H. Tributsch, *J. Phys. Chem. B*, 2000, **104**, 8920–8924.
14. G. Renger, *Photosynth. Res.*, 1993, **38**, 229; G. Renger, *Chem. Unserer Zeit*, 1994, **28**, 118.
15. D. R. Ort and W. W. Parsons, *J. Biophys. Soc.*, 1979, **25**, 355.
16. *Oxygenic Photosynthesis*, ed. D. R. Ort and C. F. Yocum, Kluwer Academic Publishers, Dordrecht, 1996.
17. M. Haumann and W. Junge, Protons and charge indicators in oxygen evolution, in *Oxygenic Photosynthesis*, ed. D. R. Ort and C. F. Yocum, Kluwer Academic Publishers, Dordrecht, 1996.
18. E. Schrodder and H. T. Witt, *J. biol. Chem.*, 1999, **274**, 30387.
19. G. W. Brudvig and R. H. Crabtree, *Progr. Inorg. Chem.*, 1988, **37**, 99.
20. Y. Umena, K. Kawakami, J.-R. Shen and N. Kamiya, *Nature*, 2011, **473**, 55–60.
21. W. Rüttiger and C. Dismukes, *Chem. Rev.*, 1997, **79**, 1.
22. D. M. Proserpio, R. Hoffmann and G. C. Dismukes, *J. Am Chem. Soc.*, 1992, **114**, 4374.
23. M. M. Najafpour, M. Z. Ghobadi, B. Haghighi, J. J. Eaton-Rye, T. Tomo, J.-R. Shen and S. I. Allakhverdiev, *Biochemistry*, 2014, **79**(4), 324–336.
24. H. Tributsch and L. Pohlmann, *Science*, 1998, **279**, 1891–1895; H. Tributsch and L. Pohlmann, *J. Electroanal. Chem.*, 1995, **395**, 53; H. Tributsch and L. Pohlmann, *J. Theor. Biol.*, 1996, **178**, 17.
25. L. Pohlmann and H. Tributsch, *J. Theor. Biol.*, 1992, **155**, 443; L. Pohlmann and H. Tributsch, 1992, **156**, 63; L. Pohlmann and H. Tributsch, *Electrochim. Acta*, 1997, **42**, 2737.
26. F. Gutmann and O. J. Murphy, *Mod. Asp. Electrochem.*, 1983, **15**, 1.
27. G. F. Nader and C. D. Tokarz, *J. Thermophys. Heat Transfer*, 2006, **20**(3), http://arc.aiaa.org/doi/abs/10.2514/1.19140?journalCode=jtht.
28. *Computation in Cellular and Molecular Biological Systems*, ed. R. Cutberthson, R. Paton, M. Holcombe, World Scientific, 1996.
29. G. Varo and J. K. Lanyi, *Biochem.*, 1991, **30**, 5016.
30. H. Tributsch and L. Pohlmann, *J. Theor. Biol.*, 1996, **178**, 17–28.
31. D. Walker, *New Phytol.*, 1992, **121**, 325.
32. R. A. Marcus, *J. Chem. Phys.*, 1956, **24**, 966.
33. R. A. Marcus, *Trans. N. Y. Acad. Sci.*, 1957, **19**, 423.
34. A. Fujishima and K. Honda, *Nature*, 1972, **238**, 37.
35. T. Jafari, E. Moharreri, A. S. Amin, R. Miao, W. Song and S. L. Suib, *Molecules*, 2016, **21**, 900.

36. A. Fujishima, K. Hashimoto and T. Watanabe, *TiO_2 Photocatalysis: Fundamentals and Applications, BKC Tokyo*, 1999.
37. K. Hashimoto, H. Irie and A. Fujishima, *Jap. J. Appl. Phys.*, 2005, **44**.
38. *Photoelectrochemical Water Splitting: Materials, Processes and Architectures, Royal Society of Chemistry Book*, ed. H.-J. Lewerenz and L. Peter, 2013, DOI: 10.1039/9781849737739, Print ISBN: 978-1-84973-647-3, http://pubs.rsc.org/en/content/ebook/978-1-84973-647-3#!divbookcontent.
39. *Photoelectrochemical Water Splitting, Springer Briefs in Energy*, ed. Z. Chen, E. Miller and H. N. Dinh, 2013, DOI: 10.1007/978-1-4614-8298-7_3.
40. H. Tributsch, Photovoltaic hydrogen generation, *Int. J. Hydrogen Energy*, 2008, **33**, 5911–5930.
41. T. Jafari, E. Moharreri, A. S. Amin, R. Miao, W. Song and S. L. Suib, *Molecules*, 2016, **21**, 900.
42. X. Chen and S. S. Mao, Titaniumdioxide nanomaterials: Synthesis, properties, modifications and applications, *Chem. Rev.*, 2007, **107**, 2891–2959.
43. S. K. Dutta, S. K. Mehetor and N. Pradhan, *J. Phys. Chem. Lett.*, 2015, **6**, 936–944.
44. C. Madden, M. D. Vaughn, I. Díez-Pérez, K. A. Brown, P. W. King, D. Gust, A. L. Moore and T. A. Moore, *J. Am. Chem. Soc.*, 2012, **134**(3), 1577–1582.
45. H. Tributsch, *Ber. Bunsenges. Phys. Chem.*, 1978, **82**, 169–174; H. Tributsch, *J. Electrochem. Soc.*, 1978, **125**, 1086–1093; H. Tributsch, *J. Electrochem. Soc.*, 1981, **128**(6), 1261–1268; H. Tributsch, *Faraday Discussion Roy. Soc. Chem.*, 1980, **70**, 189–205.
46. H. M. Kühne, W. Jaegermann and H. Tributsch, *Chem. Phys. Lett.*, 1984, **112**, 160–164.
47. H. M. Kühne and H. Tributsch, *J. Electroanal. Soc.*, 1983, **130**(6), 1448–1450.
48. N. Mehtougui, D. Rached, R. Khenata, H. Rached, M. Rabah and S. Bin-Omran, *Mater. Sci. Semicond. Process.*, 2012, **15**, 331–339.
49. Y. Ze-Jin, G. Yun-Dong, L. Jin, L. Jin-Chao, D. Wei, C. Xin-Lu and Y. Xiang-Dong, *Chin. Phys. B*, 2010, **19**, 077102.
50. F. A. Armstrong, *Philos. Trans. Roy. Soc. B*, 2008. DOI: 10.1098/rstb.2007.2223.
51. L. I. Krishtalik, *Biochim. Biophys. Acta.*, 1986, **849**, 162–171.
52. J. Kim, Xi Yin, K.-C. Tsao, S. Fang and H. Yang, *J. Am. Chem. Soc.*, 2014, **136**(42), 14646–14649.
53. M. E. Abrishami, M. Risch, J. Scholz, V. Roddatis, N. Osterthun and C. Jooss, *Materials*, 2016, **9**, 921.
54. J. Suntivich, K. J. May, H. A. Gasteiger, J. B. Goodenough and Y. Shao-Horn, *Science*, 2011, **334**, 1383–1385.
55. J. D. Johnson, http://www.chm.bris.ac.uk/motm/oec/motm.htm, retrieved 14/01/2017.
56. M. Huynh, C. Shi, S. J. L. Billinge and D. G. Nocera, *J. Am. Chem. Soc.*, 2015, **137**(47), 14887–14904.

57. J. H. Jung, K. H. Kim, D. J. Eom, T. W. Noh, E. J. Choi, J. Yu, Y. S. Kwon and Y. Chung, *Phys. Rev. B*, 1997, **55**, 15489.
58. S. Mukhopadhyay, S. K. Mandal, S. Bhaduri and W. H. Armstrong, *Chem. Rev.*, 2004, **104**, 3981–4026.
59. H. Tributsch, *J. Mod. Phys.*, 2016, 7, 365–374.
60. H. Tributsch, *J. Mod. Phys.*, 2016, 7, 1455–1482.
61. H. Tributsch, *Irrationality in Nature or in Science? Probing a Rational Energy and Mind World*, CreateSpace, a Company of Amazon, 2015, ISBN-13:978-1514724859.
62. Belousov–Zjabotinsky reaction, Wikipedia.

CHAPTER 2

Theoretical Design of PEC Materials

JUNJIE WANG,*[a] PAKPOOM REUNCHAN,[b] WEI ZHOU[c] AND NAOTO UMEZAWA*[d,†]

[a] Tokyo Institute of Technology, Materials Research Center for Element Strategy, 4259 Nagatsuta-cho, Midori-ku, Yokohama, Kanagawa 226-8503, Japan; [b] Kasetsart University, Department of Physics, Faculty of Science, Bangkok 10900, Thailand; [c] Department of Applied Physics, Tianjin Key Laboratory of Low Dimensional Materials Physics and Preparing Technology, Faculty of Science, Tianjin University, Tianjin 300072, China; [d] International Center for Materials Nanoarchitectonics (MANA), National Institute for Materials Science, 1-1 Namiki, Tsukuba, Ibaraki 305-0044, Japan
*Email: wang.junjie0810@gmail.com; umezawa.naoto@gmail.com

2.1 Introduction

Photoelectrochemical water splitting has attracted great attention for its potential impact on solar energy conversion for a hydrogen energy society. Since the discovery of the Honda–Fujishima effect of TiO_2,[1] a great deal of effort has been made to develop novel semiconductor photocatalysts that exhibit better performance than TiO_2 under irradiation with sun light. One serious drawback of TiO_2 is that it has a too wide band gap to absorb visible light and, therefore, is unable to utilise the majority of sun light. Engineering photoabsorption edges is therefore an important task for the

†Present address: Samsung Electronics, Semiconductor R&D Center, 1, Samsungjeonja-ro, Hwaseong-si, Gyeonggi-do 18448, Korea

Energy and Environment Series No. 20
Advances in Photoelectrochemical Water Splitting: Theory, Experiment and Systems Analysis
Edited by S. David Tilley, Stephan Lany and Roel van de Krol

Published by the Royal Society of Chemistry, www.rsc.org

improvement of photocatalytic efficiency. The band 'gap' engineering alone is not sufficient for the design of semiconductor photocatalysts. Band 'edge' positions, *i.e.* ionisation potential and electron affinity must be also tailored for guaranteeing its capability of reducing and oxidising water. To satisfy this requirement, the valence band maximum (VBM) must be lower (more positive) than the oxidation potential of water for an anode material, while the conduction band minimum (CBM) must be higher (more negative) than the reduction potentials of water for a cathode material as shown in the illustration in Figure 2.1. Modern *ab initio* computations are a powerful tool for finding promising semiconductors that meet the aforementioned criteria. In fact, the band alignments of semiconductor materials have been extensively studied by DFT calculations for perovskite materials,[2] oxynitrides,[3] and two-dimensional materials.[4–8] The effects of molecular adsorption,[9] water solution,[10–12] and formation of heterostructures[13] on band edge positions have been also studied in detail. Computational details for estimating ionisation potential and electron affinity of semiconductor materials are well described in ref. 14.

Doping is a viable approach for sensitising wide gap semiconductors under visible light irradiation. Several theoretical studies of anion doping in TiO_2 showed that N doping gives rise to impurity state in the band gap,[15–18] consistent with the pioneering work of Asahi *et al.*,[19] which demonstrated that N-doped TiO_2 is a visible-light sensitised photocatalyst. Co-doping with donor–acceptor pair is even more effective since it can increase solubility limit of dopants. For example, W and C act as a donor and acceptor when substituted for Ti and O, respectively, and so (W, C) co-doped TiO_2 nanowire exhibited improved photoelectrochemical performance.[20] N-doping into $BiVO_4$ was reportedly advantageous for improving electron conductive property as well as photoabsorption property.[21] Effects of metal doping

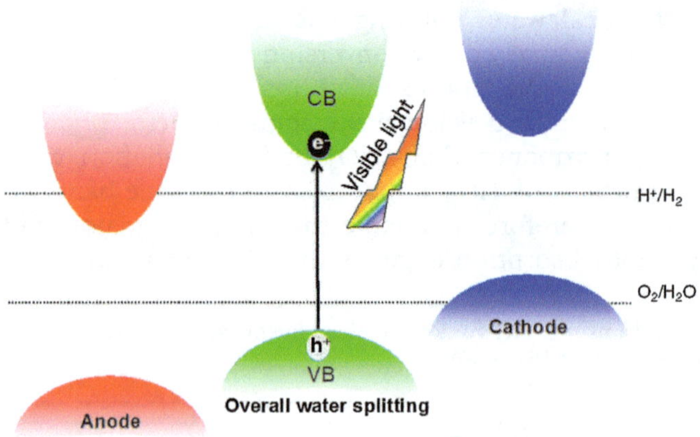

Figure 2.1 Required band edge positions for overall water splitting, and anode and cathode materials for photoelectrochemical water splitting. VB and CB denote valence band and conduction band, respectively.

on hematite surfaces were studied by LDA + U approach, where strong electronic correlation effects in the Fe d shell was described by a Hubbard U energy. It was predicted that Ni and Co doping is helpful for adjusting local charge of oxygen ions which contributes to smaller overpotentials for oxidation reaction.[22] Black TiO_2 was synthesised by hydrogenation treatment of powder-formed TiO_2 and its photocatalytic activity was experimentally demonstrated.[23] The origin of the disordered structure in black TiO_2 has been discussed using anatase nanoparticle models and DFT calculations.[24] The authors suggested that hydrogenation helps breaking Ti–O bonds on surface which facilitates forming disordered structures. $SrTiO_3$ is a popular photocatalyst for various reasons such as a higher CBM than that of TiO_2 [25] and its high stability in water. A number of dopants have been tested for $SrTiO_3$ to enhance visible light absorption in this wide gap semiconductor that is also known as a transparent conducting oxide.[26] For example, Ir-doped $SrTiO_3$ was reported as a promising cathode material for electrochemical water splitting reaction.[27] Ru or Rh-doped $SrTiO_3$ have been theoretically investigated using DFT calculations[28] although a band gap correction was not considered in this study. It is known that (semi-) local functionals such as local density approximation (LDA) or generalised gradient approximation (GGA) underestimate semiconductor band gaps. The band gap correction is crucial when we discuss impurities and defects in semiconductors since it leads to even qualitatively different results in some cases.[29] Recently, a comprehensive theoretical study has been conducted for N-doped $SrTiO_3$ with an appropriate band gap correction.[30] More references related to doping strategy for increasing visible light absorption can be found in ref. 31.

In the field of electronics, doping is usually performed for the sake of introducing conductive carriers in semiconductors by substituting aliovalent elements for a host element. In photocatalysis, however, doping plays two roles; one is the introduction of impurity states in the band gap for sensitising the host material under visible light irradiation as discussed above, the other is for the carrier doping, similarly to the case in electronics, for the design of anode (n-type) and cathode (p-type) materials. For example, we have theoretically designed n-type Ag_3PO_4 for enhancing the oxygen evolution rate from water solution based on the strategy shown in Figure 2.2. Hole accumulation on the surface is more expected in n-type Ag_3PO_4 owing to the upward band bending towards the oxide–water interface. The problem of such carrier doping in Ag_3PO_4 is discussed in the next section.

The surface structure is of great interest for photocatalysis study, especially in view of recent advances in experimental measurements for characterising catalysis reactions.[32,33] The impact of point defects on the surface reactivity were investigated by means of scanning tunnelling microscopy in combination with DFT calculations using slab models.[34,35] Surface faceting is also an important factor to improve photocatalysts performance and reactive facets were designed on the bases of surface energy analysis using DFT calculations.[36–40] The chemical dynamics of the oxidation evolution reaction was described by first-principles molecular dynamics simulation for

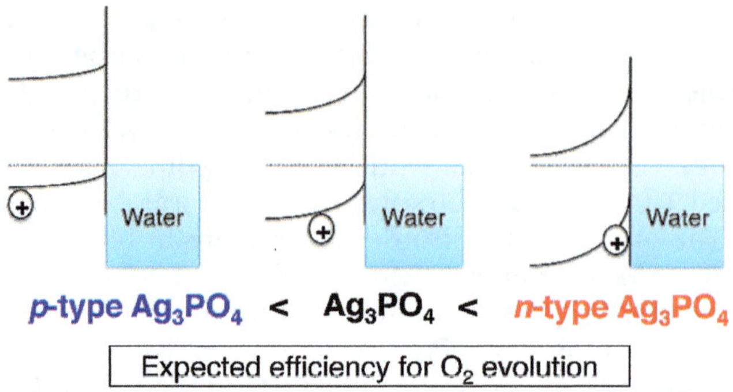

Figure 2.2 Carrier doping strategy for the improvement of O_2 evolution efficiency of Ag_3PO_4.

TiO_2–water solution interface.[41] It was revealed that Ti interstitial defect plays crucial role in stability and reactivity of TiO_2 surface[42] as well as visible light absorption.[43] Additionally, it is believed that placing co-catalysts on the surface of semiconductor photocatalysts can be effective for promoting carrier separation. The effects of various co-catalysts such as graphene or Pt nanoparticle have been studied with the help of DFT calculations.[44,45]

In this chapter, we will present our recent attempts on the band edge engineering for the design of advanced photocatalysts using DFT calculations. For the computational materials screening of photocatalysts, it is important to consider how the band structure is formed by the atomic orbitals of the constituent atoms. Specifically, it is preferable that the VBM or CBM consist of multiple orbitals originating from both cations and anions, so to enable good carrier transport properties. This guiding principle for the design of p-type oxide semiconductors was firstly pointed out in ref. 46. For example, Ag_3PO_4 is known as a highly active photocatalyst for the oxidation reaction, and the hole mobility is crucial for its performance. As shown in Figure 2.3, the atomic eigenvalues of Ag d and O p orbitals are close to each other, which results in the mixed character of these states at the VBM contributing to beneficial hole transport properties and high photocatalytic performance of Ag_3PO_4 as discussed in our previous work.[47] The same situation is observed in the other prominent photocatalyst $BiVO_4$, where Bi s orbitals are located within a similar energy range as O p, which should explain the high reactivity for oxygen evolution in aqueous solution.

As seen in the examples cited above, the quality of photocatalysts is determined not only by the bulk electronic structure but also by the chemical stability,[49–51] surface orientation and morphology, concentration of recombination centres, and the type of co-catalysts located on the surface *etc.* Nevertheless, we consider a desirable electronic structure as a necessary condition for a good photocatalyst and this criterion should be used in a first screening step of semiconductor materials. Thus, we mainly focus on the optimisation of the electronic structure in the following sections, although stability is also investigated to some extent.

Figure 2.3 Energy eigenvalues for single atoms given by Hartree–Fock method.[48]

In Section 2.2, we will present our studies on doping effects both for introducing impurity levels and conductive carriers. In Section 2.3, the effects of strain and thickness on the band edge positions of nanosheet materials are discussed. In Section 2.4, we present our recent attempts on evolutional crystal structure search for the discovery of novel photofunctional materials.

2.2 Effects of Doping in Photocatalyst

Doping is the intentional introduction of a small amount of impurities into an intrinsic semiconductor for the purpose of tuning its properties for desired applications. In semiconductors that are utilised in electronic devices, doping is essential to enhance concentration of free charge carriers. An n-type semiconductor can be achieved by doping with suitable donor impurities while a p-type semiconductor is obtained by doping with acceptor impurities. In a photoelectrochemical cell, n-type and p-type semiconductors are required as photoanode and photocathode, respectively. Solid-state photocatalysts are basically semiconductors with sufficiently large band gaps. In addition, the position of valence and conduction band edges must be suitable for the desired photocatalytic reaction. It is often found that the true performance of as-grown photocatalysts is limited by their physical properties, and the materials require further tuning. For example, large band gap of some metal oxides allows optical absorption only in the ultraviolet region, which can be altered by band gap engineering. The development of effective photocatalysts can also be accelerated by optimising the doping so to realize the desired n-type and p-type conductivities. In this section, two key important roles of doping in photocatalyst will be addressed: (1) The narrowing of the optical absorption edge in the case of chromium doping in $SrTiO_3$.[52] (2) Introducing electron and hole charge carriers by sulfur and silicon doping in Ag_3PO_4.[53]

2.2.1 Chromium Doping in $SrTiO_3$

$SrTiO_3$ has been realised as an alternative choice to TiO_2 for hydrogen production from water splitting. Its conduction band edge is 0.8 eV higher than the standard hydrogen electrode potential,[54] resulting in a stronger reduction capability than that of TiO_2 under ultraviolet irradiation without any bias voltage.[55,56] However, $SrTiO_3$ can absorb radiation only ultraviolet (UV) portion from solar spectrum because of its large band gap of 3.2 eV, which hinders the efficiency of solar-energy conversion. A strategy to improve the efficiency is to lift the valence band edge upwards into the visible range of its band gap, whereas the position of conduction band edge remains untouched in order to maintain the ability to reduce water into hydrogen (H_2). Chromium (Cr) doping is an approach that exhibited the successful enhancement of visible-light absorption in $SrTiO_3$.[57–59] Its effects on the photocatalytic water splitting and decomposition of organic molecules have been widely studied. Experiments revealed that Cr-doped samples are successful for H_2 evolution but not O_2. This suggests that incorporated Cr ions in those samples introduce occupied levels above the valence band while leave the conduction band edge untouched.

X-Ray photoelectron spectroscopy (XPS) measurements suggested that the photocatalytic activity was improved by the presence of trivalent Cr ions (Cr^{3+}), but degraded by the presence of hexavalent Cr ions (Cr^{6+}).[58,59] Experimental results additionally showed that co-doping with donor-type impurities could enhance the photocatalytic H_2 evolution on Cr-doped $SrTiO_3$, indicating that n-type condition, in other words Fermi-level is pinned near the CBM, is advantageous for stabilising Cr^{3+}.[57,59] Our recent work based on state-of-the-art density functional theory (DFT) using a hybrid functional has revealed the energetically favourable site for Cr dopant and the effect of its oxidation state on the electronic structure of Cr-doped $SrTiO_3$.[52] Based on a thermodynamic approach and formation energy calculations, we find that the Cr substitution on the Ti site (Cr_{Ti}) is energetically favourable under all equilibrium growth conditions. Thus, Cr_{Ti} should be responsible for the observed visible light absorption. Our calculations show that under n-type conditions, Cr_{Ti} occurs in the negative charge state (Cr_{Ti}^-), and introduces fully occupied t_{2g}-derived states (t'_{2g}) above the valence band maximum (VBM) (Figure 2.4(a)). Electrons in these occupied states can be excited by photons, giving rise to a redshift in photoabsorption edge. When the Fermi level moves toward the valence band electrons are removed from these t'_{2g} states, causing Cr_{Ti} occurs in neutral (Cr^0_{Ti}) and 2+ charge states (Cr_{Ti}^{2+}) in which t'_{2g} states are empty and locate in the band gap, likely acting

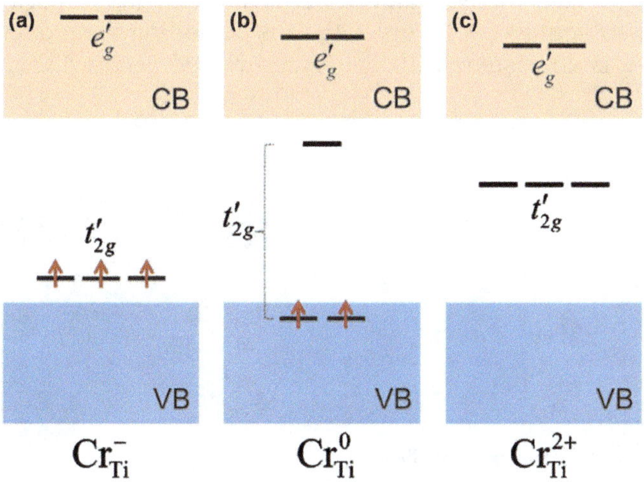

Figure 2.4 Schematic illustration of the d-states single-particle energy levels in the band gap for the majority spin channel of the Cr_{Ti} impurity in $SrTiO_3$ for (a) the 1−, (b) the neutral, and (c) the 2+ charge state. The arrows indicate the electron occupation.
Reproduced from ref. 52 with permission from the PCCP Owner Societies.

as trapping centres for photo-excited carriers. However, based on charge neutrality analysis, $Cr_{Ti}{}^{2+}$ is only stable in p-type $SrTiO_3$.

We investigate the effect of Cr_{Ti} on the optical properties by constructing the configuration coordinate diagram as shown in Figure 2.5. The optical transitions associated with $Cr_{Ti}{}^{-}$ occur through the process $Cr_{Ti}{}^{-} + h\nu \rightarrow Cr^0_{Ti} + e^-$ in which an electron is transferred from $Cr_{Ti}{}^{-}$ to the conduction band, and $Cr_{Ti}{}^{-}$ is converted to Cr^0_{Ti}. The calculated absorption energy is 2.32 eV, in good agreement with the photoabsorption edge reported in experiments at 530 nm (2.34 eV).[58–60] The photoabsorption caused by $Cr_{Ti}{}^{-}$ corresponds to the photoexcitation of an electron from a fully occupied in-gap t'_{2g} state to the CBM. In other words, this photoabsorption happens *via* an optical transition between a discrete impurity-level and the CB, rather than a band-to-band transition. The photo-excited holes that are created in the impurity level associated with $Cr_{Ti}{}^{-}$ are thus likely to be trapped, thereby decreasing their mobility. This is consistent with the results of photocatalytic experiments on Sb and Cr co-doped $SrTiO_3$ in which the O_2 evolution was found to be much smaller than the H_2 evolution.[57] Moreover, the emission energy, which results from the recombination of electron in the CB and hole trapped at $Cr_{Ti}{}^{-}$, is calculated to be 1.32 eV. The emission energy is in reasonably agreement with photoluminescence peak obtained from Sb and Cr co-doped $SrTiO_3$ in which the emission peak occurred at about 800 nm (1.55 eV).[57] Similar analysis of optical transition for the process $Cr^0_{Ti} + h\nu \rightarrow Cr_{Ti}{}^{-} + h^+$ yields the absorption energy of 1.93 eV and emission energy of 0.93 eV, suggesting that Cr^0_{Ti} can induce visible light absorption as well. However, in samples with a higher oxidation state of Cr (Cr^0_{Ti}), H_2 evolution is unlikely to be enhanced because the in-gap impurity states may act as trapping centres (*cf.* Figure 2.4), thus degrading the reduction activity. This picture is consistent with the experimental report[58] that the higher

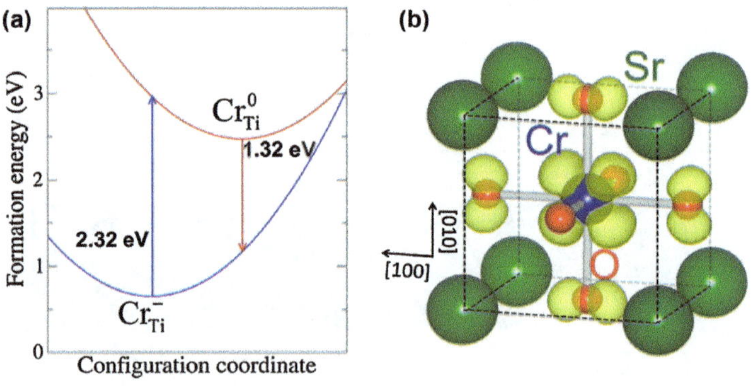

Figure 2.5 (a) Configuration-coordinate diagram for the Cr_{Ti} in $SrTiO_3$. The formation energy corresponds to the O-poor condition and to a Fermi level at the CBM. (b) The spin density of $Cr_{Ti}{}^{-}$ derived from a t'_{2g} state. Reproduced from ref. 52 with permission from the PCCP Owner Societies.

oxidation state of Cr is unfavourable for the photocatalytic activity although the visible-light absorption was observed.

The charge states of Cr_{Ti} are related to the oxidation state using Bader charge analysis. The Bader charges of Cr ions corresponding to Cr_{Ti}^{2+}, Cr_{Ti}^{0}, and Cr_{Ti}^{-} are compared with those in CrO_3, CrO_2, and Cr_2O_3. The Bader charges for Cr_{Ti}^{2+} (+2.26*e*), Cr_{Ti}^{0} (+2.05*e*) and Cr_{Ti}^{-} (+1.91*e*) are equivalent to those for CrO_3 (+2.25*e*), CrO_2 (+2.06*e*), and Cr_2O_3 (+1.82*e*), respectively. This analysis suggests that the oxidation states of Cr_{Ti}^{2+}, Cr_{Ti}^{0}, and Cr_{Ti}^{-} are likely 6+, 4+ and 3+, respectively. These results indicate that the growth conditions and the Fermi level of the system can influence the oxidation state of Cr dopants and thereby the photocatalytic performance. Based on our results, Cr_{Ti}^{-} with Cr being in the 3+ oxidation state, is responsible for the visible-light absorption, and this state is beneficial for H_2 evolution because it gives rise to the fully occupied states above the VBM without creating any trap states in the band gap. This model is consistent with recent experiments in which Cr-doped samples with a high Cr^{3+} content exhibited high photocatalytic activities under visible light, whereas the samples that contained a higher oxidation state of Cr were inactive.[58,59] These results indicate that the Fermi level should be raised toward the CBM to enhance the desirable charge state of Cr_{Ti}, Cr_{Ti}^{-}. In other words, the Cr_{Ti}^{-} will dominate in Cr-doped $SrTiO_3$ that exhibit n-type conductivity. These results also explain the enhanced photocatalytic performance of Cr-doped $SrTiO_3$ with co-doping by Sb, Nb, or Ta, which could act as donor when substituted for Ti.[57] We thus believe that co-doping with donor impurities will stabilise Cr_{Ti}^{-}, which acts as active Cr ions (Cr^{3+}). To identify the most effective donor that can act as co-dopant for stabilising Cr^{3+} in Cr-doped $SrTiO_3$, we comprehensively investigated the effects of doping $SrTiO_3$ with group-V metals tantalum (Ta), antimony (Sb), niobium (Nb) and vanadium (V), the group-III metals lanthanum (La), yttrium (Y) and the halogen (F), and reported in ref. 61. Our theoretical investigation based on hybrid density-functional predicts that La is the best donor among the candidate dopants for stabilizing Cr^{3+}. The calculated results showed that La substituted for Sr site has the highest solubility, act as single shallow donor, and does not induce trap states in the band gap. Moreover, the complex between La_{Sr}^{+} and Cr_{Ti}^{-} is unlikely to form due to its very small binding energy, thus the visible-light absorption property of Cr-doped $SrTiO_3$ should be maintained. The prediction was successfully confirmed by experiments,[61] which clearly revealed that La and Cr co-doped $SrTiO_3$ owns the best performance for H_2 evolution from water splitting under visible-light irradiation.

2.2.2 Sulfur and Silicon Doping in Ag_3PO_4

Recently, silver orthophosphate (Ag_3PO_4) has been reported to exhibit very high quantum efficiency of nearly 90% for the O_2 evolution under visible-light irradiation.[62] The combination of its band gap of 2.45 eV (direct) and an appropriate position of the valence-band (VB) edge, as well as a suitable band character is what makes Ag_3PO_4 suitable for photocatalytic oxidation reaction, and has become a prototype for visible-light responsive photocatalyst. This discovery triggered many extensive studies, both experimental[63–65] and computational.[66–69] A recent DFT study on this material suggested that the point native defects are unlikely to cause carrier trapping and degrade the photocatalytic activity.[68] It also suggested that in as-grown Ag_3PO_4, the Fermi level is located near the middle of the band gap and thus intrinsic carrier conductivity is negligible for any equilibrium growth condition. In fact, n-type and/or p-type semiconductors are required in a photoelectrochemical cell as either photoanode and/or photocathode, respectively. The use of two semiconductors with suitable band edges position as anode (n-type) and cathode (p-type) should yield a high electric field that facilitates the separation of photo-excited electron–hole pairs, thereby increasing the rate of the desired chemical reactions.[70] Because Ag_3PO_4 does not easily form native defects that would act as electron or hole killers, we expect that external doping should work readily in this semiconductor.

Here we address the effect of sulfur (S) and silicon (Si) impurities on the electronic properties of Ag_3PO_4, focusing on their efficacy for n-type and p-type doping. These dopants were selected in this study because they are adjacent to phosphorous (P) in the Periodic Table, and their atomic radii are just slightly different; Si: 1.18 Å, P: 1.10 Å, and Si: 1.03 Å. This suggests that S and Si could be readily incorporated on the P site where the substitutional S (S_P) and Si (Si_P) act as donor and acceptor, respectively. Based on a thermodynamic approach and formation energy calculations, we actually find that S and Si prefer to occupy P site over Ag or interstitial site. Sulfur substituted for O site is not considered because S and O are isovalent, and thus S_O is electrically inactive. We neglect Si_O as well because the large size mismatch between Si and O atoms should cause very high formation energy and thereby its very small concentration in the material can be ignored.

Figure 2.6(a) shows the projected density of states (PDOS) of the S, P and Si 3p orbitals and 2p orbitals of their nearest O atoms for the S_P, P_P, and Si_P configurations. The electron distributions for the bonding states associated with S–O, P–O, and Si–O bonds are shown in the insets. The bonding states, derived from the p orbitals, are located at ~8.60, ~6.84, and ~5.76 eV below the VBM. To understand the qualitative electronic structure of S_P and Si_P, the molecular orbital diagram is created and depicted in Figure 2.6(b). For simplicity, we can regard S, P and Si as ions with 6+, 5+ and 4+ oxidation state, respectively. The substitutional S on the P site, S_P, thus acts as a donor while the substitutional Si on the P site, Si_P acts as an acceptor.

We find that S_P does not create any in-gap state. Four electrons from S can fill three holes in V_P, forming complete set of S–O bonding states in $(SO_4)^{2-}$, while an extra electron which would occupy the antibonding state

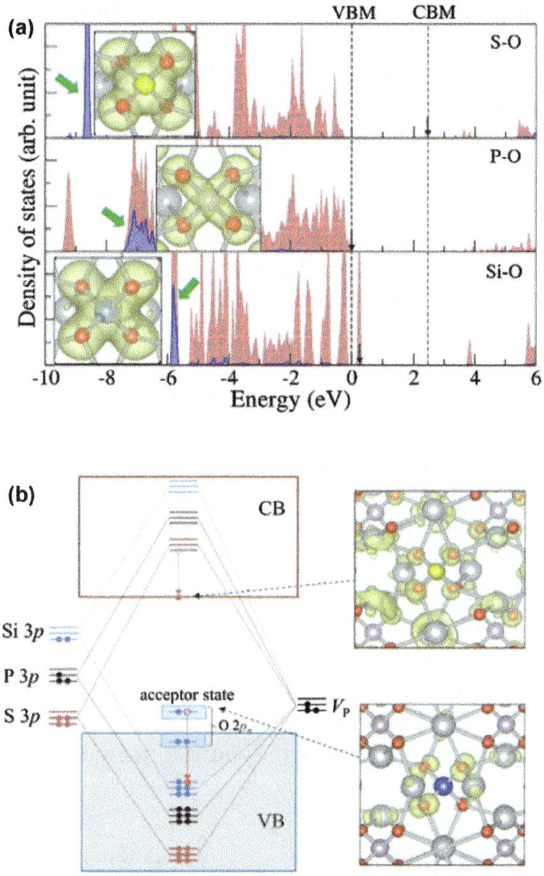

Figure 2.6 (a) Projected density of states (PDOS) of S, P, and Si 3p orbitals (blue shaded) and 2p orbitals (red shaded) of the four nearest O atoms for S_P, P_P, and Si_P in the neutral charge state. The insets show the charge distribution of the bonding state associated with S–O, P–O, and Si–O bonds, which are indicated by the solid green arrows. The vertical black arrows indicate the highest occupied state. (b) Schematic diagram for the interaction between 3p orbitals of S, P, or Si with the p orbitals of P vacancy (V_P) in S_P, P_P, and Si_P in Ag_3PO_4. The solid dots represent the electron occupation. The charge distributions of the bottom of the conduction band in S_P and of the acceptor state which is located above the valence band (VB) in Si_P are also shown.
Reprinted with permission from P. Reunchan and N. Umezawa, *J. Phys. Chem. C*, 2015, 119, 2284. Copyright (2015) American Chemical Society.

transferred to the conduction band minimum. The delocalised charge distribution of the conduction-band like state with the extra electron of S_P^0 (Figure 2.6(b)) indicates that S_P^0 can be readily ionised into $S_P{}^+$, suggesting that S_P is a shallow donor. On the other hand, we find that Si_P creates a deep acceptor state, derived from an O $2p_\pi$ state. Therefore, Si_P is a deep acceptor, which occurs in neutral and 1− charge state. The charge distribution of this acceptor state is shown in Figure 2.6(b).

Our analysis of the electronic structure above suggests that S is a shallow donor while Si is a deep acceptor when substituted at the P site. Next, their solubilities in the material under equilibrium growth conditions are investigated through the formation energy calculations.[53] Figure 2.7 shows that S_P is very easily incorporated into Ag_3PO_4 because its formation energy is very low even under P-rich conditions. This finding can be attributed to the slightly smaller atomic size of S compared to that of P, and to the formation of a strong S–O bond, as seen by the lower energy of the S–O bonding state compared to the respective P–O state (see Figure 2.6). The thermodynamic transition level from S_P^+ to S_P^0, $\varepsilon(0/+)$ is located at 3.22 eV above the VBM, which is ~0.8 eV above the CBM, suggesting that S_P^0 can be fully ionised into S_P^+even under room temperature. The combination of the very low formation energy of S_P and its shallow donor behavior, suggests that S-doped Ag_3PO_4 can give rise to the electron carrier in the conduction band, thereby shifting Fermi level upward. In contrast, the substitution of P by Si is not feasible although the atomic radius of Si is only slightly larger than that of P. The relatively higher formation energy of Si_P compared to S_P even under the P-poor growth condition suggests that the incorporation of Si is much more difficult than that of S into Ag_3PO_4. The solubility of Si_P is likely limited by the formation of SiO_2, which is energetically very stable. As shown in Figure 2.7, Si_P is stable in the neutral, Si_P^0, and negative, Si_P^-, charge states with the (0/−) thermodynamic transition level located at

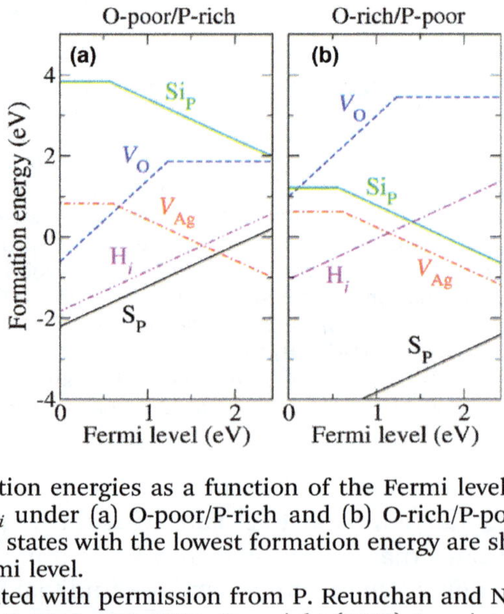

Figure 2.7 Formation energies as a function of the Fermi level for S_P, Si_P V_O, V_{Ag}, and H_i under (a) O-poor/P-rich and (b) O-rich/P-poor limits. Only the charge states with the lowest formation energy are shown for each value of Fermi level.
Reprinted with permission from P. Reunchan and N. Umezawa, *J. Phys. Chem. C*, 2015, **119**, 2284. Copyright (2015) American Chemical Society.

0.58 eV above the VBM. This indicates that Si_P is a deep acceptor; only a small number of holes is introduced in the valence band at room temperature, consistent with the hole localisation on the neighbouring O atoms as shown in Figure 2.6(b). Under the P-rich or O-poor growth condition, the formation energy of Si_P is very high even when the Fermi level is close to the CBM (n-type condition), meaning that the concentration of Si_P is negligible. As the P chemical potential decreases from the P-rich to the P-poor limit, the likelihood of Si_P formation increases by several orders of magnitude. However, our carrier concentration analysis based on charge neutrality[53] indicates that even under the P-poor (O-rich) growth condition, the density of hole carriers in the valence band introduced by Si_P is still much lower than the desired hole concentration in a good p-type material.

In summary, from hybrid-density functional calculations, we find that both S and Si are energetically favourable on P site. S_P is a shallow donor and can form in very high concentration while Si_P acts as a deep acceptor, and its solubility is very low, which results in a low concentration of hole carriers. We suggest that n-type Ag_3PO_4 can be readily achieved by S doping while p-type Ag_3PO_4 by Si doping is unlikely.

2.3 Band Structure Design of Highly Efficient Photocatalysis by Strain Engineering

Hydrogen generation and storage has received much attention for the usage as a clean fuel. Compared to conventional hydrogen generation by steam reforming, solar water splitting using semiconductor photocatalysts is a potential clean and renewable energy source. However, photocatalytic materials are often limited by their physical properties. For example, a too large band gap will not allow for visible light absorption. It is well known that the internal strain, which can be easily produced from the lattice mismatch or external stress, is an efficient approach to modify the electronic structure of materials.[71–73] Recently, strain engineering has been widely employed to design the desired physical and chemical properties of nano-materials, such as the two dimensional layered materials. In this chapter, the band gap and band edge alignment engineering by the strain effect for SnS_2 and SnO layered compounds is discussed as examples to illustrate the application of strain engineering for designing photocatalytic water splitting materials.[74,75]

2.3.1 Strain Engineering for Single-layer SnS_2

As a non-toxic, low cost and chemically stable van der Waals layered material, SnS_2 has attracted the interests of researchers in catalysis, solar cell and microelectronics areas. Especially, single-layer SnS_2 was experimentally found to have an incident photon-to-current conversion efficiency of 38.7%, much higher than in its bulk form with an efficiency of 2.33%,[76] making the single-layer SnS_2 more promising for photocatalytic water splitting than its bulk phase. Subsequently, Zhang *et al.* studied the single-layer phase theoretically and confirmed the stability in aqueous solution and the optical band-gap within the range of visible-light irradiation, thereby suggesting it as a promising photocatalytic material.[77] However, their results also indicate that the conduction band edge of single-layer SnS_2 is lower than the standard reduction potential for H^+/H_2 by 0.9 eV, which implies that hydrogen production from water splitting cannot spontaneously take place. Therefore, the electronic structure modification of the single-layer SnS_2 is crucial for improving its photocatalytic ability. Zhou *et al.* have systematically investigated the dependence of the band structure of single-layer SnS_2 on the in-plane biaxial strain with hybrid functional calculations (HSE06).[74] The strain engineering was considered as a possible way to realise visible-light overall water splitting in single-layer SnS_2.

The stable structure of single-layer SnS_2, shown in the inset of Figure 2.8(b), has a hexagonal lattice and belongs to the space group $P6_3/mmc$. As for its intrinsic electronic structure, the valence band maximum (VBM) of single-layer SnS_2 is mainly constituted by S-3p states, while both Sn-5s and S-3p orbitals give contributions to its conduction band minimum (CBM). An indirect band gap of 2.57 eV was obtained from calculations at the hybrid functional level. Similar as in the previous report,[77] the conduction band edge of un-strained single-layer SnS_2 is actually lower than the hydrogen reduction potential.

When in-plane biaxial strain is imposed on single-layer SnS_2, significant changes of the band alignment can be observed (see Figure 2.9). The band gap of single-layer SnS_2 approaches the largest value of 2.7 eV at 2% tensile strain. At this point, increasing or decreasing the biaxial strain will narrow

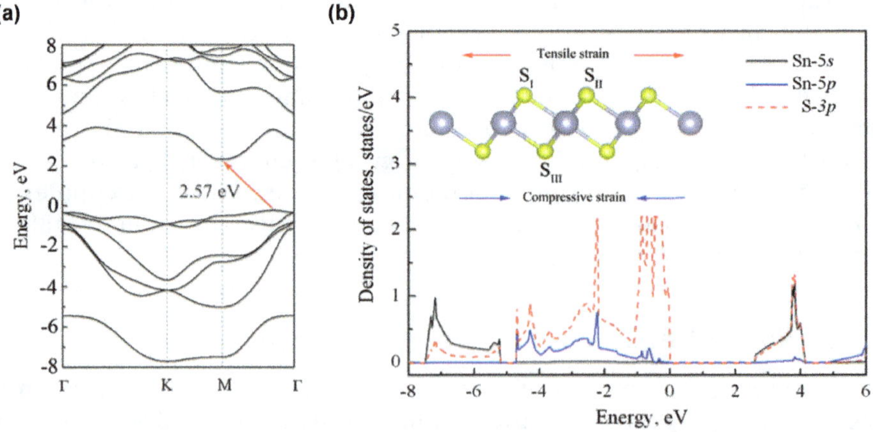

Figure 2.8 (a) Band structure, and (b) density of states of the single-layer SnS_2.[74] Reproduced from ref. 74 with permission from the PCCP Owner Societies.

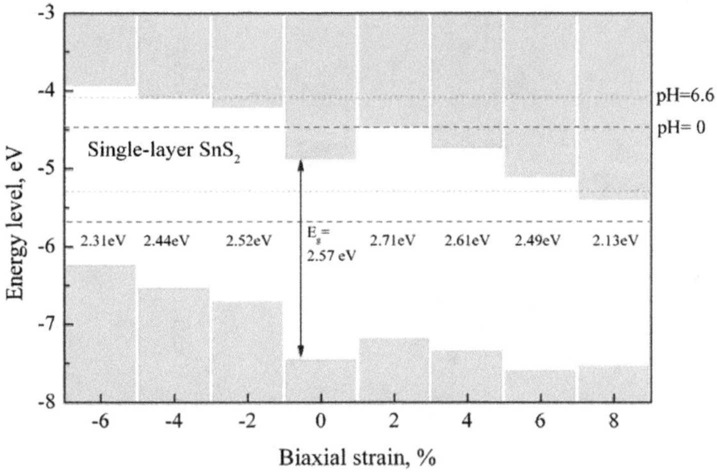

Figure 2.9 Band-edge positions of single-layer SnS_2.
Reproduced from ref. 74 with permission from the PCCP Owner Societies.

the band gap to enhance the visible-light absorption. Overall water splitting can spontaneously take place at a compressive strain of 2% and 6% at pH values of 0 and 6.6, respectively (Figure 2.9). Therefore, compressive in-plane strain could be a viable way to realise the visible-light absorbing photocatalysts for overall water splitting. For tensile strain larger than 6%, the band gap is reduced due to the shift of VBM and CBM in opposite directions. This non-linear change of the band gap is induced by the transition of *k*-point for band edge from the competing effect of different orbital interactions, which has been discussed in ref. 74 in detail. Under tensile strain, the ability of oxygen production is preserved, but the CBM become too low for hydrogen generation.

The above described change of the band edge offsets as a function of biaxial strain is explained by a bonding analysis. From the crystal orbital overlap population analysis, Zhou *et al.*[74] concluded that the CBM (VBM) of single-layer SnS_2 is dominated by the anti-bonding states between Sn-5s and S-3p_{xy} (S_I-3p_{xy} and S_{II}-3p_{xy}) interactions as shown in Figure 2.10. Bonding states associated with S_I3p_z–$S_{III}3p_z$ interactions are also formed at the bottom of the conduction band. It indicates that the band edges of single-layer SnS_2 are determined by both the Sn–S and S–S interactions. When compressive biaxial strain is introduced, the bonding and the anti-bonding splitting associated with Sn5s–S3p and S_I3p_{xy}–$S_{II}3p_{xy}$ hybridisation is increased due to the bond length shortening, which shifts up both its VBM and CBM (Figure 2.10(a)). The weakened S_I3p_z–$S_{III}3p_z$ interactions due to the Poisson effect further assists the upward shift of the CBM (Figure 2.10(b)). This analysis confirms that biaxial strain is an effective way to tune the band structure of single-layer SnS_2.

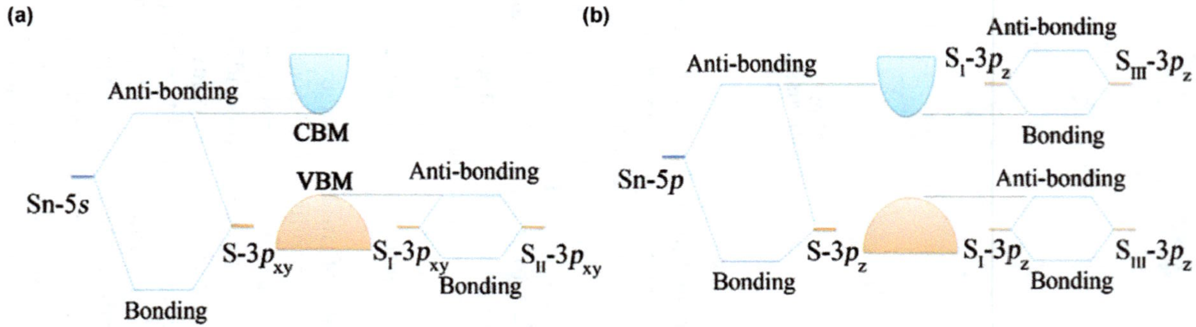

Figure 2.10 (a), (b) The bonding diagram of single-layer SnS_2.
Reproduced from ref. 74 with permission from the PCCP Owner Societies. (The S atoms at different sites are indicated as S_I, S_{II} and S_{III} in Figure 2.8(b).)

2.3.2 Strain Engineering for Layered SnO

Tin monoxide (SnO) is a layered semiconductor material promising for a wide variety of technological applications, such as Li-rechargeable batteries[78] and bipolar conductivity materials.[79] SnO has attracted particular attention for electronic and optoelectronic applications, because it is a rare example of an oxide semiconductor in which bipolar conductivity can be easily realized. However, SnO possesses a relatively small indirect band gap of only ∼0.7 eV. This narrow band gap limits the technological potential of SnO in areas such as photocatalytic water splitting and transparent conductivity. The possibility to modify the electronic structure of SnO by strain engineering to extend its application range was studied Zhou *et al.*[75] with hybrid functional calculations.

Bulk SnO is constructed with Sn–O–Sn slabs stacked along the [001] crystal direction. Most studies focus on the weak van der Waals interaction in the [001] direction of SnO.[80,81] However, the calculated band structure in Figure 2.11(a) shows a significant band dispersion along Γ–Z direction at the valence band which implies a pronounced inter-layer interaction. It was also

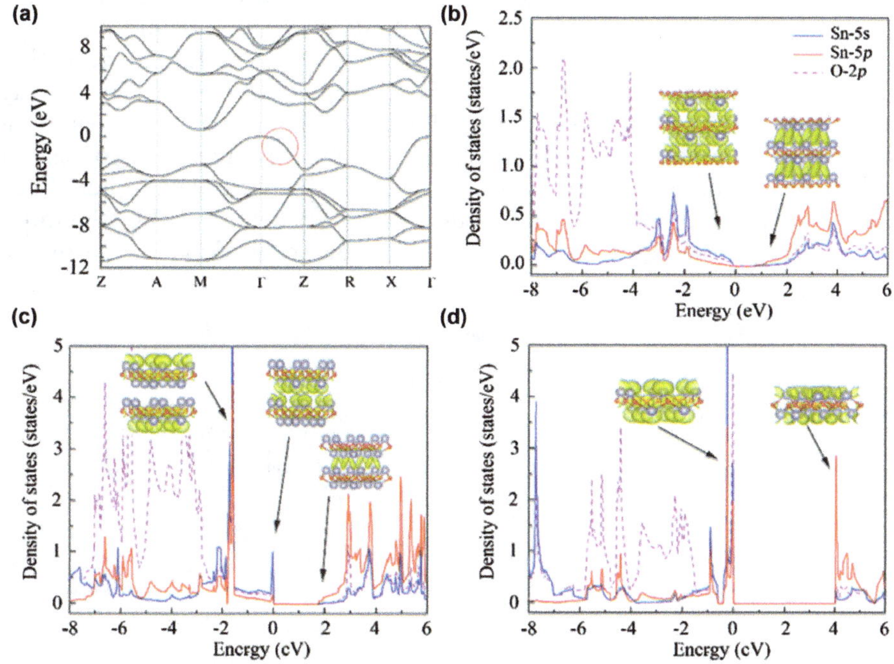

Figure 2.11 Band structures and density of states for (a), (b) bulk SnO; (c) double-layer SnO; (d) single-layer SnO. The Fermi level is set to 0 eV. The insets in (b), (c) and (d) show the partial electron density isosurfaces (0.007 e $Å^{-3}$) corresponding to the states in the regions indicted by arrows.[75]
Reproduced from ref. 75 with permission from the PCCP Owner Societies.

confirmed that the band gap of monolayer SnO is much wider than that of the bulk form.

As a typical lone pair compound, the lone pair distribution of SnO is projected away from the cation (Sn^{2+}) toward the outside void region to satisfy the orbital stabilization.[81] All the Sn-5p states contribute to the on-site hybridisation with Sn-5s through the (Sn5s–O2p)* antibonding state, resulting in a large band gap denoted as Eg1 in Figure 2.12(b). At the interface between two neighboring layers, interlayer Sn–Sn interactions are allowed and bonding-antibonding splitting occurs. These states associated with Sn–Sn interactions are responsible for the band edges of nanosheets and bulk SnO as schematically illustrated in the upper panel of Eg2 in Figure 2.12(b). The above scenario is also supported by the observations made in the partial electron density shown in the insets of Figure 2.11(b)–(d). According to this diagram, it can be expected that the bonding–antibonding splitting will be weakened with increasing the Sn(I)–Sn(II) distance from biaxial tensile strain to open the band gap of Eg2.

As shown in Figure 2.13(a), the band gap of bulk SnO increases almost linearly with strain. A semiconductor-metal transition can be observed at a strain of −5% where the band gap closes. The case of metallisation of SnO was also observed experimentally by Forthaus *et al.* at high pressure conditions.[82] The relatively low magnitude of strain required for the transition implies that in-plane strain is a good way to adjust the type of conductivity in bulk SnO. For the double-layer SnO, the band gap changes non-monotonically with the biaxial strain. This band gap change is consistent with the evolution of the inter-layer Sn–Sn distance which has been discussed in ref. 75, since the band edges of SnO is determined by the inter-layer Sn–Sn interactions. For more details, the strain dependence of the

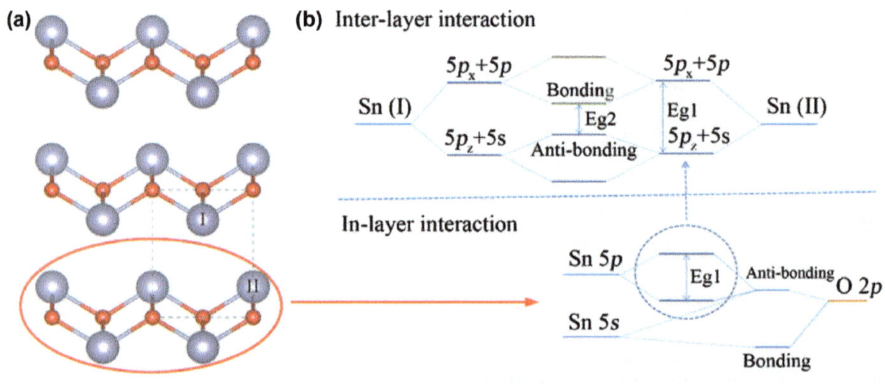

Figure 2.12 (a) Crystal structure of SnO. (b) Bond diagram associated with in-layer Sn–O interaction (lower panel) and Sn–Sn inter-layer interaction (upper panel). A band gap of isolated lone pairs on surface (Eg1) and that of inter-layer interaction (Eg2) are indicated in the figure.[75]
Reproduced from ref. 75 with permission from the PCCP Owner Societies.

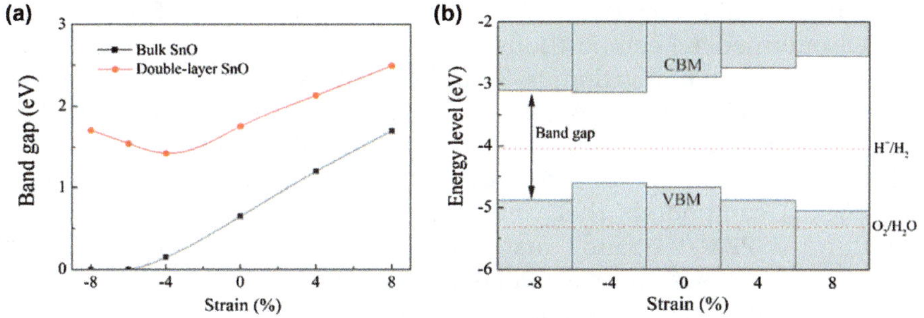

Figure 2.13 The calculated (a) band gap *vs.* biaxial strain relationship, (b) band alignment of strained double-layer SnO with respect to normal hydrogen electrode levels in electrolyte of pH = 6.6.[75]
Reproduced from ref. 75 with permission from the PCCP Owner Societies.

band-edge positions with respective to the water reduction and oxidation potentials for the double-layer SnO are shown in Figure 2.13(b). The VBM of double-layer SnO shifts downward while the CBM shifts upward when the in-plane strain is imposed. It is also apparent that the shift in the band edge is more significant for tensile strain. With 8% tensile strain, its VBM is only 0.2 eV higher than the reduction potential of O_2/H_2O. Thus, tensile strain can improve the potential of H_2 production and decrease the over-potential for oxygen evolution in the water splitting process.

2.4 Exploration of Photofunctional Materials Employing Evolutional Structure Search

Structure prediction using evolutionary algorithms has become a very useful tool in the rational design of new materials and structure identification in those situations where experimental structure refinement is difficult. Our calculations were carried out using the computer codes Universal Structure Prediction: Evolutionary Xtallography (USPEX)[83,84] and Vienna *Ab initio* Simulation Package (VASP).[85,86] Here, we present our recent research results on *ab initio* evolutionary searches for mixed valence tin oxides[87] and graphitic carbon nitride (g-C_3N_4)[88] for photocatalysis applications.

2.4.1 Mixed Valence Tin Oxides as Novel Photocatalysts

Tin dioxide (SnO_2) and tin monoxide (SnO) are the well-known binary oxides for the common 4+ and 2+ oxidation states of Sn. SnO_2 is an n-type semiconductor with a band gap of 3.6 eV, whereas SnO is a p-type semiconductor with a narrow indirect (0.7 eV) band gap. Recently, both experimental and theoretical efforts have been made to synthesise new mixed valence tin oxides (Sn_xO_y) besides SnO_2 and SnO. The structures of Sn_2O_3 and Sn_3O_4 have been predicted respectively, using the cluster expansion technique combined with density functional theory (DFT).[89,90] However, the structures of several other reported intermediate compositions such as Sn_4O_5 and Sn_5O_6 are still under debate. With this motivation, a systematic investigation[87] of the crystal structures and electronic properties of mixed valence tin oxides has been performed with a focus on their possible photocatalysis applications.

Figure 2.14(a) shows a series of stable crystal structures for Sn_xO_y ($0.5 < x/y < 1$) compounds that were predicted by combining evolutionary algorithm USPEX[83,84] and first-principles calculations of VASP.[85,86] Figure 2.14(b) shows the calculated formation enthalpies relative to the energies of α-SnO and rutile SnO_2. These data imply that the formation of Sn_2O_3, α-Sn_3O_4, Sn_4O_5, Sn_5O_6, Sn_7O_8, Sn_9O_{10} and $Sn_{11}O_{12}$ is thermodynamically stable with respect to physical mixtures of α-SnO and rutile SnO_2. Interestingly, we found that all of the Sn_xO_y structures predicted in our study consist of a combination of three basic layer types (Figure 2.14(a)): distorted SnO, Sn_3O_4, and Sn_2O_3. This suggests that the newly discovered tin oxides form a new class of van der Waals (vdW) materials. The 'Lego blocks' in Sn_xO_y consist of various mixed valence tin oxide layers, which is different from the conventional vdW heterostructures that have been fabricated *via* the layer-by-layer deposition of different nanosheet materials. Therefore, high-quality interfaces can be achieved in such Sn_xO_y vdW structures because the lattice mismatch between different forms of Sn_xO_y is small (about 1%) and can be accommodated by the flexible connection between the layers.

Figure 2.15(a) shows the linear dependence of the calculated band gap of Sn_xO_y on the interlayer distance. The band gaps of all the Sn_xO_y structures were calculated within the HSE06 + vdW-DF framework.[87] To investigate the effect of interlayer distance on the band gap, a series of calculations were performed for the well-characterized system α-SnO (Figure 2.15(b)). Here, the interlayer distance was increased from an equilibrium value while keeping the other lattice parameters fixed and electronic structure calculations were performed for a given geometry. Figure 2.15(b) shows that there is a direct relationship between the band gap and interlayer distance of α-SnO. Sn_5O_6 was selected as a representative case of Sn_xO_y to illustrate the relationship between the electronic structures and the Sn^{2+}–Sn^{2+} interaction. The band structure of Sn_5O_6 together with the partial charge densities of electrons occupying the valence-band maximum (VBM) and conduction-band minimum (CBM) are shown in Figure 2.15(c). Both the VBM and CBM mainly consist of Sn^{2+}-derived states, which is confirmed by the local density of states shown in Figure 2.15(d). One can see that the contribution of less ionized Sn_1^{2+} is more dominant at the VBM and CBM. Therefore, we can expect the interlayer Sn^{2+}–Sn^{2+} interaction is responsible for the linear dependence of the band gap on the interlayer distance.

The theoretically calculated band alignment of 11 tin oxides together with that of rutile-TiO_2 are shown in Figure 2.16. All the mixed valence tin oxides examined in the current work possess relatively small ionization potentials and thus might exhibit p-type conductivity, which is challenging to achieve in most oxide semiconductors. The red dashed line in Figure 2.16 indicates

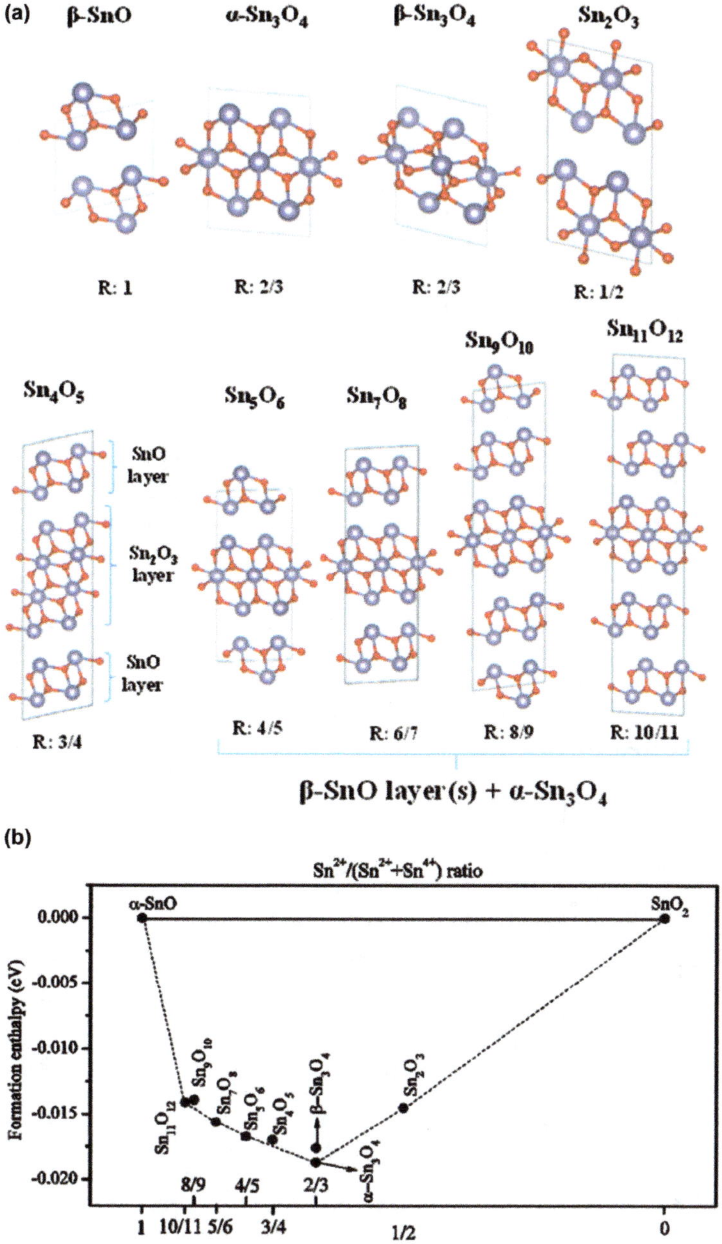

Figure 2.14 Crystal structures of predicted Sn_xO_y and SnO phases and convex hull diagram for Sn_xO_y system. (a) Grey and red spheres denote Sn and O atoms, respectively. R indicates the ratio of $[Sn^{2+}]/([Sn^{2+}]+[Sn^{4+}])$ in these compounds. (b) Enthalpies of formation per atom (ΔH) for Sn_xO_y compounds (closed circles). The dashed line indicates the convex hull of the ground state. α-SnO and SnO_2 were adopted as the limiting phases of this diagram.
Reprinted with permission from ref. 87.

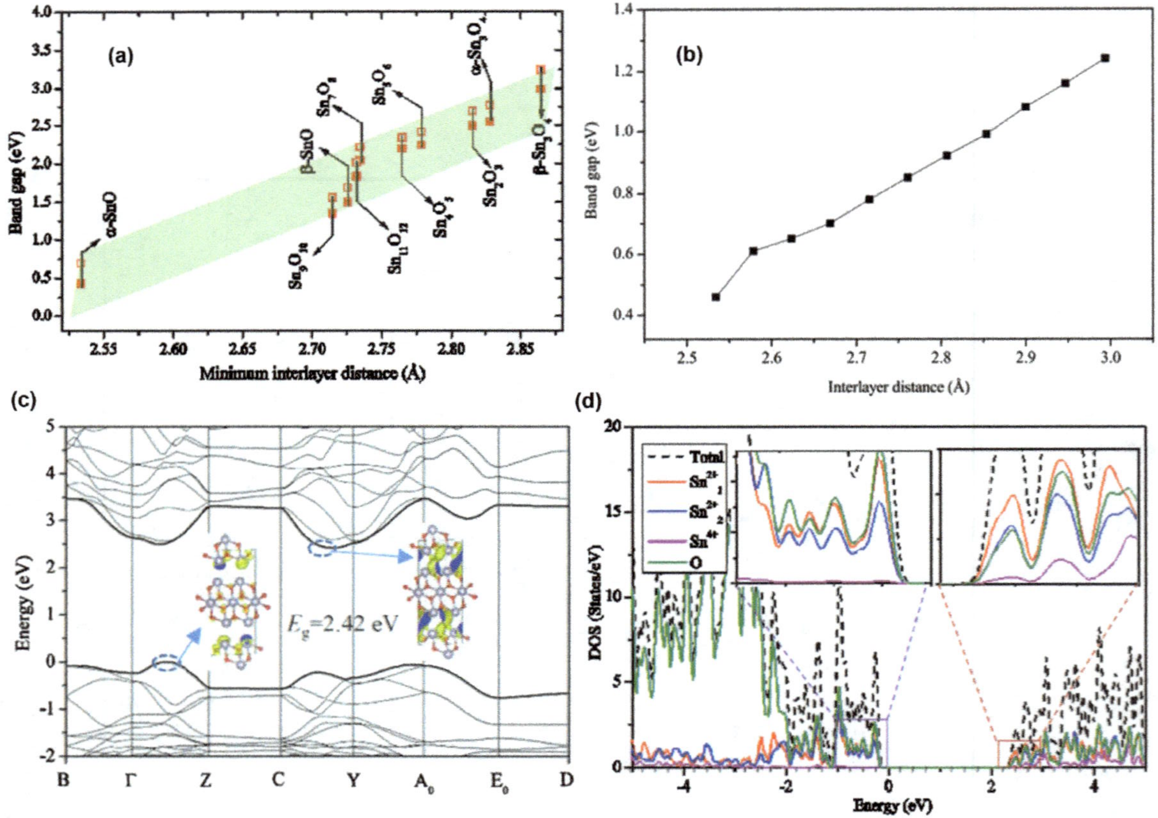

Figure 2.15 Relationship between calculated band gap and interlayer distance in Sn_xO_y structures and calculated electronic structure of Sn_5O_6. (a) Dependence of calculated band gap on minimum interlayer distance for Sn_xO_y. Solid and open symbols indicate values calculated with the HSE06 and HSE06 + vdW-DF functionals, respectively. (b) Calculated band gaps for α-SnO (litharge) as a function of interlayer distance obtained from HSE06 calculations. (c) Band structure and partial charge densities at valence band maximum (VBM) and conduction band minimum (CBM) of Sn_5O_6. The isosurfaces are at 0.003 e Å^{-3}. (d) Total and projected density of states (DOS) of Sn_5O_6. The distributions of O, Sn^{4+} and two kinds of Sn^{2+}, Sn_1^{2+} (the less ionised Sn^{2+}) and Sn_2^{2+} (the more ionised Sn^{2+}), are shown.
Reprinted with permission from Ref. 87.

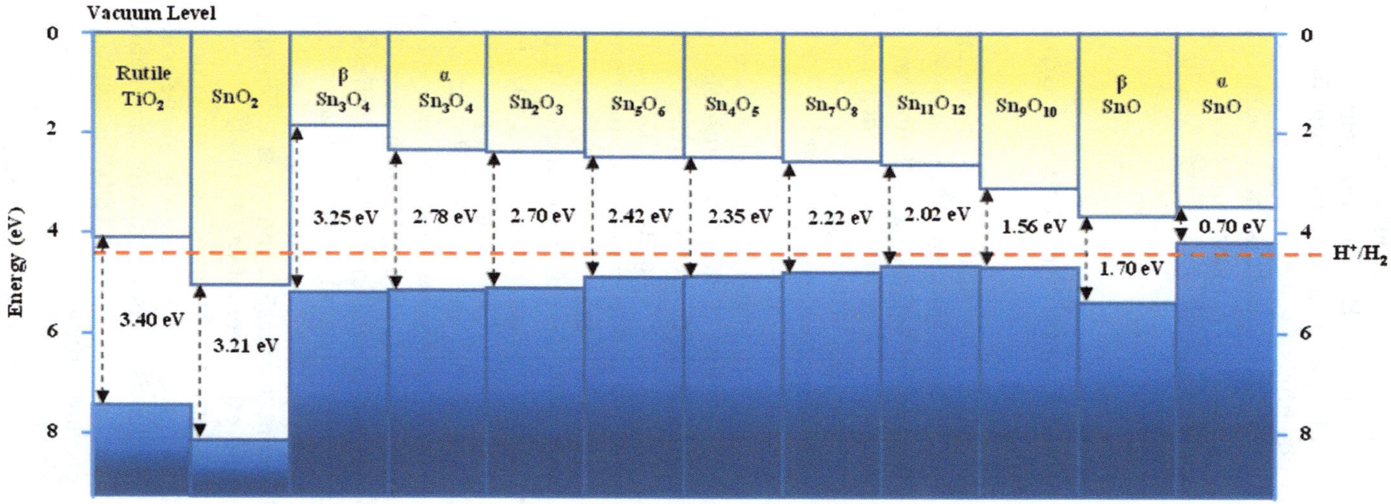

Figure 2.16 Predicted band alignment of Sn_xO_y suboxides, SnO, SnO_2 and TiO_2. The ionisation potential with respect to the common vacuum level is represented by the vertical axis. The red dashed line indicates the reduction potential of water (H^+/H_2). The band gaps and ionisation energies of SnO and Sn_xO_y were calculated within the HSE + vdW-DF framework. Reprinted with permission from ref. 87.

the standard reduction potential with respect to the vacuum level for H^+/H_2. In the present study, we adopted standard conditions of pH = 0 and $p_{H2} = 1$ atm, and found that the H^+/H_2 reduction potential is situated just below the CBM of TiO_2, which is consistent with experimental results. We note that the reduction potential of water is situated below the CBM of all the predicted Sn_xO_y structures, implying that mixed valence tin oxides have great potential for H_2 evolution from water under visible light irradiation. It is noteworthy that the photocatalytic activity of Sn_3O_4 has been validated in a recent work.[91]

2.4.2 Determination of Crystal Structures of Graphitic Carbon Nitride

Graphitic carbon nitride (g-C_3N_4) has attracted extensive interest in recent several years because it is promising as a metal-free photocatalyst.[92] However, the crystal structure of g-C_3N_4 is still unclear. Therefore, the basic characteristics of g-C_3N_4, *e.g.* electronic and optical properties, are not well understood. Two controversial structures can be obtained by different synthesis methods, poly-condensation of dicyandiamide (DCDA: $C_2N_4H_4$) in an oxidising environment and salt-melt synthesis (SMS) in an inert environment. None of the crystal structures have been solved experimentally.

To address this problem, we first studied the synthesis reaction pathway proceeding through intermediates to g-C_3N_4 from DCDA in air and inert atmospheres by using density functional theory (DFT) calculations. DFT calculations revealed that DCDA condensation in an inert atmosphere (SMS) can only proceed to an incomplete degree ($C_6N_{11}H_9$) and has a rather high concentration of ammonia (NH_3) in the final product. Because of the involvement of O_2 in the reaction, the condensation of DCDA in air will lead to the production of melem ($C_6N_{10}H_6$) quickly, a direct precursor of heptazine-based g-C_3N_4.[88] This means that the composition of product of poly-condensation in air can be C_3N_4, while SMS synthesis could introduce NH_3 contamination to the expected C_3N_4 product. Lithium and chlorine can also be included owing to the use of LiCl in the SMS synthesis. Based on the possible compositions of condensation products suggested by these DFT calculations, we then presented a systematic study of stable crystal structures of g-C_3N_4 for different synthesis conditions by combining *ab initio* evolutionary searching and experimental validation.[88]

We discovered that off-plane distortion of heptazine units is a characteristic of the most stable structure corresponding to the poly-condensation synthesis in air, which explains a known discrepancy between the lattice parameters determined by X-ray diffraction (XRD) patterns and the planar structures modelled in previous studies. For the recently suggested salt-melt synthesis of g-C_3N_4, we revealed that the synthesised sample is subject to the contamination of hydrogen, chlorine, and lithium according to our detailed analysis of the crystal structures of $C_6N_9H_3$–Li_3Cl and $C_6N_9H_3$–LiCl in comparison with the measured XRD patterns of these samples. Finally, we proposed a viable synthesis pathway for purifying high-crystallinity g-C_3N_4.

The most stable structure of predicted g-C_3N_4 corresponding to the poly-condensation synthesis in air is shown in Figure 2.17(a). The calculated lattice parameters of the predicted g-C_3N_4 is consistent with experiment. This most stable g-C_3N_4 configuration, shows a unique structure in that the

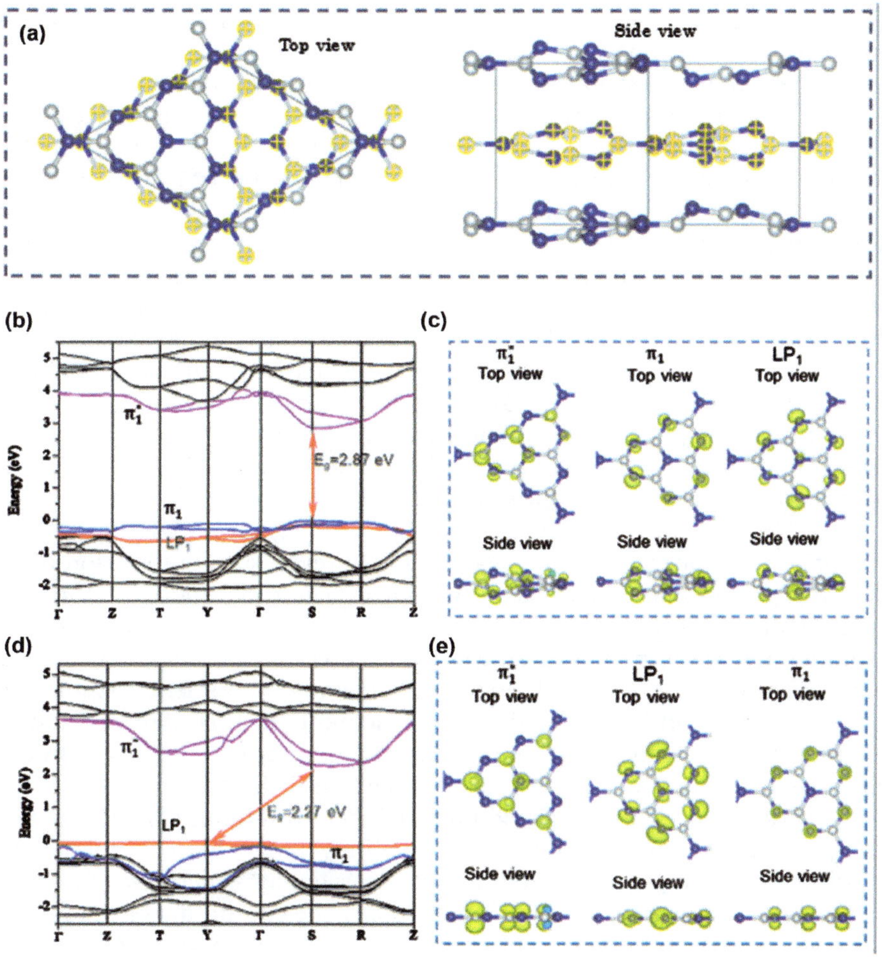

Figure 2.17 Predicted configurations and electronic structures of predicted heptazine-based g-C_3N_4. (a) Top and side views of predicted g-C_3N_4 with lowest energy; (b) Calculated band structure and (c) partial charge densities of π_1, π_1* and LP states at the VBM and the CBM of the most stable g-C_3N_4; (d) calculated band structure and (e) partial charge densities of π_1, π_1* and LP states at the VBM and the CBM of planar. Reprinted (adapted) with permission from J. Wang, D. Hao, J. Ye, N. Umezawa, *Chem. Mater.*, 2017, **29**, 2694–2707. Copyright (2017) American Chemical Society.

heptazine units in one layer are rotated by 60° with respect to the basic units in the adjacent layers. The electronic structures of the predicted g-C_3N_4 are shown in Figure 2.17(b) and (c). One can see that the predicted g-C_3N_4 possesses a direct band gap of 2.87 eV (Figure 2.17(b)) which is consistent with the experimental band gap of g-C_3N_4 synthesised by thermal condensation. Figure 2.17(b) show that the lone pair (LP) bands (LP_1 in Figure 2.17(c)) are largely shifted downward and partially hybridised with the π_1 bands (Figure 2.17(c)) compared with those of the planar configuration of g-C_3N_4 (Figure 2.17(d) and (e)). Figure 2.17(e) clearly shows that LP bands correspond to the in-plane electron distribution of peripheral nitrogen, while the electron distributions of π_1 bands are perpendicular to the C–N plane. The band hybridisation of LP_1 and π_1 bands of the most stable g-C_3N_4 shown in Figure 2.17(c) originates from the release of the LP orbitals toward the out-of-plane open space; the lobes are sticking up from the heptazine plane (Figure 2.17(c)), alleviating the Coulombic repulsive interactions among the LP orbitals. The distortion, therefore, greatly stabilises the system (~0.3 eV/C_3N_4 lower than the planar configuration), giving a clear explanation of why this structure is energetically favoured over other configurations. The on-site LP-π_1 hybridisation, which also contributes to the stabilisation of the system, is observed as resonance of the two orbitals in the band structure (Figure 2.17(b)) for the predicted g-C_3N_4. One can see that the band gap of g-C_3N_4 is dramatically opened from 2.27 eV of planar configuration (Figure 2.17(d)) to 2.87 eV of the distorted one (Figure 2.17(b)) due to the hybridisation of LP and π_1 orbitals.

The SMS high-crystalline phase has been used as an initial guess for the *ab initio* evolutionary structure searching. However, the composition of the SMS high-crystalline phase was still unclear. The experimental lattice parameters of SMS high-crystalline phase are used as input guesses and the evolutionary searching based on the adjusted compositions was continued until the criterion is satisfied. Following previous experimental studies, we assume that C and N are the dominant elements in this highly-crystalline phase in the first step, and introduce potential impurities such as hydrogen, chlorine, or lithium in the subsequent steps until the criterion is satisfied. Through a series of high-throughput *ab initio* evolutionary structure searches, we confirmed that the high-crystallinity SMS phase is not g-C_3N_4. Instead, we finally approached the most reliable structures (shown in Figure 2.18(a) and (b)) by gradually increasing the complexity of the system.

An *ab initio* evolutionary structure searching was carried out for the composition of $C_6N_9H_3 \cdot LiCl$ (40 atoms per cell), and finally the most stable configuration with space group of *Cmcm* (shown in Figure 2.18(a)) was obtained. The simulated XRD pattern of $C_6N_9H_3 \cdot LiCl$ (Figure 2.18(c)) also shows good consistency with the experimental XRD pattern of the SMS high-crystallinity sample except for the peak at 18.02°. This inconsistency may be attributed to the different symmetries of the predicted structure (*Cmcm*) and the experimental one ($P6_3cm$). In a previous research, Wirnhier *et al.*[93] found by elemental analysis that the composition of the SMS highly crystalline

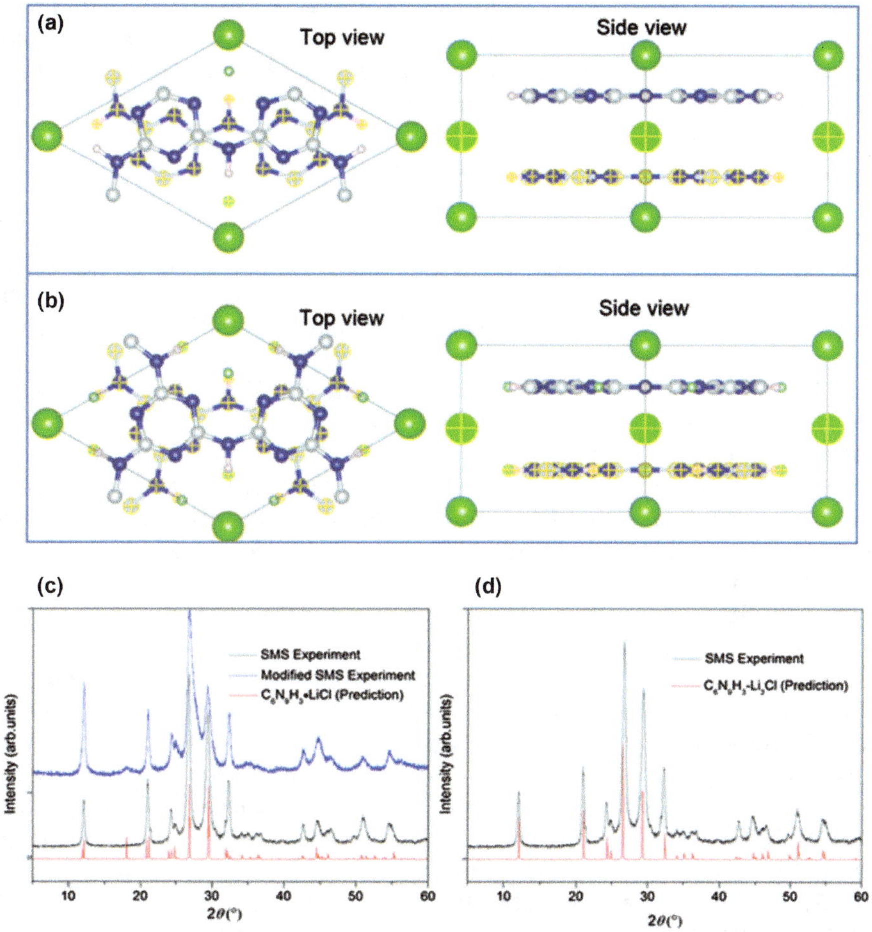

Figure 2.18 Predicted structures corresponding to the SMS condition and comparisons of the simulated and experimental XRD patterns. Predicted structures of (a) $C_6N_9H_3 \cdot LiCl$ and (b) $C_6N_9H_3 \cdot Li_3Cl$, comparison of simulated XRD patterns of (c) $C_6N_9H_3 \cdot LiCl$ and (d) $C_6N_9H_3 \cdot Li_3Cl$ with the experimental XRD pattern of the highly crystalline samples synthesised by SMS and modified SMS.
Reprinted (adapted) with permission from J. Wang, D. Hao, J. Ye, N. Umezawa, *Chem. Mater.* 2017, **29**, 2694–2707. Copyright (2017) American Chemical Society.

phase was $C_{12}N_{17.5}H_{6.3}Cl_{1.5}Li_{3.2}$. To compare with their result, our formula for the predicted structure can be rewritten as $C_{12}N_{18}H_6Cl_2Li_2$, which shows a lower Li content than the experimentally determined formula. This indicates that extra Li atoms could exist in the lattice cell and might be helpful to stabilise the $P6_3cm$ symmetry. Therefore, a series of new structure searches was carried out for the systems with the composition of $C_6N_9H_3 \cdot Li_{1+x}Cl$ ($x = 1, 2 \ldots$). Finally, we arrived at a new structure (Figure 2.18(b)) with the composition of $C_6N_9H_3 \cdot Li_3Cl$ and space group of *P*6$_3$*cm*; its simulated XRD pattern shows perfect agreement with the experimentally observed XRD pattern (Figure 2.18(d)) of the highly crystalline SMS sample. For the first time, we explicitly proved the composition of the high-crystallinity SMS sample and obtained its crystal structure.

Figure 2.18(a) and (b) show that the structures of $C_6N_9H_3 \cdot LiCl$ and $C_6N_9H_3 \cdot Li_3Cl$ are quite similar. However, because of the removal of the extra Li atoms, the overlap of the triazine rings in the $C_6N_9H_3 \cdot LiCl$ cell (Figure 2.18(a)) is not as perfect as that in the $C_6N_9H_3 \cdot Li_3Cl$ cell (Figure 2.18(b)). This deviation can be attributed to the transition from the 'micro' A–B to the 'macro' A–B stacking,[88] which can stabilise the structure by decreasing the interlayer repulsion, and lead to the change in the symmetry from *P*6$_3$*cm* to *Cmcm*. Since the concentration of Li in $C_6N_9H_3 \cdot Li_3Cl$ is three times that of Cl, it is not so stable from the point of view of charge balance. Therefore, we suspected that $C_6N_9H_3 \cdot Li_3Cl$ might be a metastable phase in comparison with $C_6N_9H_3 \cdot LiCl$. We designed a modified experimental procedure based on the method reported by Bojdys *et al.*[94] to confirm our speculation. In the modified procedure, we boiled our final sample for 5 min instead of just washing with boiling water, and then centrifuged it at 16 000 rpm for 10 min. This procedure was repeated at least 15 times. The XRD pattern of this modified SMS sample is presented in Figure 2.18(c) together with those of the SMS high-crystallinity phase and the predicted $C_6N_9H_3 \cdot LiCl$ result. We can see that a new peak appears at about 18°, which is consistent with the simulated XRD pattern of the $C_6N_9H_3 \cdot LiCl$ structure. This indicates the transition of the symmetry from *P*6$_3$*cm* to *Cmcm* and our hypothesis is validated. Therefore, we can conclude that a more stable new phase of $C_6N_9H_3 \cdot LiCl$ has been synthesised experimentally for the first time.

In addition to the determination of the most stable structure $C_6N_9H_3 \cdot LiCl$ in SMS, a series of new structures with composition of $C_6N_9H_3 \cdot HCl$, $C_6N_9H_3$ and g-C_3N_4 (triazine-based) (Figure 2.19) were identified by *ab initio* evolutionary structure searches. Therefore, we proposed a possible experimental procedure to produce highly crystalline g-C_3N_4 from $C_6N_9H_3 \cdot LiCl$ as illustrated in Figure 2.19. In this proposed procedure, the first step is to replace Li with H from $C_6N_9H_3 \cdot LiCl$. The experiment revealed that the Li ions cannot be removed by further boiling and centrifuging. Therefore, we suggest dispersing $C_6N_9H_3 \cdot LiCl$ in an aqueous HCl solution to replace Li^+ with H^+. Two possible solutions are suggested for the removal of Cl ions. The first solution is to peel the bulk $C_6N_9H_3 \cdot HCl$ into nano-sheets by soft chemistry synthesis; the Cl ions can then be easily removed by a chemical method to obtain C_2N_3H nanosheets. Graphitic carbon nitride can be synthesised through the oxidation of the C_2N_3H nanosheets. Another option is to heat $C_6N_9H_3 \cdot HCl$ in a H_2 gas flow at a medium to high temperature. The small H_2 molecules can diffuse into the $C_6N_9H_3 \cdot HCl$ to drag the Cl out in the form of HCl. The continuous gas flow can remove the HCl from the sample surface to perpetuate the reaction. Finally, highly crystalline g-C_3N_4 can be obtained by heating C_2N_3H in a proper oxidising environment.

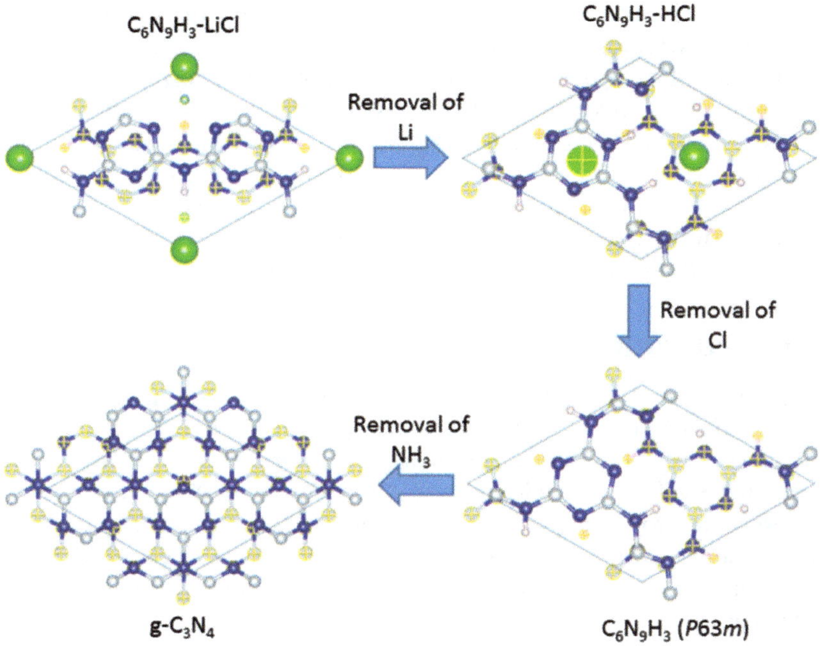

Figure 2.19 Proposed reaction pathway to synthesise highly crystalline g-C_3N_4 starting from $C_6N_9H_3 \cdot$ LiCl.

2.5 Conclusions

We present our recent researches of employing state-of-the-art computational methods to understand and search promising PEC materials. Using density functional theory (DFT) calculations, we revealed the mechanism of doping on visible light absorption, introduction of conductive carriers and band edge engineering by biaxial strain of semiconductor photocatalysts. To extend the family of photo-functional materials, high-throughput *ab initio* evolutionary structure searches were carried out and successfully predicted the structures of a series of new van der Waals materials Sn_xO_y. Moreover, the crystal structures of the well-known photo-functional materials g-C_3N_4 were finally revealed by combining evolutionary structure search and experiment validation. For further discovery of novel PEC materials from computational methods, one potential direction should be an inverse materials design for desired properties by means of machine learning technique applied on training set database. Such an approach should allow us to carry out systematic materials screening for excellent PEC materials. In addition to the development of such an automated materials discovery tool, deepening physical insight into photocatalysis phenomena is also important. For example, the role of exciton in photocatalysis process is insufficiently understood and the clarification of its dynamics should offer an insight into the carrier separation mechanism, leading to a new design principle. In particular, excitons are relatively stable in low dimensional systems such as nanosheets due to the limited electronic screening effect and thus should significantly affect their photocatalytic performance. Finally, more focused research on surface chemistry should be promoted for understanding overall photocatalysis reactions. A comprehensive study that reveals the correlation between surface morphology and reactivity under the supply of photo-excited carriers remains elusive. A theory based on atomistic scale simulation should provide us with new concepts that go beyond the d-band theory widely used in the catalysis field.

Acknowledgements

Junjie Wang is an International Research Fellow of the Japan Society for the Promotion of Science (JSPS). JJW acknowledge financial support from JSPS through project P14207. This work was partly supported by the Core Research for Evolutional Science and Technology (CREST) program, Precursory Research for Embryonic Science and Technology (PRESTO) program of Japan Science and Technology Agency (JST), and by the World Premier International Research Center Initiative on Materials Nanoarchitectonics (MANA), MEXT.

References

1. A. Fujishima and K. Honda, *Nature*, 1972, **238**, 37–38.
2. I. E. Castelli, D. D. Landis, K. S. Thygesen, S. Dahl, I. Chorkendorff, T. F. Jaramillo and K. W. Jacobsen, *Energy Environ. Sci.*, 2012, **5**, 9034–9043.
3. Y. Wu, P. Lazic, G. Hautier, K. Persson and G. Ceder, *Energy Environ. Sci.*, 2013, **6**, 157–168.
4. H. L. Zhuang and R. G. Henning, *Chem. Mater.*, 2013, **25**, 3232–3238.
5. H. L. Zhuang and R. G. Henning, *J. Phys. Chem. C*, 2013, **117**, 20440–20445.
6. Y. Li, Y.-L. Li, C. M. Araujo, W. Luo and R. Ahuja, *Catal. Sci. Technol.*, 2013, **3**, 2214–2220.
7. A. K. Singh, K. Mathew, H. L. Zhuang and R. G. Henning, *J. Phys. Chem. Lett.*, 2015, **6**, 1087–1098.
8. Y. Li, Y.-L. Li, B. Sa and R. Ahuja, *Catal. Sci. Technol.*, 2017, **7**, 545–559.
9. S. Yang, D. Prendergast and J. B. Neaton, *Nano Lett.*, 2012, **12**, 383–388.
10. J. Cheng and M. Sprik, *Phys. Chem. Chem. Phys.*, 2012, **14**, 11245–11267.
11. J. Chang, X. Liu, J. A. Kattirtzi, J. VandeVondele and M. Sprik, *Angew. Chem., Int. Ed.*, 2014, **53**, 12046–12050.
12. R. Crespo-Otero and A. Walsh, *J. Phys. Chem. Lett.*, 2015, **6**, 2379–2383.
13. J. Liao, B. Sa, J. Zhou, R. Ahuja and Z. Sun, *J. Phys. Chem. C*, 2014, **118**, 17594–17599.
14. V. Stevanovic, S. Lany, D. S. Ginley, W. Tumas and A. Zunger, *Phys. Chem. Chem. Phys.*, 2014, **16**, 3706–3714.
15. C. D. Valentin and G. Pacchioni, *Catal. Today*, 2013, **206**, 12–18.
16. N. Umezawa and J. Ye, *Phys. Chem. Chem. Phys.*, 2012, **14**, 5924–5934.

17. J. B. Varley, A. Janotti and C. G. Van de Walle, *Adv. Mater.*, 2011, **23**, 2343–2347.
18. W.-J. Yin, S.-H. Wei, M. M. Al-Jassim and Y. Yan, *Phys. Rev. Lett.*, 2011, **106**, 066801.
19. R. Asahi, T. Morikawa, T. Ohwaki, K. Aoki and Y. Taga, *Science*, 2001, **293**, 269–271.
20. I. S. Cho, C. H. Lee, Y. Feng, M. Logar, P. M. Rao, L. Cai, D. R. Kim, R. Sinclair and X. Zheng, *Nat. Commun.*, DOI: 10.1038/ncomms2729.
21. T. W. Kim, Y. Ping, G. A. Galli and K.-S. Choi, *Nat. Commun.*, DOI: 10.1038/ncomms9769.
22. P. Liao, J. A. Keith and E. A. Cater, *J. Am. Chem. Soc.*, 2012, **134**, 13296–13309.
23. X. Chen, L. Liu, P. Y. Yu and S. S. Mao, *Science*, 2011, **331**, 746.
24. L. Liu, P. Y. Yu, X. Chen, S. S. Mao and D. Z. Shen, *Phys. Rev. Lett.*, 2013, **111**, 065505.
25. A. Kudo and Y. Miseki, *Chem. Soc. Rev.*, 2009, **38**, 253–278.
26. J. Ravichandra, W. Siemons, H. Heijmerikx, M. Huijben, A. Majumadar and R. Ramesh, *Chem. Mater.*, 2010, **22**, 3983–3987.
27. S. Kawasaki, R. Takahashi, K. Akagi, J. Yoshinobu, F. Komori, K. Horiba, H. Kumigashira, K. Iwashita, A. Kudo and M. Lippmaa, *J. Phys. Chem. C*, 2014, **118**, 20222–20228.
28. H.-C. Chen, C.-W. Huang, J. C. S. Wu and S.-T. Lin, *J. Phys. Chem. C*, 2012, **116**, 7897–7903.
29. A. Janotti, J. B. Varley, P. Rinke, N. Umezawa, G. Kresse and C. G. Van de Walle, *Phys. Rev. B*, 2010, **81**, 085212-1–085212-7.
30. P. Reunchan, N. Umezawa, A. Janotti, J. T-Thienprasert and S. Limpijumnong, *Phys Rev. B*, 2017, **95**, 205204-1–205204-8.
31. H. Tong, S. Ouyang, Y. Bi, N. Umezawa, M. Oshikiri and J. Ye, *Adv. Mater.*, 2012, **24**, 229–251.
32. M. A. Henderson and I. Lyubinetsky, *Chem. Rev.*, 2013, **113**, 4428–4455.
33. M. Plaza, X. Huang, J. Y. Peter Ko, M. Shen, B. H. Simpson, J. Rodriguez-Lopez, N. L. Ritzert, K. Letchworth-Weaver, D. Gunceler, D. G. Schlom, T. A. Arias, J. D. Brock and H. D. Abruna, *J. Am. Chem. Soc.*, 2016, **138**, 7816–7819.
34. Y. Wang, H. Sun, S. Tan, H. Feng, Z. Cheng, J. Zhao, A. Zhao, B. Wang, Y. Luo, J. Yang and J. G. Hou, *Nat. Commun.*, DOI: 10.1038/ncomms3214.
35. M. Setvin, U. Aschauer, P. Scheiber, Y.-F. Li, W. Hou, M. Schmid, A. Selloni and U. Diebold, *Science*, 2013, **341**, 988–991.
36. C. Li, C. Koenigsmann, W. Ding, B. Rudshteyn, K. R. Yang, K. P. Regan, S. J. Konezny, V. S. Batista, G. W. Brudvig, C. A. Schmuttenmaer and J.-H. Kim, *J. Am. Chem. Soc.*, 2015, **137**, 1520–1529.
37. H. Xu, P. Reunchan, S. Ouyang, H. Tong, N. Umezawa, T. Kako and J. Ye, *Chem. Mater.*, 2013, **25**, 405–411.
38. Y. Bi, S. Ouyang, N. Umezawa, J. Cao and J. Ye, *J. Am. Chem. Soc.*, 2011, **133**, 6490–6492.
39. D. J. Martin, N. Umezawa, X. Chen, J. Ye and J. Tang, *Energy Environ. Sci.*, 2013, **6**, 3380–3386.

40. L. I. Bendavid and E. A. Carter, *J. Phys. Chem. C*, 2013, **117**, 15750–15760.
41. J. Chen, Y.-F. Li, P. Sit and A. Selloni, *J. Am. Chem. Soc.*, 2013, **135**, 18774–18777.
42. A. Tilocca and A. Selloni, *J. Phys. Chem. C*, 2012, **116**, 9114–9121.
43. S. Wendt, P. T. Sprunger, E. Lira, G. K. H. Madsen, Z. Li, J. Ø. Hansen, J. Matthiesen, A. Blekinge-Rasmussen, E. Lægsgaard, B. Hammer and F. Besenbacher, *Science*, 2008, **320**, 1755–1759.
44. N. Yang, Y. Liu, H. Wen, Z. Tang, H. Zhao, Y. Li and D. Wang, *ACS Nano*, 2013, **7**, 1504–1512.
45. N. Umezawa, H. H. Kristoffersen, L. B. Vilhelmsen and B. Hammer, *J. Phys. Chem. C*, 2016, **120**, 9160–9164.
46. H. Kawazoe, H. Yanagi, K. Ueda and H. Hosono, *MRS Bull.*, 2000, **25**, 28–36.
47. P. Li, N. Umezawa, H. Abe and J. Ye, *J. Mater. Chem. A*, 2015, **3**, 10720–10723.
48. C. F. Fischer, *The Hartree-Fock Method for Atoms, A Numerical Approach*, A Wiley-Interscience Publication John Wiley & Sons, New York, 1977.
49. K. A. Persson, B. Waldwick, P. Lazic and G. Ceder, *Phys. Rev. B*, 2012, **85**, 235438-1–235438-11.
50. Á. Valdés, J. Brillet, M. Grätzel, H. Gudmundsdóttir, H. A. Hansen, H. Jónsson, P. Klüpfel, G.-J. Kroes, F. L. Formal, I. C. Man, R. S. Martins, J. K. Nørskov, J. Rossmeisl, K. Sivula, A. Vojvodic and M. Zäch, *Phys. Chem. Chem. Phys.*, 2012, **14**, 49–70.
51. S. Chen and L.-W. Wang, *Chem. Mater.*, 2012, **24**, 3659–3666.
52. P. Reunchan, N. Umezawa, S. Ouyang and J. Ye, *Phys. Chem. Chem. Phys.*, 2012, **14**, 1876.
53. P. Reunchan and N. Umezawa, *J. Phys. Chem. C*, 2015, **119**, 2284.
54. Y. Xu and M. A. A. Schoonen, *Am. Mineral.*, 2000, **85**, 543.
55. M. S. Wrighton, *et al.*, *J. Am. Chem. Soc.*, 1976, **98**, 2774.
56. A. K. Ghosh and H. P. Maruska, *J. Electrochem. Soc.*, 1977, **124**, 1516.
57. H. Kato and A. Kudo, *J. Phys. Chem. B*, 2002, **106**, 5029.
58. D. Wang, J. Ye, T. Kako and T. Kimura, *J. Phys. Chem. B*, 2006, **110**, 15824.
59. T. Ishii, H. Kato and A. Kudo, *J. Photochem. Photobiol., A*, 2004, **163**, 181.
60. J. W. Liu, G. Chen, Z. H. Li and Z. G. Zhang, *J. Solid State Chem.*, 2006, **179**, 3704.
61. P. Reunchan, *et al.*, *J. Mater. Chem. A*, 2013, **1**, 4221.
62. Z. Yi, *et al.*, *Nat. Mater.*, 2010, **9**, 559.
63. Y. Bi, S. Ouyang, N. Umezawa, J. Cao and J. Ye, *J. Am. Chem. Soc.*, 2011, **133**, 6490.
64. W. Yao, *et al.*, *J. Mater. Chem.*, 2012, **22**, 4050.
65. G. Botelho, J. Andres, L. Gracia, L. S. Matos and E. Longo, *ChemPlusChem*, 2016, **81**, 202.
66. N. Umezawa, O. Shuxin and J. Ye, *Phys. Rev. B*, 2011, **83**, 035202.
67. J. J. Liu, X. L. Fu, S. F. Chen and Y. F. Zhu, *Appl. Phys. Lett.*, 2011, **99**, 191903.
68. P. Reunchan and N. Umezawa, *Phys. Rev. B*, 2013, **87**, 245205.
69. J. M. Kahk, *et al.*, *J. Mater. Chem. A*, 2014, **2**, 6092.

70. A. J. Nozik and R. Memming, *J. Phys. Chem.*, 1996, **100**, 13061.
71. E. Ungersboeck, S. Dhar, G. Karlowatz, V. Sverdlov, H. Kosina and S. Selberherr, *IEEE Trans. Electron Devices*, 2007, **54**, 2183–2190.
72. V. Swamy, A. Y. Kuznetsov, L. S. Dubrovinsky, A. Kurnosov and V. B. Prakapenka, *Phys. Rev. Lett.*, 2009, **103**, 075505.
73. A. Janotti, D. Steiauf and C. G. Van de Walle, *Phys. Rev. B*, 2011, **84**, 201304.
74. W. Zhou and N. Umezawa, *Phys. Chem. Chem. Phys.*, 2016, **18**, 7860.
75. W. Zhou and N. Umezawa, *Phys. Chem. Chem. Phys.*, 2015, **17**, 17816.
76. Y. Sun, H. Cheng, S. Gao, Z. Sun, Q. Liu, Q. Liu, F. Lei, T. Yao, J. He, S. Wei and Y. Xie, *Angew. Chem., Int. Ed.*, 2012, **51**, 8727–8731.
77. H. L. Zhang and R. G. Henning, *Phys. Rev. B*, 2013, **88**, 115314.
78. Y. Idota, T. Kubota, A. Matsufuji, Y. Mackawa and T. Miyasaka, *Science*, 1997, **176**, 1395.
79. H. Hosono, Y. Ogo, H. Yanagi and T. Kamiya, *Electrochem. Solid-State Lett.*, 2011, **14**, H13.
80. K. Govaerts, R. Saniz, B. Partoens and D. Lamoen, *Phys. Rev. B*, 2013, **87**, 235210.
81. J. P. Allen, D. O. Scanlon, S. C. Parker and G. W. Watson, *J. Phys. Chem. C*, 2011, **115**, 19916.
82. M. K. Forthaus, K. Sengupta, O. Heyer, N. E. Christensen, A. Svane, K. Syassen, D. I. Khomskii, T. Lorenz and M. M. Abd-Elmeguid, *Phys. Rev. Lett.*, 2010, **105**, 157001.
83. A. R. Oganov and C. W. Glass, *J. Chem. Phys.*, 2006, **124**, 244704.
84. A. R. Oganov, A. O. Lyakhov and M. Valle, *Acc. Chem. Res.*, 2011, **44**, 227.
85. G. Kresse and J. Hafner, *Phys. Rev. B*, 1994, **49**, 14251.
86. G. Kresse and J. Furthmuller, *Comput. Mater. Sci.*, 1996, **6**, 15.
87. J. Wang, N. Umezawa and H. Hosono, *Adv. Energy Mater.*, 2016, **6**, 1501190.
88. J. Wang, D. Hao, J. Ye and N. Umezawa, *Chem. Mater.*, 2017, **29**(7), 2694–2707.
89. M. A. Mäki-Jaskari and T. T. Rantala, *Modell. Simul. Mater. Sci. Eng.*, 2004, **12**, 33.
90. A. Seko, A. Togo, F. Oba and I. Tanaka, *Phys. Rev. Lett.*, 2008, **100**, 045702.
91. M. Manikandan, T. Tanabe, P. Li, S. Ueda, G. V. Ramesh, R. Kodiyath, J. Wang, T. Hara, A. Dakshanamoorthy, S. Ishihara, K. Ariga, J. Ye, N. Umezawa and H. Abe, *ACS Appl. Mater. Interfaces*, 2014, **6**, 3790–3793.
92. X. C. Wang, K. Maeda, A. Thomas, K. Takanabe, G. Xin, J. M. Carlsson, K. Domen and M. Antonietti, *Nat. Mater.*, 2009, **8**, 76–80.
93. E. Wirnhier, M. Döblinger, D. Gunzelmann, J. Senker, B. V. Lotsch and W. Schnick, *Chem. - Eur. J.*, 2011, **17**, 3213–3221.
94. M. J. Bojdys, J.-O. Müller, M. Antonietti and A. Thomas, *Chem. - Eur. J.*, 2008, **14**, 8177–8182.

CHAPTER 3

Computational Screening of Light-absorbing Materials for Photoelectrochemical Water Splitting

IVANO E. CASTELLI,*[a,†] KORINA KUHAR,[b] MOHNISH PANDEY[b] AND KARSTEN W. JACOBSEN[b]

[a] Department of Chemistry, University of Copenhagen, DK-2100 Copenhagen Ø, Denmark; [b] CAMD, Department of Physics, Technical University of Denmark, DK-2800 Kgs. Lyngby, Denmark
*Email: ivca@dtu.dk

3.1 Introduction

The production of green energy is one of the necessary means to address the issues of increased energy demand, finite fossil fuel resources, and increase in pollution. Photovoltaics (PV) and photoelectrochemical (PEC) water splitting are two ways for converting the energy from the sun into electricity or chemical energy, respectively. Both technologies rely on the absorption of a large fraction of the photons coming from the Sun and the generation of electron–hole pairs that can be used directly as electricity or to induce chemical processes. In the former case, storage of the electrical energy

[†]Present address: Department of Energy Conversion and Storage, Technical University of Denmark, DK-2800 Kgs. Lyngby, Denmark

Energy and Environment Series No. 20
Advances in Photoelectrochemical Water Splitting: Theory, Experiment and Systems Analysis
Edited by S. David Tilley, Stephan Lany and Roel van de Krol

Published by the Royal Society of Chemistry, www.rsc.org

requires the use of batteries where capacity and cost pose challenges. In the latter case, the energy is directly stored in chemical bonds in fuels like for example hydrogen, which can then be transported and used as required.

A fundamental requirement for both technologies is the efficient absorption of a significant fraction of the solar spectrum. The search for new materials that can push the limits and achieve higher efficiencies has therefore been the subject of investigation for the last half century leading to very significant advances. However, the search has mostly been based on trial-and-error combined, of course, with chemical knowledge and intuition about material properties. The focus here is to discuss how computational screening may contribute to the advancement of a more systematic approach to identifying new energy materials in particular for water splitting.

The basic principle of a photoelectrochemical water splitting device is the generation of electron–hole pairs using photons. Once the charges are generated, they have to reach the surface of the light-absorbing materials without recombining and induce hydrogen and oxygen evolution leading to the net splitting of water into its constituents. The water splitting process using solar light was achieved for the first time by Fujishima and Honda in 1972 using TiO_2.[1]

The bare free energy to split water is 1.23 eV per molecule, but with the current technology a number of losses have to be added to this. The hydrogen and oxygen evolution processes have overpotentials associated with them resulting in energy losses (around 0.1 and 0.4 eV, respectively[2]). Furthermore, the photo-excited electrons and holes are not in equilibrium, and therefore the Fermi level is split into electron and hole quasi-Fermi levels giving rise to an additional loss of about 0.25 eV for each band edge.[3] The net result is that the band gap of the material used to generate the charges should be significantly larger than the 1.23 eV, rather of the order of 2.2 eV. Because of the size of the band gap, the maximum theoretical efficiency that can be achieved by this type of device is only around 7%.[4]

One way to increase the efficiency is to build a device composed of more than one light-harvesting material, so that two or more photons are used to build-up the energy required to split water. In the two-photon or tandem device, the first material absorbs the high energy part of the solar spectrum while the remaining photons go through to be absorbed by the second photoactive material with a smaller band gap. Usually the hydrogen evolution process is taking place on the side with the small band gap (SBG) material and the oxygen evolution on the side with the large band gap (LBG) material as shown at Design 1 in Figure 3.1(a). It has been shown,[6–8] that an efficiency as high as 29% in principle can be obtained by combining two gaps of around 1.0 and 1.7 eV.[4,7,9–11] Silicon can work as a small band gap material but, as of today, materials with an optimal matching between the size of the band gap and the position of the band edges, *i.e.* the potential of the photogenerated charges, which should straddle the redox levels of water, have not been found yet. It has recently been discussed,[5] that an alternative design (Design 2 in Figure 3.1(b)), where the hydrogen evolution takes place on the large band gap side and the oxygen is formed on the small band gap side should also be considered. In fact, a computational screening study based on the Inorganic Crystal Structure Database (ICSD)[12] and the Materials Project database[13] indicates that the material requirements for Design 2 seem to be considerably easier to fulfill.[5,14]

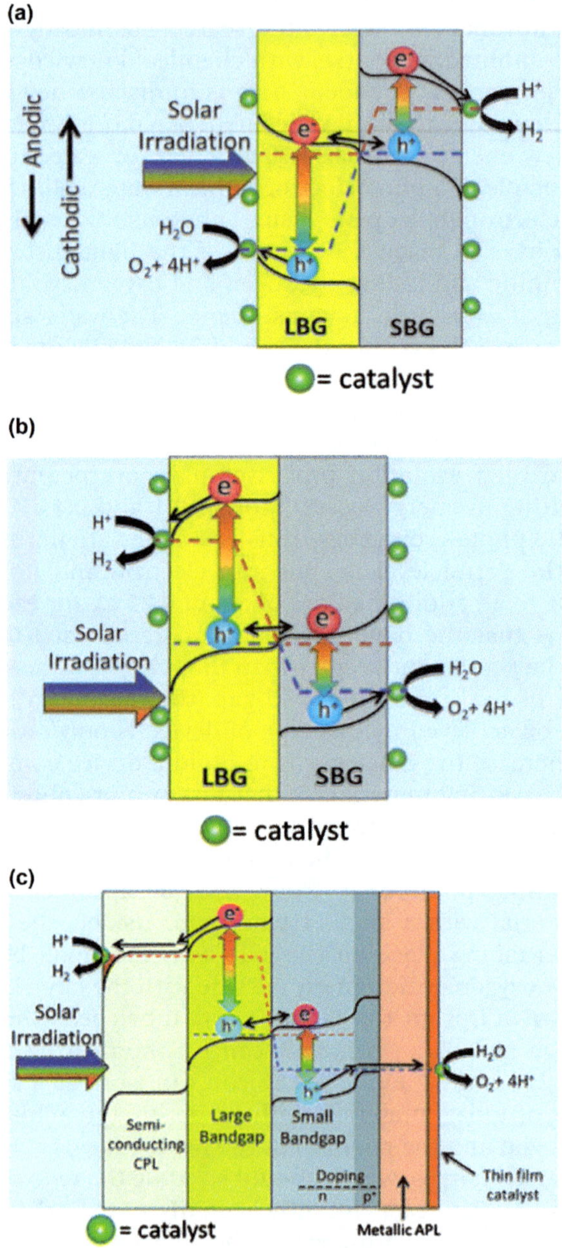

However, a more realistic device, as sketched in Figure 3.1(c), may not only involve the light absorbing materials but also protection layers for the anode and cathode, so that the semiconductors are shielded from the water. Furthermore, catalyst layers or particles will be necessary to lower the overpotentials for the hydrogen and oxygen evolution processes. The device may be fully integrated as shown in Figure 3.1(c) or split into photovoltaic and electrolysis components.[15] In any case, such a device involves many different types of materials each of which should possess appropriate photoelectric, chemical, and/or mechanical properties. Computational screening is likely to play an important role in the identification of relevant new materials with tuned properties. This we shall discuss in the following with particular emphasis on light absorbing semiconductors.

3.2 Density Functional Theory and High-throughput Screening

The main approach to calculating material properties from quantum mechanical principles is Density Functional Theory (DFT).[16,17] In particular for computational screening studies, DFT realises a convenient compromise between reliability and computational speed. The theory was developed in the 1960s as a formally exact approach to calculating total energies of electronic systems. However, a key quantity in the theory, the so-called exchange-correlation energy, is not known exactly. Since then, numerous approximations to the exchange-correlation energy have been suggested, providing better accuracy and improved predictive power of the approach.[18,19] At the same time the numerical techniques are continuously improved,[20] and the computers have grown tremendously in memory and speed allowing for the treatment of more complex and realistic atomic systems.

A standard DFT calculation for a material takes as input the chemical element types and real space positions of all the atoms, often conveniently described with a periodically repeated unit cell. Based on this information the total energy can be calculated and, for example, minimised with respect to the atomic positions to find a low-temperature stable structure. A number of (opto-)electronic, mechanical, and chemical properties can then be calculated.

Direct application of DFT calculations therefore requires knowledge of which chemical elements are present in the system and also information

Figure 3.1 Different designs for a two-photon water splitting device. (a) The large band gap material (LBG) is the photoanode and the small band gap material (SBG) the photocathode. *Vice versa* in (b), LBG is the photocathode and SBG the photoanode. (c) Complete structure of a two-photon water splitting device with a transparent conductive protective layer for the cathode (CPL), a large band gap photocathode, a small band gap photoanode, (with a *p–n* homojunction), and a metallic protecting layer for the anode.
Reproduced from ref. 5 with permission from The Royal Society of Chemistry.

about relevant crystal structures. The design problem (or inverse problem), however, turns this approach around in the sense that we do not start with a known material. On the contrary, we want to find a material with some particular properties without knowing the building blocks—the atoms and their positions—of the material in advance. This means that we need a search or screening strategy to identify relevant materials as we shall return to below.

More importantly, we need to know what to calculate. For a water splitting device a clear aim is to get as high a solar-to-hydrogen conversion efficiency as possible, however, this is a complicated quantity which cannot directly be computed. We, therefore, need to break down the challenge into smaller problems and identify quantities that we are able to calculate at the atomic scale. These so-called *descriptors* bridge the calculated microscopic quantities with the macroscopic properties measured in experiments.[21,22] One example of a descriptor could be the electron effective mass. It can be computed at the atomic level since it can be obtained from a band structure calculation, and it plays an important role for the charge mobility. We shall discuss descriptors for water splitting devices in more detail in the next section. The descriptor-based search has been used to suggest novel materials for batteries,[23] for carbon capture and storage,[24] stable binary and ternary alloys,[25] electrocatalysts,[26] transparent conductors,[27] dye sensitised solar cells,[28] photovoltaics,[29,30] and water splitting materials.[10,31,32]

Once the descriptors have been defined, a material design or high-throughput screening project can proceed in different ways. On the most simple level, a crystal structure is defined as a template and different chemical elements are tried in the various crystallographic positions. This 'brute force' approach does not show a particularly high rate of success because each calculation is performed without learning from the data collected until that point. Despite this, it allows to study trends because it generates a complete set of data and it is a good starting point for more sophisticated tools that can help in achieving a higher rate of success. The trends can be generalised to design structural or chemical rules that reduce the space of investigation, for example excluding combinations of chemical elements which cannot produce interesting materials. For example, for the ABO_3 perovskite structures that will be discussed later in Section 3.4.1, investigation of a large number of systems lead to the recognition that stable semiconductors in the perovskite structure are only formed if the A-cation is larger than the B-anion and if the valences add up to match the oxygen valence.[33–35]

Given a set of descriptors the screening process can also be viewed as a high-dimensional optimisation problem where a—possibly multi-dimensional—fitness function is optimised.[35] Optimisation schemes, like minima hopping and metadynamics, can be used to investigate different stable structures at a fixed composition.[36,37] More generally genetic algorithms can be applied to explore the materials space. A genetic algorithm requires definition of mutation and crossover operators, which define how new materials for consideration are being generated. Therefore, the efficiency of the approach depends sensitively on the particular implementation of the mutation and crossover operations. Genetic algorithms have been used—among many other applications—to search for stable alloy compositions[38] and are extensively used for determination of crystal structures.[39]

Recent years have seen a tremendous development in the data sciences where new techniques to analyse, describe, and predict data have been introduced. These techniques are currently being taken up and used by the materials science community, so that the properties of new materials can be predicted based on correlations in existing materials data.[40] This development goes hand in hand with the establishment of large computational materials databases like the NOMAD repository,[41] the Materials Project database,[13] the AFLOWLIB,[42] the Open Quantum Materials Database,[43] the NREL Materials Database[44] and the Computational Materials Repository (CMR).[45,46] In the future these databases can be expected to be the starting point for many computational screening studies.

Before going into the descriptors of particular interest for light absorption and water splitting we now discuss an example of a *screening funnel*. As mentioned above a computational screening approach typically involve a number of requirements that the materials will have to obey as expressed through appropriate descriptors. However, the different descriptors may not be equally important for the performance of the device and, what is equally important in this context, they may take very different amounts of computer resources to calculate. This means that it is beneficial to organise the screening so that the number of materials, which should be taken into consideration, is gradually reduced by considering the most easily computed descriptors first.

An example of how the number of structures investigated gets reduced step by step is shown in the funnel in Figure 3.2 which could be applied to the problem of finding light-absorbing materials for photovoltaic and water splitting applications, by exploring compounds of a particular, say $A_xB_yC_z$, stoichiometry. At each step a new criterion is introduced, the properties of the candidates are calculated and the candidates that do not satisfy the given criterion are removed from the study. The first couple of criteria do not require DFT calculations and are therefore very fast. In the first step a number of chemical elements are excluded, because they posses unfavourable properties like being toxic or very rare. Second, chemical rules are applied. For example all the compositions where the valences do not add up to zero, including all known possible valences for the elements, are removed. This step excludes metals and narrows down the group of possible materials significantly.

The next step could be to investigate with DFT the remaining compositions in a number of different crystal structures of relevance to the particular composition. The structures are checked for stability and only the stable candidates are kept in the pool of potentially interesting materials.

Further down the funnel, we find criteria related to the electronic properties of the materials like the band gap and the effective masses. These properties require detailed DFT calculations or even higher-level calculations performed with many-body perturbation theory. The last item in the sketched funnel involves DFT calculations of defect properties. Such

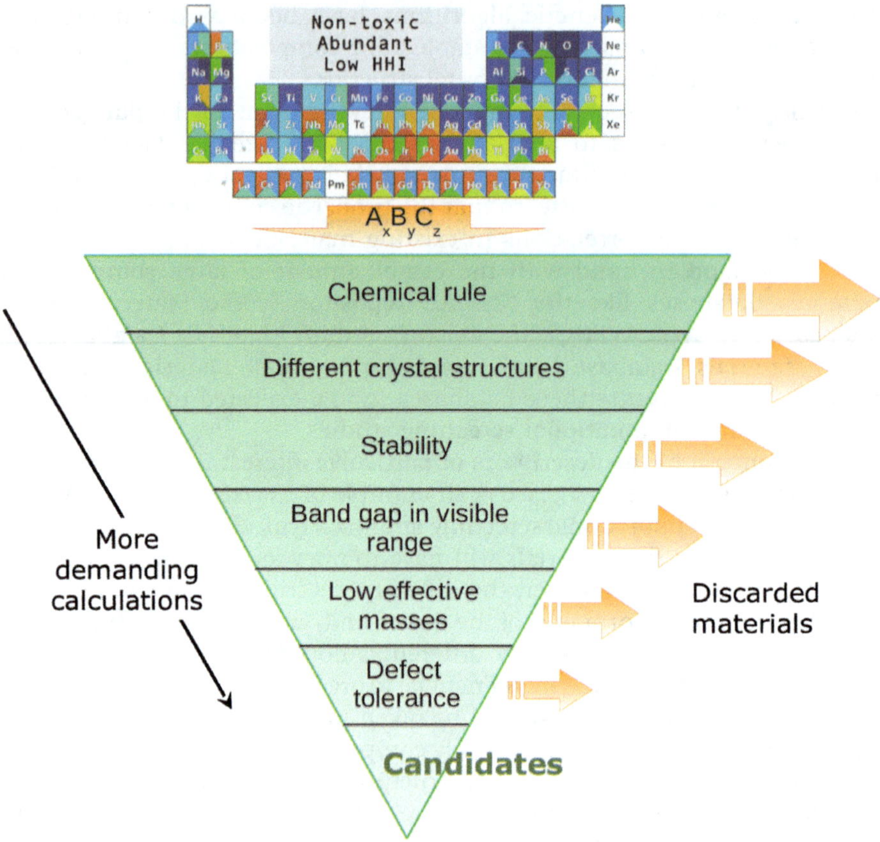

Figure 3.2 Illustration of a funnel screening approach: starting with a large number of candidates, the selection is reduced every time a new property is investigated. As moving towards the bottom of the funnel, the number of candidates decreases, and the cost of the DFT calculations increases.

calculations require consideration of large unit cells with many atoms and are feasible only for a small subset of materials. Therefore they are considered only late in the screening process.

In the end the funnel leads to identification of a small number of materials with desirable properties, which can be considered as interesting candidates for further theoretical or experimental investigation.

Funnels, like the one showed here, can be constructed for many computational screening challenges and provide a convenient way to organise the screening in an optimal fashion.

3.3 Screening Descriptors and Criteria

Defining descriptors and criteria is a key step in any high-throughput screening study. In this section, we describe some of the most important ones. We can distinguish between different categories of criteria, whether they depend on (i) the properties of the constituent atoms (abundance/cost, and toxicity), (ii) the crystal structure (stability), (iii) the electronic properties, (iv) if they are related with the direct calculation of the light absorption, and (v) with the properties of interfaces.

3.3.1 Abundance, Cost and Herfindahl–Hirschman Index

One of the first things to consider is which kind of chemical elements to include in a screening. This choice can, for example, be based on the cost of the pure elements, their abundance, and/or their toxicity. This criterion also allows us to reduce the chemical space of investigation by removing some elements that might not be easily treated experimentally. If we aim to produce cheap energy, a material for which the pure elements are too expensive, will not be feasible. To a first approximation, the cost of a compound can be estimated as the sum of the costs of its pure constituents weighted by the stoichiometry. Usually elements in their pure form are more expensive than compounds, like their oxide form, but this somehow takes into consideration also the manufacture cost. Prices are in general subjected to the market and they might change dramatically as it happened for the price of rhodium which dropped from more than 300 USD g^{-1} at the end of may 2008 to less than 50 USD g^{-1} three months later.[47]

If we want to avoid the fluctuations due to the market, we can consider the estimated abundance of chemicals in the Earth's crust.[48] 99% of the crust is composed of only a few elements (oxygen 46%, silicon 28%, aluminum 8.2%, iron 5.6%, calcium 4.2%, sodium 2.5%, magnesium 2.4%, potassium 2.0%, and titanium 0.61%) and the other elements contribute for less than 0.15% each. The composition of the crust does not always correlate with the cost of the element. Copper or tungsten, for example, show an abundance of 0.0066% and 0.0001%, respectively, but their cost is rather low (around 6 and 25 USD kg^{-1}).[47] However, the more abundant metals barium or strontium (0.034% and 0.036%) have higher costs (550 USD kg^{-1} and 1000 USD kg^{-1}).[49] Another example is the rare-earth metals that even though they are not too scarce, they have a high cost due to their relatively low production.

Considering the production of a pure element is, indeed, a better indicator whether an element can be used on a large scale or not. Vesborg and Jaramillo[50] have investigated the global annual production of elements and suggest that only elements with a production larger than 33 kt per year are feasible for large scale application. The yearly production of the different elements is indicated with colours in the x- and y-axis of Figure 3.6.

Another possible indicator is the Herfindahl-Hirschman index (HHI), that determines how unevenly distributed the production is. Economically speaking, it is a measure of the market concentration. In case of chemical elements, the HHI is determined not only by the production and cost, but also the geographical provenance, the current political situation and other factors.[51]

Figure 3.3 shows the HHI for production and reserve for most of the elements in the periodic table as well as their abundance (or scarcity). We

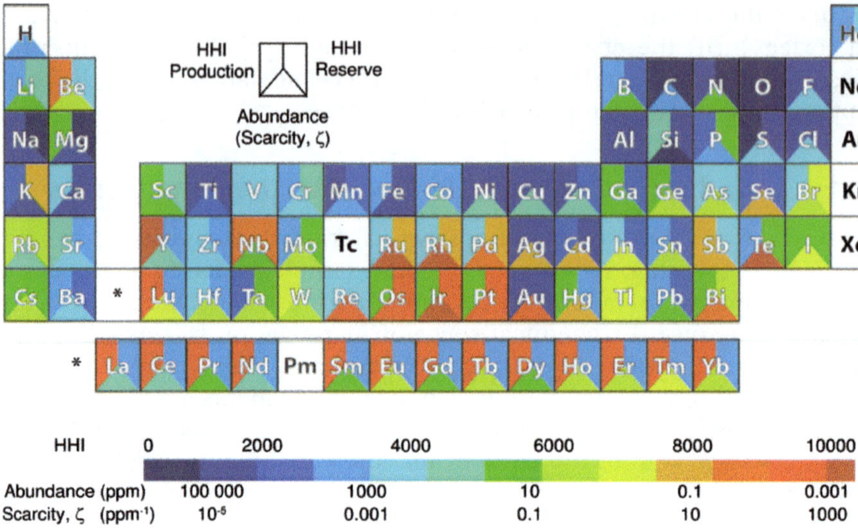

Figure 3.3 HHI for productions and reserves and abundance for most of the elements in the periodic table. Optimal elements for a 'green' device are indicated in tones of blue.
Reprinted with permission from Gaultois *et al.*, *Chemistry of Materials*, 2013, **25**, 2911.[51] Copyright 2013 American Chemical Society.

can identify some elements in different tones of blue, that have a low HHI and are abundant. They are thus optimal to be used in a 'green' device. There are elements, instead, for which the HHIs are low, but they are rather scarce, like Ta, Ru, and Rh. These elements can be used, for example, to make high-impact discoveries of novel materials, understand trends and mechanisms, but they should be replaced by more abundant elements if the device has to be up-scaled. There are elements like K, Os, Ir, and Pt, that have a reasonably low HHI for the production, but a very high HHI for the reserves. This means that these elements are in 'danger', so a technology based only on them might not be sustainable in the long run.

These and other considerations should be taken into account when designing a screening protocol, because the choice of the chemical elements involved largely depends on the aim of the project. For example, in the screening of novel cubic perovskites, we have considered all non-radioactive elements (excluding rare-earth) because one of the goals was to study trends on top of searching for novel compounds (Section 3.4.1). In a later screening, in which we calculated the band gaps of known materials from databases (Section 3.4.2), we have selected only abundant or low HHI elements because the focus was on identifying optimal compounds.

3.3.2 Toxicity

To achieve the goal of clean/green energy, and also in general for mass production of a device, it is important to consider issues about toxicity of the constituent chemical elements. For example it may be preferable to remove radioactive and highly toxic chemicals from the set of elements to use in a screening. One interesting case is the element lead. Hybrid lead halide perovskites have recently shown extremely high performances when used in a photovoltaic device.[52] However, the presence of lead may be unwanted and the current regulation is forbidding their use in a urban setting.[53] Yet, careful engineering of a device (encapsulation, protection layers *etc.*) might be able to solve some of the issues related to safety.

3.3.3 Stability

Independent of the application, stability is one of the most important requirements for a candidate compound. Here, we distinguish between different levels of stability: (i) stability against phase changes of the material, (ii) stability against segregation into different materials, and (iii) stability against dissolution in water. These can all be investigated in high-throughput screenings using DFT, and for the dissolution case using a combination of DFT and experimental dissolution energies. Convenient tools for such studies are for example implemented in the ASE module |ase.phasediagram|.[54,55] The investigation of formation energies and stability of defects (vacancies or dopants) requires more computational power and cannot presently be carried out for large classes of materials.

- *Segregation/decomposition*: the stability of a given material against segregation or decomposition into other materials or phases at low temperatures can be investigated by construction of the so-called convex hull. The competing materials can typically be taken from experimental or theoretical databases, like the Inorganic Crystal Structure Database (ICSD),[12] Material Project,[13] or Open Quantum Materials Database (OQMD).[43] The convex hull defines which material or combination of materials, at a given composition, has the lowest energy, *i.e.* is the most stable, at 0 K. A new candidate compound can then be tested against the already existing convex hull and the material can be considered stable and thus becomes a new point of the hull if its energy is below the hull.
- *Stability in water—Pourbaix diagrams*: it is also possible to investigate the stability of a material against decomposition where some parts are dissolved in water. This is done by means of the so-called Pourbaix diagrams in which the candidate material is now tested not only against solid phases but also against dissolved ions and compounds. This can be generalised in a chemical reaction of the form: $r\mathrm{R} + w\mathrm{H_2}O = p\mathrm{P} + h\mathrm{H}^+ + ne^-$, where reagents and products are indicated by R and P with the coefficients r and p, respectively, and $h\mathrm{H}^+$ and ne^- indicate h dissolved protons and n free electrons. Using the Nernst

equation (at 300 K), we can obtain the equilibrium potential of this reaction (eqn (3.1)):

$$nU = \Delta G + 0.0591 \log \frac{[a_\mathrm{P}]^p}{[a_\mathrm{R}]^r} - 0.0591\ h\ \mathrm{p}H, \tag{3.1}$$

where ΔG is the standard free energy of the reaction (in eV), and $[a_\mathrm{P}]$ ($[a_\mathrm{R}]$) is the concentration of the products (reagents). By convention, the concentrations of the ions involved are usually taken equal to 10^{-6} M. We can then plot the Pourbaix diagram by solving the Nernst equation at different pHs and potentials. A Pourbaix diagram is formed by regions separated by straight lines each of them indicating a different equilibrium reaction.[56] Following the scheme described in ref. 57 and 58, the DFT total energies of the solid phases can be combined with the experimental dissolution energies for ions and other species in aqueous solution from ref. 59 and 60. The Pourbaix diagrams take into account only the thermodynamics of reactions and not their kinetics. It might thus be, that a reaction is thermodynamically possible but in reality kinetically hindered. This method has been applied, for example, as a stability descriptor to select novel stable light-harvesting materials[14] and to explain the lack of stability of the $SrRuO_3$ as a catalyst for the oxygen evolution reaction.[61]

3.3.4 Electronic Properties

So far, we have discussed how elemental and structural properties can be used as criteria for a screening. We now describe some of the main electronic descriptors used in computational screening of light-absorbing materials.

- *Band gap*: if we assume that all photons with an energy higher than the gap are absorbed and the ones with a lower energy are not, the band gap is the simplest descriptor to quantify the efficiency of a light-absorber. One of the issues with standard DFT is, however, the underestimation of the band gap due to the self-interaction error for semi-local exchange-correlation functionals[62] and the derivative discontinuity.[63] A possible way to overcome these issues is to include a fraction of exact exchange (hybrid functionals like PBE0 or HSE06[64]) or to use many-body methods (GW approximation, for example) which give a better estimate of the band gap but with a significant increase of the computational cost. Another possibility is to use an exchange-correlation functional which explicitly includes the calculation of the derivative discontinuity, like the GLLB-SC,[65,66] which is a good compromise between accuracy and computational cost for screening purposes. The GLLB-SC band gaps have been benchmarked against hybrid functionals, different levels of the GW approximation, and experiments for binary and up to quaternary semiconductors,[14,67] single metal and perovskite oxides,[31,34] and halide perovskites[68] resulting in an error of the order 0.5 eV with respect to the G_0W_0 values and of around 15% with respect to the eigenvalue-self-consistent GW. The GLLB-SC functional is thus accurate enough to be used in high-throughput calculations without having a dramatic increase in the computational cost over conventional gradient corrected functionals.

- *Direct/indirect gap*: the first improvement of the band gap criterion described above, is to consider the character of the transition, *i.e.* whether the band gap is direct or indirect. If our device has to be of reduced dimensions in the direction of the absorption of light (thin films or nanoparticles, for example), only the direct transitions will contribute to the absorption, since phonons are required to absorb the indirect ones. This is of course less relevant in the case of thicker materials where the phonon density of states is sufficiently high. The simplest way to determine the character of the gap, is to identify the position of the transition in reciprocal space. Depending on the application, one can then decide if to consider only materials with a direct gap or also the ones with an indirect gap. A more sophisticated method has been proposed by Yu and Zunger[69] in which the character of the gap is combined with the shape of the absorption at the band gap edges and with the nonradiative recombination losses into a single metric to search for improved chalcopyrite materials for photovoltaics. In Section 3.3.5, we are going to improve this descriptor even further by including the direct calculation of the matrix elements.
- *Effective masses*: the effective masses of electrons and holes can be convenient descriptors for the mobility of photo-excited carriers in semiconductors. The mobility, μ, is in the simple Drude model expressed as a function of the relaxation time τ and the effective mass, m^*, $\mu = \frac{q \times \tau}{m^*}$, where q is the electron/hole charge. The relaxation time can come from several different scattering processes and is in general difficult to calculate, but the inverse proportionality of the mobility with the mass can be used as a rough estimate: the lower the mass, the better the chances are that the semiconductor will show a high mobility of the carriers. In the simplest approximation, the energy bands at the valence and conduction band edges are parabolic and the effective masses are isotropic, while in general the effective masses are described by 3×3 tensors or the three eigenvalues. Carrier effective masses can thus be calculated by fitting parabolas to the valence band and the conduction band edges, for the hole and electron effective masses, respectively. A small region close to the band edges should be considered, to assure a good quality of the fit, which means that a dense k-point sampling is necessary. If we consider isotropic effective masses, we can limit the k-point sampling to the major directions in the Brillouin zone by

defining a path which visits the high-symmetry points, as discussed in details by Setyawan and Curtarolo.[70] In this way, a denser k-point grid is achieved without an extensive usage of computational power.

- *Life times and carrier separation*: one important aspect of a photoexcited process is the lifetime of the charge carriers (the generated electron–hole pairs). The lifetime of the excited states gives the time scale of the recombination process in which an electron and a hole reannihilates. This process leads to lowering the efficiency of a PEC device and therefore, a large lifetime of the charge carriers is required for efficient energy harvesting. However, several pathways for the recombination adversely affect the lifetime. For example, localised states available at interfaces or surfaces or at defects present in the crystal may act as charge recombination centres. Additionally, a lower mobility of the charge carriers also increases the probability of recombination.

  From an engineering point of view, the separation of the charge carriers can be enhanced by various means, for example, gradient doping or controlling the shape of nanocrystals of the material.[71,72] However, taking the crystal phase engineering or doping into account in a screening study is very demanding, and there is a need for descriptors related to the lifetime and carrier separation of bulk materials, which qualitatively if not quantitatively can predict materials with longer lifetimes and efficient carrier separation.

  One such descriptor may be the detailed nature of the band structure around the gap. It has been suggested in recent theoretical and experimental studies that a slightly indirect band gap due to spin-orbit Rashba-splitting may lead to lower recombination rates and higher diffusion lengths of the charge carriers in the case of organic-inorganic hybrid perovskites.[73,74] Rappe *et. al*[74] proposed that the forbidden recombination of excited charge carries due to different spins induced by the spin-orbit splitting enhances the lifetime thereby increasing the diffusion length of the charge carriers. It would be interesting if such considerations could be taken into account in future screening projects.
- *Defects*: defects play a crucial role as donors and acceptors in semiconductor technology. However, defects can also play a negative role and for example deteriorate the opto-electronic performance of semiconductors. The presence of defects may induce electronic states in the band gap, and if these states are not close to the band edges (like donor and acceptor states) they may be localised and act as recombination centres decreasing the mobility and/or lifetime of the charge carriers. Therefore, their presence is highly undesirable for the performance of a PEC device. There are two main aspects which need to be considered in studying defects in semiconductors: (i) the thermodynamics of defect formation and (ii) the nature of defect-induced states. The thermodynamics refer to formation energy of defects and therefore the possibility of their presence in a sample. However, the

latter aspect refers to the nature of states induced in the presence of defects, for example, mid-gap states or shallow states which are closer to the band edges.

- *Thermodynamics of defects*: the description of the thermodynamics of point defect formation based on DFT is well developed and quantities like the formation energy, entropy, and excess volume can be derived.[76] However, these calculations rely on the use of large supercells where careful consideration of convergence with cell size is required. This also means that the calculations often are very time consuming and—at least at present—cannot be used for high-throughput screenings of thousands or tens of thousands of materials. However, the calculation of defect thermodynamics may be of great importance for further analysis of materials that appear promising from large-scale screening studies.
- *Defect-induced gap states*: as mentioned above, the presence of defects may possibly induce mid-gap states in the band gap of significant importance for the opto-electronic performance. However, calculating the accurate positions of defect-induced states is very computationally demanding and requires a supercell approach. Furthermore, standard (semi-)local exchange-correlation functionals may also fall short of a correct determination.[76] Qualitative information about the presence of mid-gap states may be obtained by more approximate calculations, but here we shall just discuss a descriptor-based approach based on a simple physical picture behind the defect-state formation.[75,77,78] The scheme has been used to treat defect formation in 2D transition metal dichalcogenides (TMDC).[75]

The idea behind the approach is sketched in Figure 3.4, where the character of the bonding in the valence and conductance bands are indicated. (a) corresponds to the situation in which the top of the valence band is of antibonding nature whereas the bottom of the conduction band is of bonding nature. (The labels 'd' and 'p' refer to the orbital character of the states in TMDCs). In contrast, (b) represents the situation in which the top of valence band is of bonding nature whereas the bottom of the conduction band is of antibonding nature. Breaking of bonds for example through the creation of vacancy defects will in the case (a) lead to new states between the bonding and antibonding states within each band. Therefore, this will not give rise to any new states in the band gap or at least only to shallow states close to the band edges as shown in (c). In the case (b) the breaking of bonds will again induce states between the antibonding and bonding states, but these will now be dangling bonds in the mid-gap region as shown in (d). This feature of the band structure can be assessed *via* a descriptor which estimates the bonding and antibonding nature at the CBM and the VBM. The advantage of such an approach is that the determination of whether a given material is

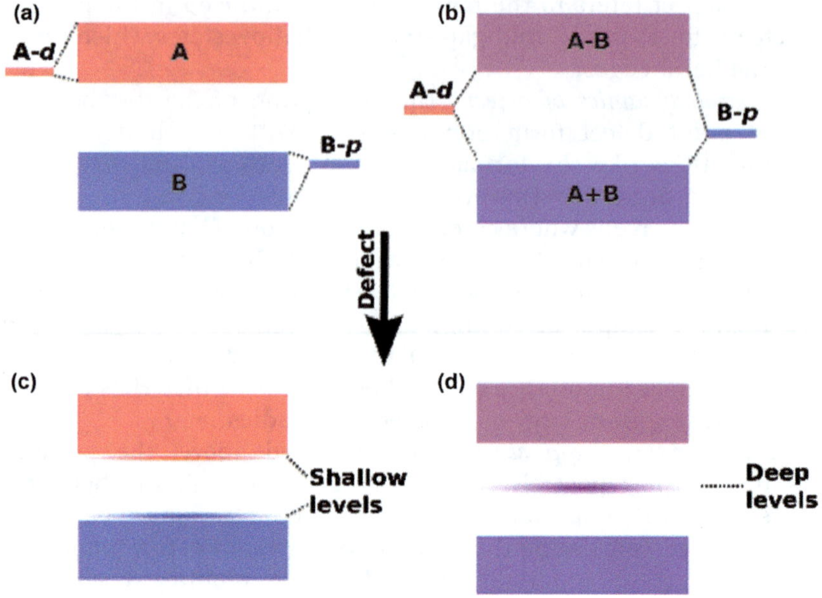

Figure 3.4 Nature of states induced depending on the orbital character of the valence and conduction band.
Reprinted with permission from Pandey *et al.*, *Nano Letters*, 2016, **16**, 2234.[75] Copyright 2016 American Chemical Society.

defect tolerant (no mid-gap states) or defect sensitive (mid-gap states) can be drawn from the calculations on the primitive cell only.[75,77,78]

3.3.5 Direct Calculation of Light Absorption

The band gap gives a simple estimation of light absorption efficiency, however, it does not consider the absorption strengths or effects due to excitons. We are now going to describe some of these effects and how to, at least partially, take them into account when screening for novel materials.

- *Matrix Elements and Absorption Spectrum*: the band gap by itself does not contain important information for light absorption like forbidden transitions due to symmetry or lack of spatial overlap between states in the conduction and valence bands. One way to access such information is by a direct calculation of the absorption spectrum in which both the character of the gap (direct/indirect) and the strengths of the dipole-transition matrix elements are considered. The standard way to do this is by means of Time Dependent Density Functional Theory (TDDFT) within the Adiabatic Local Density Approximation (ALDA). The problem of the underestimated band gap in (semi-)local DFT can be addressed by for example using the band structure from a GLLB-SC calculation which includes the derivative discontinuity. The absorption efficiency, η, is calculated as the fraction of photons absorbed (eqn (3.2)):

$$\eta = \frac{1}{n_{\text{tot}}} \int_{\text{gap}_\text{d}}^{\infty} ph_{\text{abs}}(E) n_{\text{ph}}(E) dE, \tag{3.2}$$

where the total number of photon emitted by the sun at AM1.5 is indicated with n_{tot}, the direct band gap with gap_d, the photon absorptivity, which is a characteristic of the material, with $ph_{\text{abs}}(E)$, and the number of sun photons as a function of the energy, E in eV, with $n_{\text{ph}}(E)$. The photon absorptivity is a function of the absorption coefficient $\alpha(E) = \frac{2Ek(E)}{\hbar c}$ ($\hbar$ and c are the Planck constant and the speed of light, and k is obtained from the absorption spectrum as $k^2 = \frac{1}{2}(-\text{Re}\varepsilon + \sqrt{\text{Re}^2\varepsilon + \text{Im}^2\varepsilon}))$, and on the thickness of the material, L: $ph_{\text{abs}}(E) = 1 - e^{-2\alpha(E)L}$. This model assumes that no photons with energy below the direct band gap are absorbed, since absorption through an indirect transition requires the participation of phonons because of the change in the momentum and, at this stage, phonons are not included in the calculation. Details of an implementation can be found in ref. 79 and 80.

- *Density of states*: a complete calculation of the absorption spectrum is rather computationally time-consuming and it is therefore of interest to consider simpler approaches which take more of the band structure into account than just the band gap without requiring a full TDDFT calculation. The electronic excitations of relevance for light absorption involve excitations from states near the VBM to states close to the CBM. The absorption will be more intense if many states are available for the excitation and a high joint density of states (JDOS) near the VBM and CBM is therefore preferential. Under the assumption that the velocity matrix element between the conduction and valence band states is constant the absorption simply becomes proportional to the JDOS.

  A high JDOS is typically seen where many flat bands, as for example, *d*-bands are located close to the band edges. Recently, Keszler and co-workers explored $CuTaS_3$ experimentally and theoretically as a candidate material for PV.[81] The observed strong absorption onset in this material is explained by the high JDOS. Similarly a high JDOS also gives rise to strong absorption onset in FeS_2 and MoS_2.[82,83] Therefore, this special feature in the DOS is a promising descriptor for strong absorption.

  It might be noted that there is a potential conflict between requiring high JDOS and at the same time low effective masses. Many flat bands around the band edges will give rise to a high JDOS but at the same time the masses will be high and therefore the mobility low. One way out of this could be to have many flat bands with additionally some low-mass bands taking care of the charge transport.

- *Excitons*: the fundamental quasi-particle gaps calculated with the GLLB-SC functional, or other methods like GW, do not include excitonic effects of relevance to the light absorption process. The optical generation of an electron–hole pair may result in the formation of a bound electron-hole pair, *i.e.* an exciton, rather than free charges. Due to the large effective masses and screened Coulomb interaction, the exciton binding energies for 3D inorganic semiconductors are small, of the order of 0.01 eV.[84] When the dimensionality is reduced, *i.e.* for 2D and 1D materials, the excitons binding energies increase because of an increase in the dielectric constant due to reduced screening. Monolayer MoS_2, for example, shows an excitons binding energy of 0.6 eV, and the corresponding energy for bulk MoS_2 is around 0.1 eV.[85] Organic materials also show larger exciton binding energies compared with the inorganic 3D compounds. For example, the electron–hole interaction in hybrid halide perovskites has been calculated to be around 0.15 eV.[68]

  Excitonic effects can be approximately included in TDDFT by moving beyond the time-local approximation and use the so-called bootstrap kernel.[86] However, a more accurate and transparent treatment is obtained using the Bethe-Salpeter equation.[87,88] The latter approach is, however, very time consuming and can only be performed for few systems of limited size.

3.3.6 Interfaces

So far, we have discussed light absorbing materials in their bulk form. However, in a functioning water splitting device interfaces will play a crucial role. Returning to the protected Design 2 of Figure 3.1 we see that there may be interfaces between light harvesting materials, protection layers, catalysts, and the surrounding electrolyte.

Interfaces are very difficult to study because the interface or surface dipole depends sensitively on the atomic structure. The line up of electronic levels at the two sides of an interface is therefore difficult to determine accurately.

Of particular interest for water splitting is the interface to the electrolyte. The charge carriers generated in the light absorbing material(s) need sufficient energy to perform the water splitting processes and in the simple case where a light absorber is directly in contact with water the band edges need to be appropriately positioned relative to the water redox levels.

A number of methods for determining the band edge positions relative to the water redox levels based on DFT and extensions have been proposed.[32,89–91] In these approaches the level alignment is determined by direct DFT calculations of the semiconductor–water interface, to determine for example the shift in electrostatic potential between the solid and the water. The DFT results can be further improved by applying GW corrections to the determination of the band edges in the semiconductor. Such calculations are computationally intensive, but it was recently pointed out, that the positions of the band edges can be estimated reasonably from the ionisation potentials and electron affinities alone avoiding explicit modeling of the semiconductor–water interface.[89]

In some of our material screening studies, we have used a simple estimate of the band edge positions based on an empirical formula suggested by Butler and Ginley.[92,93] The band edge energies, $E_{VB,\ CB}$, are here expressed using the electronegativities in the Mulliken scale of the constituent atoms, χ_i, and the band gap, E_{gap} (eqn (3.3)):

$$E_{VB,\,CB} = E_0 + \left(\prod_i^N \chi_i\right)^{1/N} \pm E_{gap}/2, \tag{3.3}$$

where E_0 corresponds to the difference between the normal hydrogen electrode (NHE) and the vacuum ($E_0 = -4.5$ eV) and N is the total number of atoms. Even though, this is an approximate method, the calculation of the position of the band edges in this way is in good agreement with experiments.[93]

An effect that is usually not considered in screening approaches is the band bending at the interface with the electrolyte. This effect is due to an accumulation of charges at the surface and of opposite charge in the electrolyte. pH and potential influence the charges at the interface and thus the band bending. Depending on the direction where the bands bend, the extraction of the photogenerated charges can become easier, more difficult, or even impossible. The optimal band bending for the two-photon device is shown in Figure 3.1. Different methods, like combining Density Functional Theory and Green's function or Joint Density Functional Theory,[94,95] have been proposed to model the interfaces and can be used to model the band bending. For example, Jaio *et al.*[96] have investigated the formation of Schottky barrier and band bending in niobium doped rutile titania in contact with gold.

3.4 Materials Investigated

In the following we shall go through some of the computational screening studies of water splitting materials performed in the past, with a maybe unavoidable bias towards the ones that some of the current authors have been involved in.

For this discussion, we have selected three classes of materials: perovskites, experimentally known structures from various databases, and 2D materials mainly in the 2H (MoS_2 type), 1T (CdI_2 type), and hybrid perovskite structures. A significant part of data presented in the coming sections have been collected, together with scripts to analyse the results, in the CMR database.[46]

The first PEC device ever built used TiO_2, which is a binary semiconductor.[1] Unfortunately, the band gap of TiO_2 is too large to efficiently harvest the visible part of the solar spectrum. To improve the light absorption, numerous

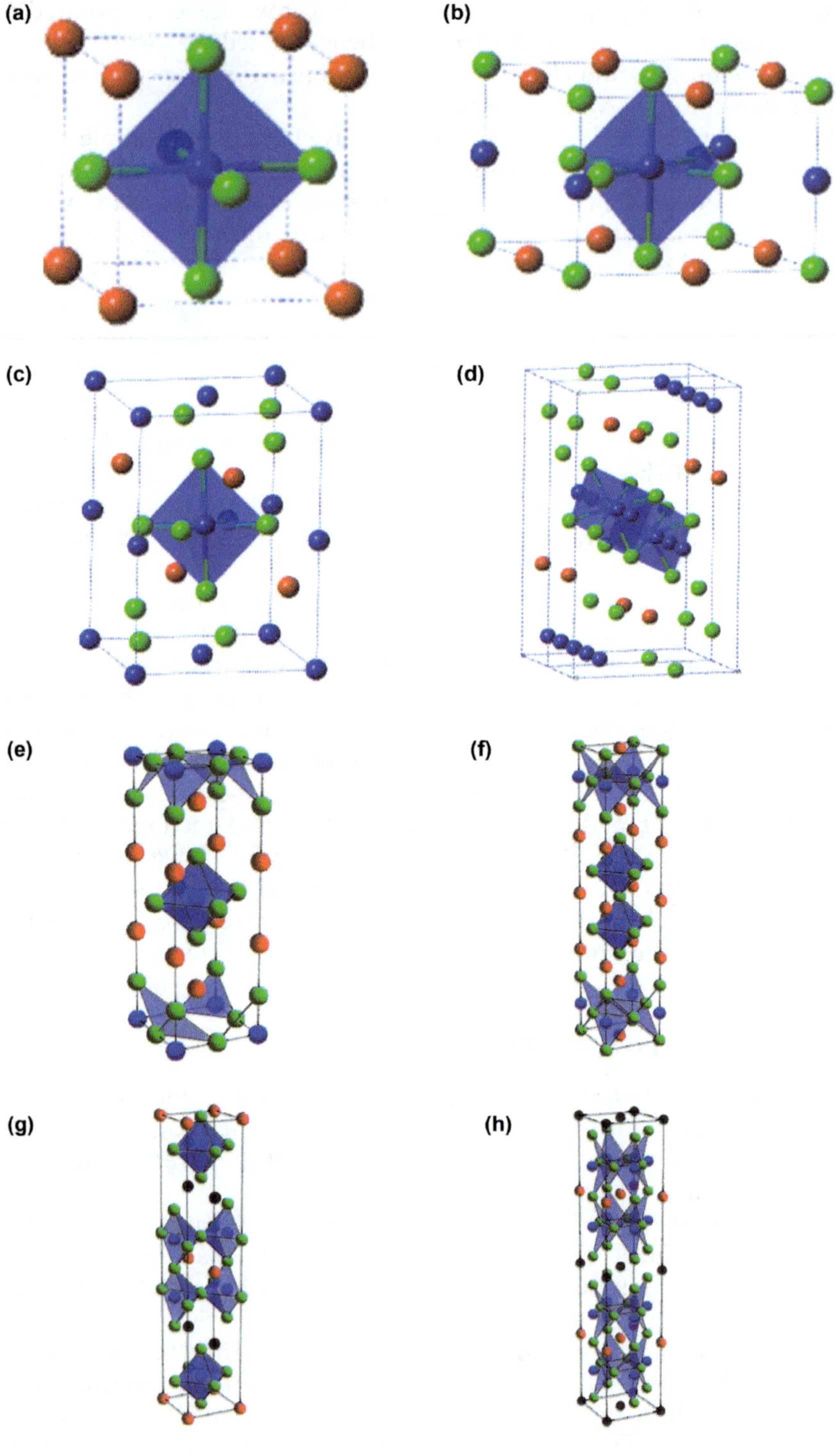
(a)
(b)
(c)
(d)
(e)
(f)
(g)
(h)

common binary and ternary semiconductors, like ZnO, GaN, Fe_2O_3, GaP, Cu_2O, Ta_2O_5, and WO_3, have been explored, but an optimal material has not been found.[97–103] In most of the binary compounds, a certain level of nanostructuring or doping is required for the material to work. Additionally, most of the binary oxides tend to have the valence band quite deep as compared to the water redox level, whereas substituting the oxygen with other non-metal like sulfur, phorphorous or arsenic, pushes the valence band and conduction band upwards at the same time.[104,105] Therefore, there are not many possibilities of controlling the band edges independently. Because of a relative limited number of polymorphs and combinations of two chemical elements, it has been natural to investigate ternary compounds that allow for more prototypes as well as a larger amount of possible combinations rather than relying on nanostructuring and doping.

3.4.1 Perovskites

Ternary compounds offer a much broader space of investigation than binaries. Because of the relatively high stability, we early on focused on oxide materials and among all the possible classes of materials the ones with perovskite symmetry because it is so common. Most of the chemical elements can be used to form stable perovskites with interesting properties such as ferroelectricity, superconductivity, magnetism, and catalysis.[106] Few perovskite materials are also known to be able to split water, often in presence of sacrificial reagents, like the oxo-perovskite $AgNbO_3$[107] and the oxynitrides $LaTiO_2N$, $BaTaO_2N$, $SrTaO_2N$, $CaTaO_2N$, and $LaTaON_2$.[108]

The perovskite structure shows different prototypes (some of them are reported in Figure 3.5). The most simple perovskite has a cubic symmetry (Figure 3.5(a)) and can be described with a five atoms unit cell. It is composed of two cations: a larger A-cation, which usually have oxidation states equal to +1, +2, or +3 and is 12-folded coordinated, and a smaller B-cation, which has oxidation states equal to +3, +4, or +6, is 6-folded coordinated and is surrounded by an octahedron of anions. It has been shown that different anions can be used: oxygen, sulfur, nitrogen, halides and their combinations. Lower symmetry perovskites are then obtained by rotation or tilting of the octahedra: the tetragonal symmetry (Figure 3.5(b)) is obtained by a planar rotation of the octahedra, and the orthorhombic symmetry (Figures 3.5(c)

Figure 3.5 Perovskite crystal structures: cubic (a, general formula ABO_3), tetragonal (b), orthorhombic (c and d), Ruddlesden–Popper layered phase with $n = 1$ (e, A_2BO_4) and $n = 2$ (f, $A_3B_2O_7$), and Dion–Jacobson layered phases (g and h, MAB_2O_7). The A-cations are indicated in red, the B- in blue, the M- in black, and the anions in green.
(a)–(d) Reproduced from I. E. Castelli, J. M. García-Lastra, K. S. Thygesen and K. W. Jacobsen, *APL Materials*, 2014, **2**, 081514. ©The Authors. Published under the terms of the CC BY 4.0 licence, https://creativecommons.org/licences/by/4.0/.[68] (e)–(h) Reproduced from ref. 80 with permission from The Royal Society of Chemistry.

and (d)) by tilting the octahedra. Both the tetragonal and orthorhombic symmetries require a more complex description of the unit cell composed of 20 atoms and are thus more computationally expensive. The octahedra can also be separated by different atomic motifs to form the so-called layered perovskite. Two examples are the Ruddlesden-Popper phases (Figures 3.5(e) and (f)) and the Dion-Jacobson phases (Figures 3.5(g) and (h)).

We first consider the simple cubic symmetry where we combine 52 elements in the A- and B-cation position together with oxygen, sulfur, nitrogen, fluorine, and some of their combinations as anions, for a total of a little less than 20 000 different combinations. The complete dataset for the oxo-perovskites is shown in Figure 3.6. Most of the investigated combinations are unstable when compared with competing phases consisting of pure and binary compounds as explained in Section 3.3.3. The main reason for this is that we consider all possible combinations of A- and B-cations without taking into consideration the size of the cations or their oxidation states. In addition, most of the compositions do not show a band gap. There are elements that are more likely to produce stable structures with or without a band gap, as also shown in Figure 3.7. Alkali and alkaline-earth elements prefers to occupy the A-ion position because of their ionic radii and small oxidation states. However, transition metals have a preference to occupy the B-site. Post transition metals can instead be used in both positions. These considerations will be used in later studies to reduce the space of investigation when more complex perovskite structures are used as prototypes.

In an earlier study the following screening criteria relevant for a one-photon light-harvesting material for PEC were applied to the dataset above:

- stability: heat of formation <0.2 eV per atom;
- harvest of visible light: $1.5<$ band gap <3.0 eV;
- photogenerated charges at the right potential: valence band edge >1.23 V *vs.* the Normal Hydrogen Electrode (NHE) and conduction band edge <0 V *vs.* NHE. (At this stage, we have not included the overpotentials and the quasi-Fermi levels neither in the size nore in the position of the gap.)

It can be noted that the band gap bracket also covers the range of interest for large gap semiconductors in a tandem device.

The 23 materials of Figure 3.8 fulfil these criteria. Among the candidates, there are ten oxides, nine oxynitrides, and four oxyfluorides. A few of these have already been investigated for water splitting, in particular $AgNbO_3$ and the oxynitrides $LaTiO_2N$, $BaTaO_2N$, $SrTaO_2N$, $CaTaO_2N$, and $LaTaON_2$. Some other, like $BaSnO_3$ or $YTaO_2N$, have been tried as light-harvesting materials, but they have bad performances, showing instabilities or phase transitions towards lower symmetry phases. Many of the cubic perovskites are in fact unstable towards symmetry breaking at low temperatures and this symmetry breaking is often accompanied by an increase of the band gap and, in some

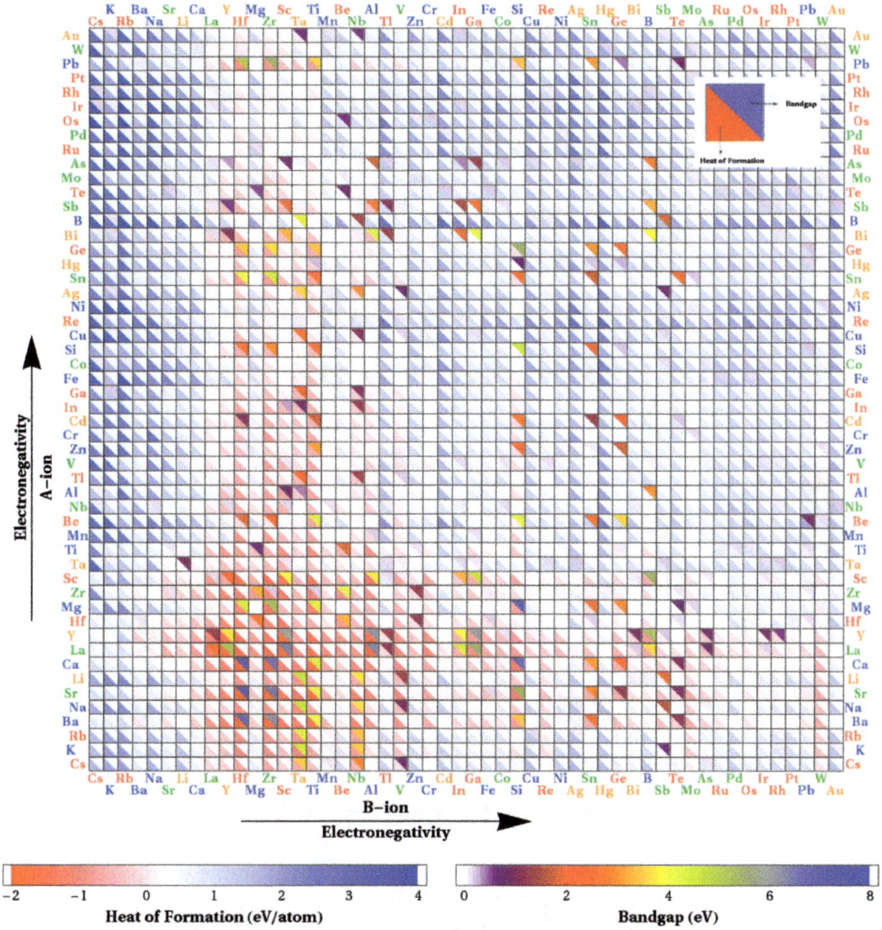

Figure 3.6 Each square represent a ABO_3 cubic perovskite, the lower, left triangle reporting the heat of formation from stable (in red) to unstable (in blue) and the upper, right triangle the band gap (a gap between 1.5 and 3.0 eV is indicated in red, a too small gap is indicated from white, corresponding to a metal, to purple and a too large gap from yellow to blue. The chemical elements are sorted for increasing electronegativity and their colours indicate the global production: >1 Mt per year in blue, >33 kt per year in green, >1 kt per year in orange, and <1 Mt per year in red.[109]
Adapted from ref. 31 with permission from The Royal Society of Chemistry.

cases, a change in the character of the gap from direct to indirect. The consideration of only the cubic phase is therefore questionable.

The selected materials have been further investigated with a stability analysis in water and a more realistic estimation of the light absorption by using TDDFT. Regarding the stability in water, we see from the colour bars in Figure 3.8 that all the materials show some instability in water,

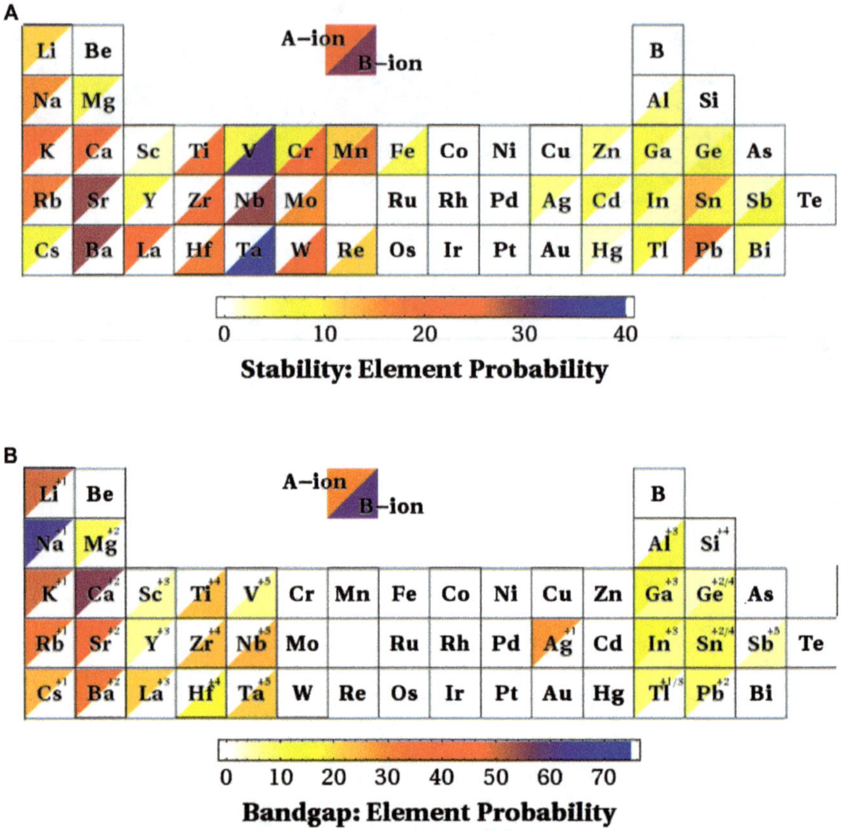

Figure 3.7 Probability (in %) for the A- and B-ion to form stable compounds (on the left) or stable combinations showing also a band gap (on the right). Figures from ref. 35, with permission of IOP Publishing.

i.e. $\Delta E > 0$ eV/atom and the oxynitrides are very likely to dissolve at high potential. As discussed before, one of the possible solutions to the water instability issue is the development of protecting shields which protect the photo-active materials from the electrochemical environment.

The light-harvesting efficiency calculated using TDDFT (Section 3.3.5) is shown in Figure 3.9(a). The oxides and oxyfluorides are characterised by having an indirect band gap, most often between the Γ- and *R*-points, and thus a large light-harvesting efficiency at the band gap for, say, a 100 nm thick material cannot be expected. In addition, the matrix elements at the transition are not particularly intense and thus, even for thick materials, we cannot achieve the maximum theoretical efficiency. Some of the oxynitrides can achieve a higher efficiency because of the direct band gap and larger matrix elements at the transition. Experiments have shown these materials to be currently among the best perovskite materials for water splitting, however, the relatively poor stability combined with difficulties in the

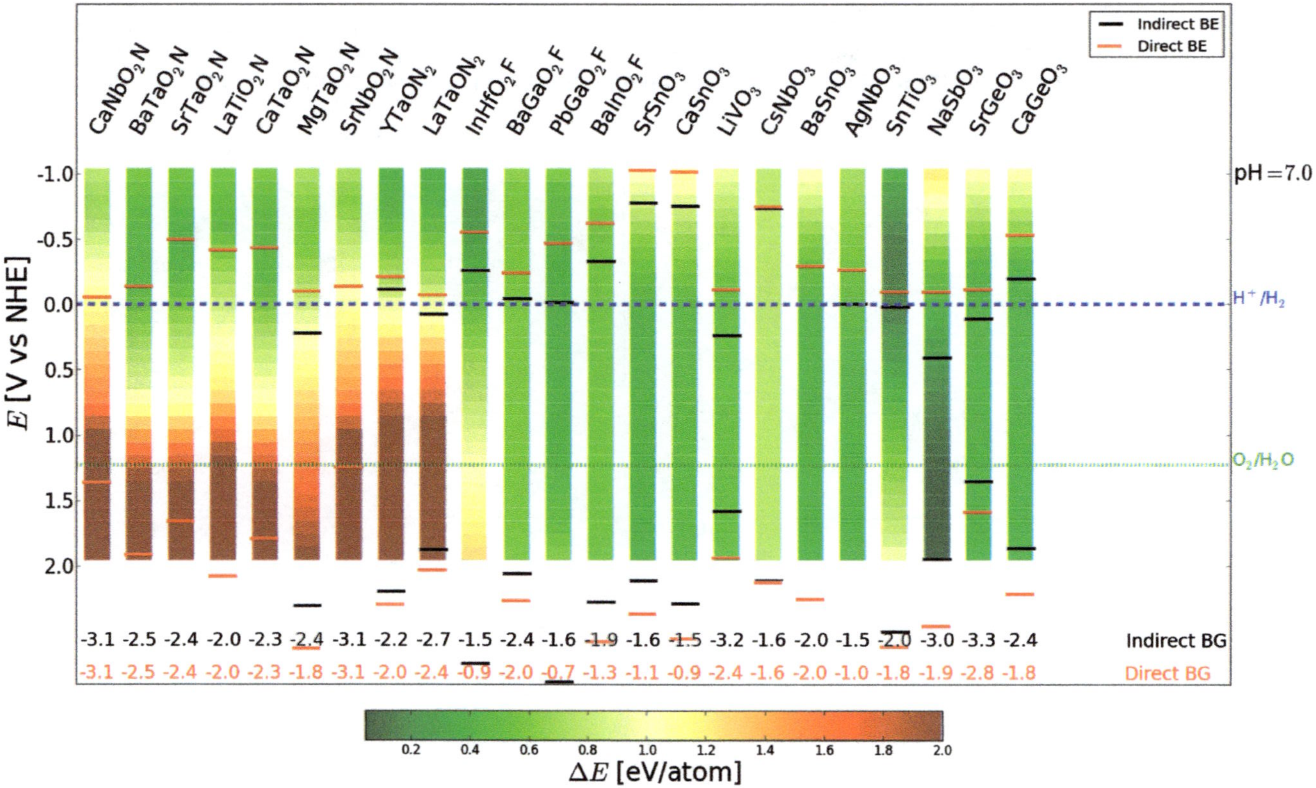

Figure 3.8 The cubic perovskite candidates as light-harvesting materials. The stability in water is indicated by the colour scale (from green, *i.e.* stable, to red, unstable compounds) and is calculated by means of Pourbaix diagrams in a potential range between −1 and 2 V *vs.* NHE and at neutral pH. The indirect and direct band gap (BG) is indicated in black and red as well as the indirect and direct positions of the valence and conduction band edges (BE). Scripts to reproduce this figure and the figure itself can be found in the 'cubic perovskites' page of the CMR database, hosted at the department of Physics at the Technical University of Denmark.[46]

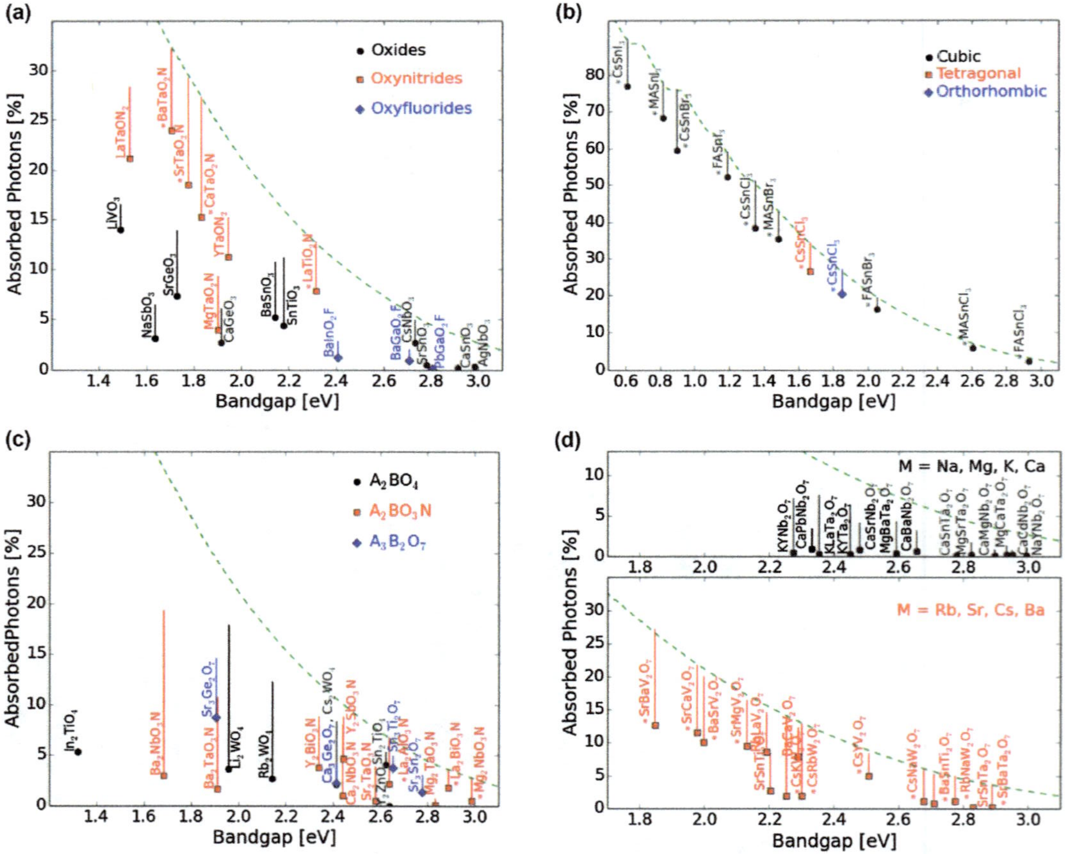

Figure 3.9 Calculated light-harvesting efficiency of the oxides, oxynitrides, and oxyfluorides (a) and halides (b) cubic perovskites and for the candidates in the Ruddlesden-Popper (c) and Dion-Jacobson (d) phases. The efficiency for a 100 nm thick material is indicated by a marker, and the vertical line goes up to the limit of infinite thickness. The direct band gap materials are indicated with a *. The green line indicates the maximum theoretical efficiency as a function of the band gap. Reproduced/adapted from ref. 80 with permission from The Royal Society of Chemistry.

extraction of the photogenerated charges make these materials still far from being practically useful as light-harvesting materials in a realistic device.

We note that an extensive computational screening of oxynitrides for water splitting has been performed by Wu *et al.*[32] based on descriptors for stability, band gap, and band edge positions. Sixteen interesting compounds are identified some of which also appear in the list of 23 perovskites discussed here. However, the screening by Wu *et al.* also considers non-perovskite structures.

In the last five years, another class of perovskites have shown extremely good performances for PV devices, making them competitive with photovoltaic thin films. Organometal, or hybrid, lead halide perovskites like $CH_3NH_3PbI_3$ (MAPI) have a direct band gap, large light absorption coefficients due to a direct transition at the band gap that involves Pb *s*-states and Pb *p*-states and high carrier mobility.[52,110] Cation mixtures of methylammonium ($MA = CH_3NH_3$), formamidinium ($FA = CH(NH_2)_2$) and some Cs together with anion mixtures of halides have now increased the power conversion efficiency of these devices to more than 21% in a photovoltaic device[111] and a solar-to-hydrogen efficiency of more than 12% when combined with Earth-abundant catalysts to split water.[112]

Computational studies can reveal some of the trends when substituting anions and cations of MAPI.[68] For example the band gaps systematically increase with the electronegativity of the anions ($I_3 < Br_3 < Cl_3$) also when these are mixed. Structural changes when lowering the symmetry (orthorhombic < tetragonal < cubic) also increases the gap. One important effect of the cation is to control the volume so that for larger volumes (Cs > MA > FA) the gap decreases.

The calculated light absorption efficiencies of some of the organometallic perovskites are shown in Figure 3.9(b). The efficiency of a 100 nm thick material is already rather close to the maximum theoretical efficiency which is reached in the limit of infinite thickness. In these materials Cs or a monovalent organic molecule, such as methylammonium ($MA = CH_3NH_3$) or formamidinium ($FA = CH(NH_2)_2$) is used as A-cation together with Pb (or Sn) as B-ions and halides as anions and they arrange in different symmetries: cubic, tetragonal, and orthorhombic. The light absorption properties of the halides are clearly superior to the ones calculated for the oxides, oxynitrides, and oxyfluorides. One of the main issues with the organometallic perovskites is their stability combined with the fact that lead is poisonous. Some efforts have been made to replace lead with other elements without losing the excellent light-harvesting properties,[113,114] but so far an optimal material has not been found.

The perovskite structure also forms the basis for more complicated crystal structures like for example, the double perovskite, in which two cubic perovskites are combined in a chessboard-like manner, or layered structures, in which the octahedra are separated by some motifs, for example an AO layer. Due to the larger unit cells and also more different chemical compositions a direct 'brute force' study of these compounds cannot be carried through. However, using some chemical or structural rules as discussed in Section 3.2 the space can be reduced and at least partly explored.[33,35] The layered perovskites shown in Figures 3.9(c) and (d) are the ones identified by a screening of stability and band gap. Only a few of the oxynitrides in the Ruddlesden-Popper (RP) phase show a direct band gap and the light absorption efficiency is rather low. A low efficiency has also been found for the Dion-Jacobson phase with M = Na, Mg, K, or Ca (DJ_1). With M = Rb, Sr, Cs, or Ba (DJ_2), instead, the materials exhibit direct band gaps with a rather large matrix element at the transition, which allows the material to reach the maximum theoretical efficiency in the limit of infinite thickness. The reason for the lower efficiency of the RP and DJ_1 phases can be traced to the relative displacement of the octahedra in neighboring units as seen in Figure 3.5. This leads to smaller matrix elements and thus reduced light absorption.

Double perovskites are obtained by combining two cubic perovskites and the number of anion and cation combinations therefore becomes huge. A dataset of calculated band gaps obtained when two stable, cubic oxo-perovskites with a non-zero gap are combined has been established.[115] The dataset has been used together with machine learning to more generally predict band gaps of double perovskites based on the band edges and the electronegativities of the constituents.[116] The double perovskites are of interest due to the large number of compositions, and some of these materials have shown good performances in energy related applications, like as catalysts for the oxygen evolution reaction.[117,118]

Another way of possibly improving the light-harvesting efficiency of the inorganic perovskites is by replacing oxygen with sulfur, which will reduce the size of band gap. Both binary and ternary sulfide semiconductors, like CdS, ZnS, CuS, SnS, Sb_2S_3, Ag_2S, Cu_2SnS_3, as well as $SrZrS_3$ in the perovskite structure, have been investigated for solar energy conversion application and have shown considerable potential.[119–123] Due to significantly lower electronegativity of sulfur compared to oxygen, sulfide perovskites prefer lower symmetry structures of corner or edge sharing BS_6 octahedra.[124] A detailed investigation on stability of different phases has been performed on a class of compounds with d^0 electronic configuration for the B-cation by Brehm *et al.*[125] In both cases, authors agree that there are no known sulfide perovskites in a perfect cubic phase, instead, they typically prefer orthorhombic (space groups *Pna*21 and *Pnma*), which is in a good agreement with the structures in the ICSD database or structures reported experimentally.[126–128]

Sun *et al.*[129] have investigated a subset of chalcogenide perovskites for photovoltaic applications. They investigated compounds in four different crystal structures: cubic perovskite, distorted cubic, hexagonal and orthorhombic (so called needle-like) structures, shown in Figures 3.10(a)–(d). The band gaps for 18 sulfide and selenide compounds in four structures are calculated using the HSE06 hybrid exchange-correlation functional. The green region in Figure 3.10(e) shows the optimal band gap region for PV light-harvesting materials. Finally, they reported $CaTiS_3$, $BaZrS_3$, $CaZrSe_3$,

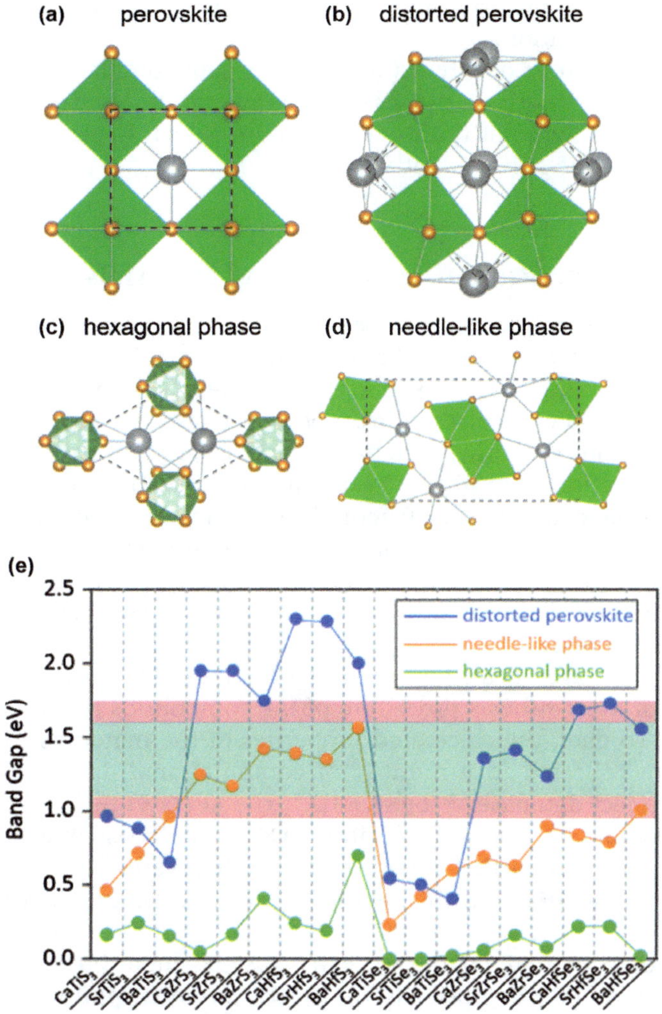

Figure 3.10 Different phases for the ABX_3 compounds: (a) ideal perovskite phase without distortion; (b) distorted perovskite phase with the *Pnma* space group (No. 62); (c) hexagonal phase with the *P63/mmc* space group (no. 194); (d) needle-like phase also with the *Pnma* space group. Grey and brown spheres represent A cations and X anions, respectively. The B cations are located at the centers of the green-coloured octahedra. (e) Band gap of 18 ABX_3 compounds in the distorted perovskite, needle-like, and hexagonal phases calculated with the HSE06 functional. The optimal band gap region for solar-cell materials is highlighted in light green. An extended region with theoretical efficiency above 28% is highlighted in light red.

Reprinted with permission from Sun *et al.*, *Nano Letters*, 2015, **15**, 581.[129] Copyright 2015 American Chemical Society.

and $CaHfSe_3$ in the distorted perovskite phase to be promising candidates for solar cell applications.

Sulfides may also be of interest as large-gap semiconductors in a tandem device for water splitting. In that case the band gap should be around 1.7 eV or slightly above. Some of the present authors currently perform computational screening of sulfides with that aim, following a screening funnel much as the one outlined in Figure 3.2.

3.4.2 Electronic Properties of Existing Materials

The screenings described above were based on the definition of some structural prototypes or templates and a following investigation of the properties of both existing and unknown materials in these structures. Another approach is to calculate the electronic properties of already known compounds as listed in various databases. The ICSD database, for example, contains information for a large number of experimentally synthesised compounds. However, it only contains the structural properties of the compounds, like XRD pattern, atomic positions, lattice vectors and synthesis conditions. Electronic, thermodynamic, vibrationals properties are often missing and, therefore, first-principle calculations can be used to calculate some of the missing information.

A major advantage of calculating the properties of known materials is the availability of the synthesis procedure of the compounds and their stability conditions. To date, the calculated properties of the materials in the ICSD database are collected in a number of computational databases, like the Materials Project database,[13] the OQMD,[43] the AFLOWLIB,[42] and the NREL Materials Database.[44] For example, in the OQMD database one can find the ICSD entries without partial site occupancy or duplicates. All the geometries are relaxed using DFT, and the details about the calculations are reported online.[130,131]

The provided computed databases can be used in an initial screening step. To take an example, at April 2017, OQMD contains around 30 000 non-duplicate structures from ICSD. If we are interested in compounds which contain only abundant, non-toxic elements with a low HHI (Figure 3.3), this number is reduced to about 9700 compounds. Furthermore, the databases contain band gap values calculated at the PBE level. Although these values are underestimated, they give enough information to allow us to eliminate metallic compounds and very large gap semiconductors, which are not suitable as visible light absorbers. In this way the number of relevant materials is reduced to only around 1500. These materials can not be investigated following a screening funnel as in Figure 3.2. However, since these materials have already been synthesised some of the investigations of stability can be avoided and the screening can quickly proceed to determination of band gaps, effective masses, and defect properties. A preliminary screening along these lines lead to identification of around 200 materials of particular interest for light absorption in PV or PEC applications.

It can be noted that the above mentioned screening study of oxynitrides by Wu *et al.*[32] also takes its starting point in the ICSD database. However, they also generate new materials using compound and prediction tools to identify favorable ionic substitutions.

3.4.3 2D Materials

In the past few years considerable efforts have gone into exploring structures of reduced dimensionality, like 2D or 1D compounds.[132] The reduced dimensionality offers both challenges and opportunities. One of the main challenges lies in synthesising materials with controlled layer thickness which may significantly affects the structural and electronic properties as well as the stability both thermodynamically and in the presence of water. For example, one of the most studied transition metal dichalcogenides (TMDC), MoS_2, is a direct band gap material in the 2D monolayer form, however, if the number of layers is increased to two or more, it becomes an indirect band gap semiconductor.[133] However, there are also major potential advantages in using 2D materials for light harvesting. The electronic properties can be controlled through heterostructuring, defect engineering, and alloying. The materials also naturally have a large surface area and depending on the architecture the charge carriers only have a small distance to move to reach the surface.[132,134,135]

Experiments have recently been carried out to explore the visible light activity of 2D materials for PEC water splitting. The main efforts until now have been centered around transition metal mono/di-chalcogenides like MoS_2, WSe_2, and SnS.[136–138] However, Luo *et al.*[132] have recently demonstrated the potential of graphitic-C_3N_4 for visible light PEC water splitting. These results show that there is a large potential to find good PEC absorbers in the different classes of 2D materials. This prospect has also been explored *via* computational screening by Hennig and co-workers in ref. 135 and 139 and collected in the Materials Web database.[140]

As for the 3D materials, a range of relevant properties of the 2D compounds can be calculated both with DFT and with many-body perturbation theory in the form of GW calculations, and some of these have been compiled in the CMR database.[46] The database includes calculations of optical properties like band gaps, effect of defects on the band structure, carrier effective masses, and dielectric engineering *via* heterostructuring.[134,141,142] The CMR 2D database contains properties of novel materials from theoretical studies as well as of already synthesised 2D materials. Figure 3.11 shows an overview of the 2D materials in the CMR database. Around a hundred materials have a band gap in the range between 1.0 and 2.5 eV of relevance for visible light absorption of both single-component and tandem devices. Additionally, the metallic 2D materials might have the possibility of being employed as co-catalysts for the hydrogen evolution or as protective shields of the dark side in a PEC device.[5,143]

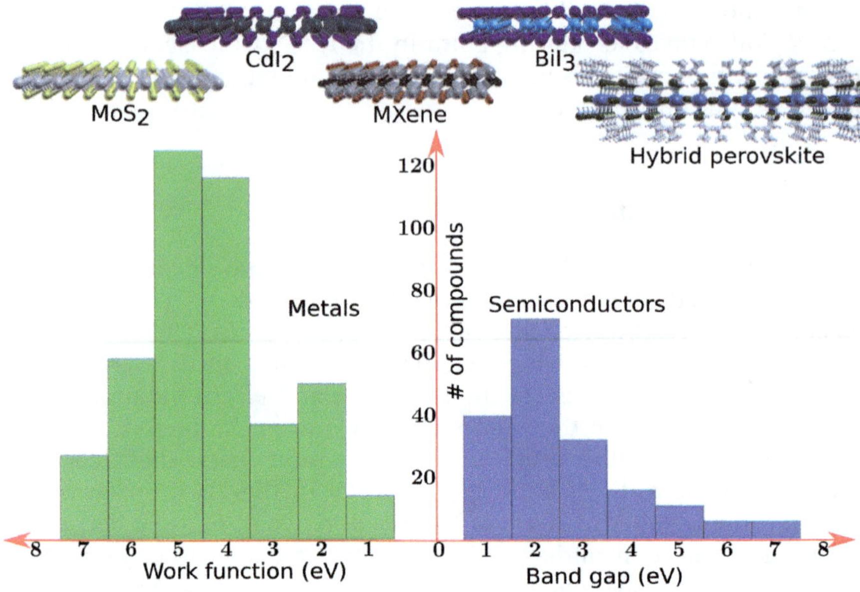

Figure 3.11 The figure shows an overview of the 2D materials in the CMR database. The database contains both semiconducting and metallic 2D materials. The histogram shows the distribution of the band gap for the semiconductors and of the workfunctions for the metals. The protypical structures are also shown.

3.5 Conclusions and Perspectives

In this chapter, we have illustrated how computational screening based on *ab initio* quantum mechanical calculations can be applied to find novel materials with potential for light harvesting in water splitting devices. The approach requires first of all a set of relevant descriptors or criteria to apply, and secondly a fast search strategy to suggest and test new materials. We considered different classes of materials including perovskites, synthesised materials from databases, and 2D materials. A few of the identified compounds are currently under experimental investigation.

Computational screenings can be expected to play a major role in the identification of new materials in the future for several reasons. Computers still become faster with increasing capacity and large-scale parallelisation. Not to mention a possible future transition to quantum computers with vastly increased capacities. The first principle calculations also go through rapid improvements both within the framework of DFT and of many-body perturbation theory, and within highly accurate wave function based quantum chemistry approaches. Last but not least we seem to be only at the beginning of a new era, where large databases with computed properties of millions of materials will be combined with data mining and machine learning techniques to extract new materials correlations, which can be used for improved searching and identification of materials with tailored properties.

This being said, the complexity of the various materials and their interfaces needed to obtain functioning PV or PEC devices should not be underestimated. Suggested new materials will require careful experimental characterisation probably leading to unexpected behaviour. Significant material and device improvement will therefore only be achieved with close collaboration between theory, synthesis, characterisation, and testing.

Acknowledgements

The authors acknowledge VILLUM FONDEN for financial support with research grant (9455).

References

1. A. Fujishima and K. Honda, *Nature*, 1972, **238**, 37–38.
2. H. Suffredini, *Int. J. Hydrogen Energy*, 2000, **25**, 415–423.
3. *Photoelectrochemical Hydrogen Production*, ed. R. van de Krol and M. Grätzel, Springer, USA, 2012.
4. M. Weber and M. Dignam, *Int. J. Hydrogen Energy*, 1986, **11**, 225–232.
5. B. Seger, I. E. Castelli, P. C. K. Vesborg, K. W. Jacobsen, O. Hansen and I. Chorkendorff, *Energy Environ. Sci.*, 2014, **7**, 2397–2413.
6. S. Hu, C. Xiang, S. Haussener, A. D. Berger and N. S. Lewis, *Energy Environ. Sci.*, 2013, **6**, 2984.
7. M. C. Hanna and A. J. Nozik, *J. Appl. Phys.*, 2006, **100**, 074510.
8. L. C. Seitz, Z. Chen, A. J. Forman, B. A. Pinaud, J. D. Benck and T. F. Jaramillo, *ChemSusChem*, 2014, **7**, 1372–1385.
9. M. D. Archer and J. R. Bolton, *J. Phys. Chem.*, 1990, **94**, 8028–8036.
10. I. E. Castelli, D. D. Landis, K. S. Thygesen, S. Dahl, I. Chorkendorff, T. F. Jaramillo and K. W. Jacobsen, *Energy Environ. Sci.*, 2012, **5**, 9034.
11. S. Haussener, S. Hu, C. Xiang, A. Z. Weber and N. S. Lewis, *Energy Environ. Sci.*, 2013, **6**, 3605.
12. *ICSDWeb*, http://www.fiz-karlsruhe.de/icsd_web.html.
13. *Materials Project – A Materials Genome Approach*, http://materialsproject.org/.
14. I. E. Castelli, F. Hüser, M. Pandey, H. Li, K. S. Thygesen, B. Seger, A. Jain, K. A. Persson, G. Ceder and K. W. Jacobsen, *Adv. Energy Mater.*, 2014, **5**, 1400915.
15. J. Jia, L. C. Seitz, J. D. Benck, Y. Huo, Y. Chen, J. W. D. Ng, T. Bilir, J. S. Harris and T. F. Jaramillo, *Nat. Commun.*, 2016, **7**, 13237.
16. P. Hohenberg and W. Kohn, *Phys. Rev.*, 1964, **136**, 864.
17. W. Kohn and L. J. Sham, *Phys. Rev.*, 1965, **140**, 1133.
18. N. Marzari, *Nat. Mater.*, 2016, **15**, 381–382.

19. *Recent Developments and Applications of Modern Density Functional Theory (Theoretical and Computational Chemistry)*, ed. J. Seminario, Elsevier Science, 1996.
20. K. Lejaeghere, G. Bihlmayer, T. Björkman, P. Blaha, S. Blügel, V. Blum, D. Caliste, I. E. Castelli, S. J. Clark, A. Dal Corso, S. de Gironcoli, T. Deutsch, J. K. Dewhurst, I. Di Marco, C. Draxl, M. Dułak, O. Eriksson, J. A. Flores-Livas, K. F. Garrity, L. Genovese, P. Giannozzi, M. Giantomassi, S. Goedecker, X. Gonze, O. Grånäs, E. K. U. Gross, A. Gulans, F. Gygi, D. R. Hamann, P. J. Hasnip, N. A. W. Holzwarth, D. Iuşan, D. B. Jochym, F. Jollet, D. Jones, G. Kresse, K. Koepernik, E. Küçükbenli, Y. O. Kvashnin, I. L. M. Locht, S. Lubeck, M. Marsman, N. Marzari, U. Nitzsche, L. Nordström, T. Ozaki, L. Paulatto, C. J. Pickard, W. Poelmans, M. I. J. Probert, K. Refson, M. Richter, G.-M. Rignanese, S. Saha, M. Scheffler, M. Schlipf, K. Schwarz, S. Sharma, F. Tavazza, P. Thunström, A. Tkatchenko, M. Torrent, D. Vanderbilt, M. J. van Setten, V. Van Speybroeck, J. M. Wills, J. R. Yates, G.-X. Zhang and S. Cottenier, *Science*, 2016, **351**, 1415.
21. A. Franceschetti and A. Zunger, *Nature*, 1999, **402**, 60–63.
22. S. Curtarolo, G. L. W. Hart, M. B. Nardelli, N. Mingo, S. Sanvito and O. Levy, *Nat. Mater.*, 2013, **12**, 191–201.
23. G. Ceder, Y.-M. Chiang, D. R. Sadoway, M. K. Aydinol, Y.-I. Jang and B. Huang, *Nature*, 1998, **392**, 694–696.
24. L.-C. Lin, A. H. Berger, R. L. Martin, J. Kim, J. A. Swisher, K. Jariwala, C. H. Rycroft, A. S. Bhown, M. W. Deem, M. Haranczyk and B. Smit, *Nat. Mater.*, 2012, **11**, 633–641.
25. S. Curtarolo, W. Setyawan, G. L. Hart, M. Jahnatek, R. V. Chepulskii, R. H. Taylor, S. Wang, J. Xue, K. Yang, O. Levy, M. J. Mehl, H. T. Stokes, D. O. Demchenko and D. Morgan, *Comput. Mater. Sci.*, 2012, **58**, 218–226.
26. J. Greeley, I. E. L. Stephens, A. S. Bondarenko, T. P. Johansson, H. A. Hansen, T. F. Jaramillo, J. Rossmeisl, I. Chorkendorff and J. K. Nørskov, *Nat. Chem.*, 2009, **1**, 552–556.
27. G. Hautier, A. Miglio, G. Ceder, G.-M. Rignanese and X. Gonze, *Nat. Commun.*, 2013, **4**, 2292.
28. K. B. Ørnsø, J. M. García-Lastra and K. S. Thygesen, *Phys. Chem. Chem. Phys.*, 2013, **15**, 19478.
29. M. d'Avezac, J.-W. Luo, T. Chanier and A. Zunger, *Phys. Rev. Lett.*, 2012, **108**, 027401.
30. J. Hachmann, R. Olivares-Amaya, S. Atahan-Evrenk, C. Amador-Bedolla, R. S. Sanchez-Carrera, A. Gold-Parker, L. Vogt, A. M. Brockway and A. Aspuru-Guzik, *J. Phys. Chem. Lett.*, 2011, **2**, 2241–2251.
31. I. E. Castelli, T. Olsen, S. Datta, D. D. Landis, S. Dahl, K. S. Thygesen and K. W. Jacobsen, *Energy Environ. Sci.*, 2012, **5**, 5814–5819.
32. Y. Wu, P. Lazic, G. Hautier, K. Persson and G. Ceder, *Energy Environ. Sci.*, 2013, **6**, 157–168.
33. A. Jain, I. E. Castelli, G. Hautier, D. H. Bailey and K. W. Jacobsen, *J. Mater. Sci.*, 2013, **48**, 6519–6534.

34. I. E. Castelli, J. M. García-Lastra, F. Hüser, K. S. Thygesen and K. W. Jacobsen, *New J. Phys.*, 2013, **15**, 105026.
35. I. E. Castelli and K. W. Jacobsen, *Modell. Simul. Mater. Sci. Eng.*, 2014, **22**, 055007.
36. S. Goedecker, *J. Chem. Phys.*, 2004, **120**, 9911–9917.
37. A. Laio and M. Parrinello, *Proc. Natl. Acad. Sci.*, 2002, **99**, 12562–12566.
38. G. H. Johannesson, T. Bligaard, A. V. Ruban, H. L. Skriver, K. W. Jacobsen and J. K. Nørskov, *Phys. Rev. Lett.*, 2002, **88**, 255506.
39. A. R. Oganov and C. W. Glass, *J. Chem. Phys.*, 2006, **124**, 244704.
40. S. Curtarolo, D. Morgan, K. Persson, J. Rodgers and G. Ceder, *Phys. Rev. Lett.*, 2003, **91**, 135503.
41. *The NOMAD Repository*, http://repository.nomad-coe.eu/.
42. *AFLOWLIB.ORG: a Distributed Materials Genome Properties Repository from High-throughput Ab-initio Calculation*, http://aflowlib.org/.
43. *The Open Quantum Materials Database*, http://oqmd.org/.
44. *NREL Materials Database*, https://materials.nrel.gov/.
45. D. D. Landis, J. S. Hummelshoj, S. Nestorov, J. Greeley, M. Dulak, T. Bligaard, J. K. Nørskov and K. W. Jacobsen, *Comput. Sci. Eng.*, 2012, **14**, 51–57.
46. *The Computational Materials Repository*, https://cmr.fysik.dtu.dk/.
47. *InfoMine*, http://www.infomine.com/chartsanddata/.
48. *Wikipedia – Abundance of the chemical elements*, https://en.wikipedia.org/wiki/Abundance_of_the_chemical_elements.
49. *Chemicool*, http://www.chemicool.com/elements/.
50. P. C. K. Vesborg and T. F. Jaramillo, *RSC Adv.*, 2012, **2**, 7933.
51. M. W. Gaultois, T. D. Sparks, C. K. H. Borg, R. Seshadri, W. D. Bonificio and D. R. Clarke, *Chem. Mater.*, 2013, **25**, 2911–2920.
52. F. Chiarella, A. Zappettini, F. Licci, I. Borriello, G. Cantele, D. Ninno, A. Cassinese and R. Vaglio, *Phys. Rev. B*, 2008, **77**, 045129.
53. *EPA – United States Environmental Protection Agency*, https://www.epa.gov/lead.
54. A. H. Larsen, J. J. Mortensen, J. Blomqvist, I. E. Castelli, R. Christensen, M. Dułak, J. Friis, M. N. Groves, B. Hammer, C. Hargus, E. D. Hermes, P. C. Jennings, P. B. Jensen, J. Kermode, J. R. Kitchin, E. L. Kolsbjerg, J. Kubal, K. Kaasbjerg, S. Lysgaard, J. B. Maronsson, T. Maxson, T. Olsen, L. Pastewka, A. Peterson, C. Rostgaard, J. Schiøtz, O. Schütt, M. Strange, K. S. Thygesen, T. Vegge, L. Vilhelmsen, M. Walter, Z. Zeng and K. W. Jacobsen, *J. Phys.: Condens. Matter*, 2017, **29**, 273002.
55. *Atomistic Simulation Environment – Phase Diagrams and Pourbaix Diagrams*, https://wiki.fysik.dtu.dk/ase/ase/phasediagram/phasediagram.html.
56. E. D. Verink, in *Simplified Procedure for Constructing Pourbaix Diagrams*, John Wiley & Sons, Inc., 2011, pp. 93–101.
57. K. A. Persson, B. Waldwick, P. Lazic and G. Ceder, *Phys. Rev. B*, 2012, **85**, 235438.

58. I. E. Castelli, K. S. Thygesen and K. W. Jacobsen, *Top. Catal.*, 2013, **57**, 265–272.
59. J. W. Johnson, E. H. Oelkers and H. C. Helgeson, *Comput. Geosci.*, 1992, **18**, 899–947.
60. M. Pourbaix, *Atlas of Electrochemical Equilibria in Aqueous Solutions*, Pergamon Press, 1966.
61. B.-J. Kim, D. F. Abbott, X. Cheng, E. Fabbri, M. Nachtegaal, F. Bozza, I. E. Castelli, D. Lebedev, R. Schäublin, C. Copéret, T. Graule, N. Marzari and T. J. Schmidt, *ACS Catalysis*, 2017, **7**, 3245–3256.
62. J. P. Perdew and A. Zunger, *Phys. Rev. B*, 1981, **23**, 5048–5079.
63. R. W. Godby, M. Schlüter and L. J. Sham, *Phys. Rev. Lett.*, 1986, **56**, 2415–2418.
64. J. Heyd, G. E. Scuseria and M. Ernzerhof, *J. Chem. Phys.*, 2003, **118**, 8207–8215.
65. O. Gritsenko, R. van Leeuwen, E. van Lenthe and E. J. Baerends, *Phys. Rev. A*, 1995, **51**, 1944.
66. M. Kuisma, J. Ojanen, J. Enkovaara and T. T. Rantala, *Phys. Rev. B*, 2010, **82**, 115106.
67. F. Hüser, T. Olsen and K. S. Thygesen, *Phys. Rev. B*, 2013, **87**, 235132.
68. I. E. Castelli, J. M. García-Lastra, K. S. Thygesen and K. W. Jacobsen, *APL Mater.*, 2014, **2**, 081514.
69. L. Yu and A. Zunger, *Phys. Rev. Lett.*, 2012, **108**, 068701.
70. W. Setyawan and S. Curtarolo, *Comput. Mater. Sci.*, 2010, **49**, 299–312.
71. F. F. Abdi, L. Han, A. H. M. Smets, M. Zeman, B. Dam and R. van de Krol, *Nat. Commun.*, 2013, **4**, 2195.
72. R. Chen, J. Zhu, H. An, F. Fan and C. Li, *Faraday Discussions*, 2017.
73. E. M. Hutter, M. C. Gelvez-Rueda, A. Osherov, V. Bulovic, F. C. Grozema, S. D. Stranks and T. J. Savenije, *Nat. Mater.*, 2017, **16**, 115–120.
74. F. Zheng, L. Z. Tan, S. Liu and A. M. Rappe, *Nano Lett.*, 2015, **15**, 7794–7800.
75. M. Pandey, F. A. Rasmussen, K. Kuhar, T. Olsen, K. W. Jacobsen and K. S. Thygesen, *Nano Lett.*, 2016, **16**, 2234–2239.
76. C. Freysoldt, B. Grabowski, T. Hickel, J. Neugebauer, G. Kresse, A. Janotti and C. G. Van de Walle, *Rev. Mod. Phys.*, 2014, **86**, 253–305.
77. A. Zakutayev, C. M. Caskey, A. N. Fioretti, D. S. Ginley, J. Vidal, V. Stevanović, E. Tea and S. Lany, *J. Phys. Chem. Lett.*, 2014, **5**, 1117–1125.
78. R. E. Brandt, V. Stevanović, D. S. Ginley and T. Buonassisi, *MRS Commun.*, 2015, **5**, 265–275.
79. J. Yan, J. J. Mortensen, K. W. Jacobsen and K. S. Thygesen, *Phys. Rev. B*, 2011, **83**, 245122.
80. I. E. Castelli, K. S. Thygesen and K. W. Jacobsen, *J. Mater. Chem. A*, 2015, **3**, 12343–12349.
81. J. Heo, L. Yu, E. Altschul, B. E. Waters, J. F. Wager, A. Zunger and D. A. Keszler, *Chem. Mater.*, 2017, **29**, 2594–2598.

82. L. Yu, S. Lany, R. Kykyneshi, V. Jieratum, R. Ravichandran, B. Pelatt, E. Altschul, H. A. S. Platt, J. F. Wager, D. A. Keszler and A. Zunger, *Adv. Energy Mater.*, 2011, **1**, 748–753.
83. M.-L. Tsai, S.-H. Su, J.-K. Chang, D.-S. Tsai, C.-H. Chen, C.-I. Wu, L.-J. Li, L.-J. Chen and J.-H. He, *ACS Nano*, 2014, **8**, 8317–8322.
84. G. Grosso and G. P. Parravicini, *Solid State Physics*, Academic Press, 2000.
85. S. Latini, T. Olsen and K. S. Thygesen, *Phys. Rev. B*, 2015, **92**, 245123.
86. S. Sharma, J. K. Dewhurst, A. Sanna and E. K. U. Gross, *Phys. Rev. Lett.*, 2011, **107**, 186401.
87. E. E. Salpeter and H. A. Bethe, *Phys. Rev.*, 1951, **84**, 1232–1242.
88. A. Marini, C. Hogan, M. Grüning and D. Varsano, *Comput. Phys. Commun.*, 2009, **180**, 1392–1403.
89. V. Stevanović, S. Lany, D. S. Ginley, W. Tumas and A. Zunger, *Phys. Chem. Chem. Phys.*, 2014, **16**, 3706.
90. Y. Wu, M. K. Y. Chan and G. Ceder, *Phys. Rev. B*, 2011, **83**, 235301.
91. M. H. Hansen and J. Rossmeisl, *J. Phys. Chem. C*, 2016, **120**, 29135–29143.
92. M. A. Butler and D. S. Ginley, *J. Electrochem. Soc.*, 1978, **125**, 228–232.
93. Y. Xu and M. A. Schoonen, *Am. Mineral.*, 2000, **85**, 543–556.
94. D. Stradi, U. Martinez, A. Blom, M. Brandbyge and K. Stokbro, *Phys. Rev. B*, 2016, **93**, 155302.
95. K. Letchworth-Weaver and T. A. Arias, *Phys. Rev. B*, 2012, **86**, 075140.
96. Y. Jiao, A. Hellman, Y. Fang, S. Gao and M. Käll, *Sci. Rep.*, 2015, **5**, 11374.
97. A. Kudo and Y. Miseki, *Chem. Soc. Rev.*, 2009, **38**, 253–278.
98. A. Wolcott, W. A. Smith, T. R. Kuykendall, Y. Zhao and J. Z. Zhang, *Adv. Funct. Mater.*, 2009, **19**, 1849–1856.
99. P. Liao, J. A. Keith and E. A. Carter, *J. Am. Chem. Soc.*, 2012, **134**, 13296–13309.
100. D. Wang, A. Pierre, M. G. Kibria, K. Cui, X. Han, K. H. Bevan, H. Guo, S. Paradis, A.-R. Hakima and Z. Mi, *Nano Lett.*, 2011, **11**, 2353–2357.
101. J. Luo, L. Steier, M.-K. Son, M. Schreier, M. T. Mayer and M. Grätzel, *Nano Lett.*, 2016, **16**, 1848–1857.
102. M. Stodolny and M. Laniecki, *Catal. Today*, 2009, **142**, 314–319.
103. W. Septina and S. D. Tilley, *Curr. Opin. Electrochem.*, 2017, **2**, 120–127.
104. M. Grätzel, *Nature*, 2001, **414**, 338–344.
105. S. Chen and L.-W. Wang, *Chem. Mater.*, 2012, **24**, 3659–3666.
106. T. Ishihara, *Perovskite Oxide for Solid Oxide Fuel Cells*, Springer Verlag, 2009.
107. H. Kato, H. Kobayashi and A. Kudo, *J. Phys. Chem. B*, 2002, **106**, 12441.
108. D. Yamasita, T. Takata, M. Hara, J. Kondo and K. Domen, *Solid State Ionics*, 2004, **172**, 591–595.
109. P. C. K. Vesborg and T. F. Jaramillo, *RSC Adv.*, 2012, **2**, 7933–7947.
110. I. Borriello, G. Cantele and D. Ninno, *Phys. Rev. B*, 2008, **77**, 235214.

111. M. Saliba, T. Matsui, J.-Y. Seo, K. Domanski, J.-P. Correa-Baena, M. K. Nazeeruddin, S. M. Zakeeruddin, W. Tress, A. Abate, A. Hagfeldt and M. Grätzel, *Energy Environ. Sci.*, 2016, **9**, 1989–1997.
112. J. Luo, J.-H. Im, M. T. Mayer, M. Schreier, M. K. Nazeeruddin, N.-G. Park, S. D. Tilley, H. J. Fan and M. Grätzel, *Science*, 2014, **345**, 1593–1596.
113. Z. Shi, J. Guo, Y. Chen, Q. Li, Y. Pan, H. Zhang, Y. Xia and W. Huang, *Adv. Mater.*, 2017, 1605005.
114. D. Cortecchia, H. A. Dewi, J. Yin, A. Bruno, S. Chen, T. Baikie, P. P. Boix, M. Grätzel, S. Mhaisalkar, C. Soci and N. Mathews, *Inorg. Chem.*, 2016, **55**, 1044–1052.
115. I. E. Castelli, K. S. Thygesen and K. W. Jacobsen, *MRS Proc.*, 2013, **1523**.
116. G. Pilania, A. Mannodi-Kanakkithodi, B. P. Uberuaga, R. Ramprasad, J. E. Gubernatis and T. Lookman, *Sci. Rep.*, 2016, **6**, 19375.
117. J. Suntivich, H. A. Gasteiger, N. Yabuuchi, H. Nakanishi, J. B. Goodenough and Y. Shao-Horn, *Nat. Chem.*, 2011, **3**, 546–550.
118. A. Grimaud, K. J. May, C. E. Carlton, Y.-L. Lee, M. Risch, W. T. Hong, J. Zhou and Y. Shao-Horn, *Nat. Commun.*, 2013, **4**, 2439.
119. F. Hergert and R. Hock, *Thin Solid Films*, 2007, **515**, 5953–5956.
120. H. Dittrich, A. Bieniok, U. Brendel, M. Grodzicki and D. Topa, *Thin Solid Films*, 2007, **515**, 5745–5750.
121. S. Fiechter, M. Martinez, G. Schmidt, W. Henrion and Y. Tomm, *J. Phys. Chem. Solids*, 2003, **64**, 1859–1862.
122. H. Katagiri, K. Jimbo, W. S. Maw, K. Oishi, M. Yamazaki, H. Araki and A. Takeuchi, *Thin Solid Films*, 2009, **517**, 2455–2460.
123. M. G. Panthani, V. Akhavan, B. Goodfellow, J. P. Schmidtke, L. Dunn, A. Dodabalapur, P. F. Barbara and B. A. Korgel, *J. Am. Chem. Soc.*, 2008, **130**, 16770–16777.
124. S. Körbel, M. A. L. Marques and S. Botti, *J. Mater. Chem. C*, 2016, **4**, 3157–3167.
125. J. A. Brehm, J. W. Bennett, M. R. Schoenberg, I. Grinberg and A. M. Rappe, *J. Chem. Phys.*, 2014, **140**, 224703.
126. C.-S. Lee, K. M. Kleinke and H. Kleinke, *Solid State Sci.*, 2005, **7**, 1049–1054.
127. J. Heo, L. Yu, E. Altschul, B. E. Waters, J. F. Wager, A. Zunger and D. A. Keszler, *Chem. Mater.*, 2017, **29**, 2594–2598.
128. W. Meng, B. Saparov, F. Hong, J. Wang, D. B. Mitzi and Y. Yan, *Chem. Mater.*, 2016, **28**, 821–829.
129. Y.-Y. Sun, M. L. Agiorgousis, P. Zhang and S. Zhang, *Nano Lett.*, 2015, **15**, 581–585.
130. J. E. Saal, S. Kirklin, M. Aykol, B. Meredig and C. Wolverton, *JOM*, 2013, **65**, 1501–1509.
131. S. Kirklin, J. E. Saal, B. Meredig, A. Thompson, J. W. Doak, M. Aykol, S. Rühl and C. Wolverton, *npj Comput. Mater.*, 2015, **1**, 15010.
132. B. Luo, G. Liu and L. Wang, *Nanoscale*, 2016, **8**, 6904–6920.

133. K. F. Mak, C. Lee, J. Hone, J. Shan and T. F. Heinz, *Phys. Rev. Lett.*, 2010, **105**, 136805.
134. M. Pandey, K. W. Jacobsen and K. S. Thygesen, *J. Phys. Chem. C*, 2016, **120**, 23024–23029.
135. A. K. Singh, K. Mathew, H. L. Zhuang and R. G. Hennig, *J. Phys. Chem. Lett.*, 2015, **6**, 1087–1098.
136. D. Deng, K. S. Novoselov, Q. Fu, N. Zheng, Z. Tian and X. Bao, *Nat. Nanotechnol.*, 2016, **11**, 218–230.
137. X. Yu, M. S. Prévot, N. Guijarro and K. Sivula, *Nat. Commun.*, 2015, **6**, 7596.
138. Y. Sun, H. Cheng, S. Gao, Z. Sun, Q. Liu, Q. Liu, F. Lei, T. Yao, J. He, S. Wei and Y. Xie, *Angew. Chem., Int. Ed.*, 2012, **51**, 8727–8731.
139. M. Ashton, J. Paul, S. B. Sinnott and R. G. Hennig, *Phys. Rev. Lett.*, 2017, **118**, 106101.
140. *Materials Web*, https://materialsweb.org/.
141. K. Andersen, S. Latini and K. S. Thygesen, *Nano Lett.*, 2015, **15**, 4616–4621.
142. M. Pandey, K. W. Jacobsen and K. S. Thygesen, *J. Phys. Chem. Lett.*, 2016, **7**, 4346–4352.
143. M. Pandey, A. Vojvodic, K. S. Thygesen and K. W. Jacobsen, *J. Phys. Chem. Lett.*, 2015, **6**, 1577–1585.

CHAPTER 4

Unravelling the Charge Transfer Mechanism in Water Splitting Hematite Photoanodes

HAMED HAJIBABAEI, YUAN GAO AND THOMAS W. HAMANN*

Michigan State University, Department of Chemistry, 578 Shaw Lane, East Lansing, Michigan 48824-1322, USA
*Email: hamann@chemistry.msu.edu

4.1 Introduction

In response to the urgent need to meet the increasing energy demand and to eliminate the environmental issues associated with combustion of fossil fuels, research has been carried out to investigate the photoelectrochemical (PEC) water splitting with semiconducting light absorbers as a viable strategy toward a carbon-free fuel cycle.[1–6] In practice, any given photoelectrode must meet several criteria to be suitable for efficient PEC water splitting: (1) band edge positions that straddle both hydrogen and oxygen evolution potentials, (2) narrow band gap that allows for substantial light absorption, (3) favourable bulk and surface characteristics that allows for efficient charge separation and charge transfer at the interface, respectively, (4) stability under PEC conditions, and (5) earth abundant composition for large scale production.[3,7,8] So far, no single semiconductor material has fulfilled all these requirements.[3,7,8] An alternative strategy is to separate the two half

Energy and Environment Series No. 20
Advances in Photoelectrochemical Water Splitting: Theory, Experiment and Systems Analysis
Edited by S. David Tilley, Stephan Lany and Roel van de Krol

Published by the Royal Society of Chemistry, www.rsc.org

reactions by coupling a photoanode with a suitable photocathode in a tandem cell configuration.[3] More effort has been put into optimising photoanodes for water oxidation because of the kinetic bottleneck of the four-hole process that has to compete with the unfavourable recombination processes. Many transition metal oxides, oxynitrides and nitrides such as $CuWO_4$,[9–11] TiO_2,[12] $BiVO_4$,[13–15] Fe_2O_3,[16–19] Ta_3N_5[20–23] and TaON[24] have been explored as photoanode materials for PEC water oxidation. Among these materials, α-Fe_2O_3 (hematite), especially, stands out as a promising photoanode material for PEC water oxidation.

Hematite exhibits outstanding stability in a wide pH range from pH 7 to ~pH 14, which is favorable for oxygen evolution reactions. In addition, Fe is one of the most abundant elements in Earth's upper continental crust after O, Si and Al.[25] It also possesses a reasonably small band gap of ~2.1 eV which has a large overlap with the solar spectrum. With the absorption edge extended to 590 nm it can produce a theoretical maximum photocurrent density of 12.5 $mA\,cm^{-2}$, which corresponds to 16.8% solar-to-hydrogen (STH) efficiency. In addition, the valence band edge position of α-Fe_2O_3 is sufficiently positive which makes it suitable for water oxidation. The first example of photolysis of water with hematite was introduced in 1976.[26] Since then, a progressively large number of studies were conducted to identify and improve the bottleneck processes limiting the efficiency of water oxidation on hematite. Very good reviews on challenges, limitations, and progress of hematite as a photoanode for PEC water oxidation are provided in these references.[27–30]

Despite the remarkable improvements of the PEC performance of hematite made over the past decade, the solar energy conversion efficiency of hematite is still far from its theoretical values. In general, the solar energy conversion efficiency of a photoanode is controlled by three processes: (1) light harvesting efficiency (LHE), (2) charge separation efficiency related to the transport of photogenerated charge carriers, and (3) hole collection efficiency (HCE) at the electrode–electrolyte interface. Hematite has a relatively long absorption depth but a short minority carrier collection length.[31–36] In order to realise a highly efficient hematite photoanode, therefore, both the LHE and charge separation efficiencies need to be simultaneously maximised. Nanostructuring and doping with foreign elements are the two well-studied strategies to optimise LHE and charge separation efficiency. The reader is encouraged to visit these references for more information.[29,37–41] Although these strategies are highly effective to improve the performance of hematite,[42–44] they are found to have less effect on HCE at the surface of the electrode that is mainly controlled by the surface properties of the semiconductor and it depends on the surface terminal atoms and surface crystal orientations.[16,45,46]

The hematite surface has a low catalytic activity toward water oxidation, thus hole transfer to the acceptor species in solution (injection) competes with the parallel electron–hole recombination reaction at the surface which substantially suppress the efficiency of the electrode. It is now well documented that water oxidation at α-Fe_2O_3 surface occurs *via* surface states.

In the case of slow water oxidation kinetics, these surface states can adversely act as the recombination centers which accelerate the electron (conduction band) and the trapped-hole (surface states) recombination reaction. To suppress this recombination, a large overpotential with respect to the flat band potential is necessary to achieve a sustainable water oxidation.[41] Thus, to realise a high-performance hematite photoanode, it is important to minimise this overpotential by promoting the kinetics of water oxidation and/or suppressing the unfavourable surface electron–hole recombination. Surface modification with various catalytic and non-catalytic materials were shown to be an effective strategy to improve the performance of hematite. The interested reader is encouraged to review these references.[17,47–52] Although the true mechanisms with which these surface modifications reduce the overpotential are not fully understood and remain an active area of study, the common observation among these studies indicates a strong correlation between the concentration of photogenerated holes at the surface, *i.e.* reduced electron–hole recombination, and water oxidation overpotential.[48,51,53,54] The chemical nature of the surface states on hematite and understanding how photogenerated holes at the surface couple with the formation of chemical bonds is, therefore, a crucial and important step towards improving the HCE. Insights gained from unraveling the interfacial charge transfer processes and how they couple with the chemical transformation at the surface of electrode are the key factors of elaborate modification and design of an efficient surface toward water oxidation reaction. The knowledge learned from α-Fe_2O_3 is not only helpful for optimising its performance, but also can be applied to many other promising water oxidation systems. In this chapter, we aim to discuss the recent advances on the mechanism of water oxidation at the surface of hematite photoelectrode. Moreover, its resemblance to the mechanism of water oxidation on the iron-based homogeneous catalysts systems and other relevant materials is reviewed.

4.2 Photoelectrochemical Methods

4.2.1 Current Density—Voltage (*J–V*) Curve Measurements

The water oxidation kinetics at the hematite (α-Fe_2O_3) surface are generally reported to be sluggish, which contributes to increased recombination at surface and therefore loss of photovoltage.[33,36,55–57] By comparing the current density–voltage (*J–V*) curve of water oxidation with α-Fe_2O_3 electrodes with that of oxidizing a suitable sacrificial hole scavenger at the same surface, one should be able to determine the extent to which surface recombination limits the water oxidation efficiency. H_2O_2 has been widely used as a sacrificial hole scavenger to study water oxidation efficiency with α-Fe_2O_3 electrodes.[36,55,58] Many research groups compared the *J–V* curves measured with and without H_2O_2 and consistently found that the addition of H_2O_2 greatly enhances the photocurrent and lowers the photocurrent onset potential for α-Fe_2O_3

electrodes prepared using various methods.[36,56] Bockris and co-worker showed that the charge transfer rate for H_2O_2 oxidation at α-Fe_2O_3 surface is ~10 times faster than that for water oxidation.[53] Hamann and co-workers also measured the photocurrent generated by α-Fe_2O_3 in contact with a one-electron redox couple, $[Fe(CN)_6]^{3-/4-}$, and showed that the photocurrent initially increases with concentration of the hole acceptor, *i.e.* $[Fe(CN)_6]^{4-}$, and quickly saturates at relatively low concentration (20 mM).[43] This indicates that when diffusion limitation is eliminated, the charge transfer efficiency of $[Fe(CN)_6]^{3-/4-}$ at hematite surface is essentially unity. The *J*–*V* curve of the hematite electrode measured in contact with H_2O shows ~600 mV positive shift compared with that measured in a $[Fe(CN)_6]^{3-/4-}$ electrolyte (Figure 4.1), which agrees with the result from comparing with oxidation of H_2O_2.[36,56] The loss in photocurrent and photovoltage in water oxidation compared with this redox couple indicates a non-optimal HCE for water oxidation with α-Fe_2O_3 electrodes. These results collectively suggest that the slow kinetics of the four-hole water oxidation reaction at the α-Fe_2O_3 surface result in enhanced surface recombination.

4.2.2 Current Transient Measurements

Photocurrent transient measurements have also been performed to investigate the kinetics of water oxidation at the hematite surface. Several groups have measured anodic spikes of current upon illumination, which quickly decay to a lower steady state water oxidation photocurrent. When the illumination is interrupted, a cathodic current spike appears and eventually decays to the steady state dark current.[60,61] A detailed mechanism of this

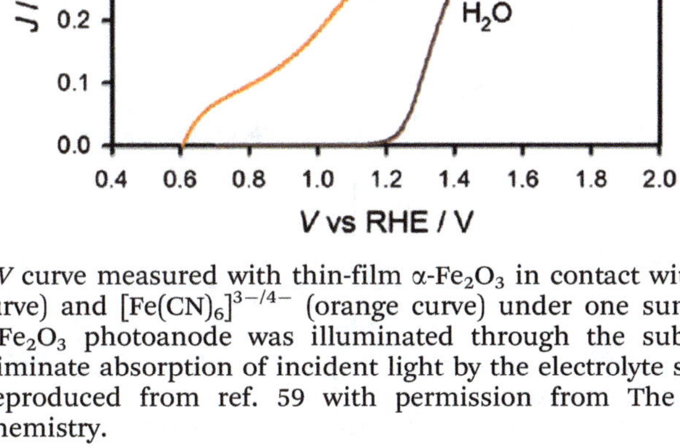

Figure 4.1 *J*–*V* curve measured with thin-film α-Fe_2O_3 in contact with H_2O (brown curve) and $[Fe(CN)_6]^{3-/4-}$ (orange curve) under one sun illumination. α-Fe_2O_3 photoanode was illuminated through the substrate side to eliminate absorption of incident light by the electrolyte solution. Reproduced from ref. 59 with permission from The Royal Society Chemistry.

process was explained by Peter *et al.*,[57] and the origin of the characteristic shape of the photocurrent response to square wave illumination is depicted in Figure 4.2. Upon illumination, the trapping of photogenerated valence band holes at surface states leads to an accumulation of surface holes, which contributes to the initial photocurrent. The hole current remains constant as long as the illumination is not interrupted. Meanwhile, an electron current is induced as the surface trapped holes recombine with conduction band electrons, and it increases towards its steady state value as the surface hole concentration builds up. Therefore, the net photocurrent that contributes to water oxidation decays from the initial value to the steady state photocurrent. When the illumination is interrupted, the hole current instantaneously drops to zero since the photogeneration of holes is ceased. The holes trapped in the surface state recombine with electrons from conduction band, resulting in a cathodic instantaneous current which eventually decays to the steady state dark current.

According to the theory described above, the anodic and cathodic current transients are attributed to trapping of holes (oxidation) and recombination of electrons (reduction) with surface species. A generalised model that describes the water oxidation process at α-Fe_2O_3 electrode surface was therefore proposed and widely recognised as shown in Figure 4.3. After charge

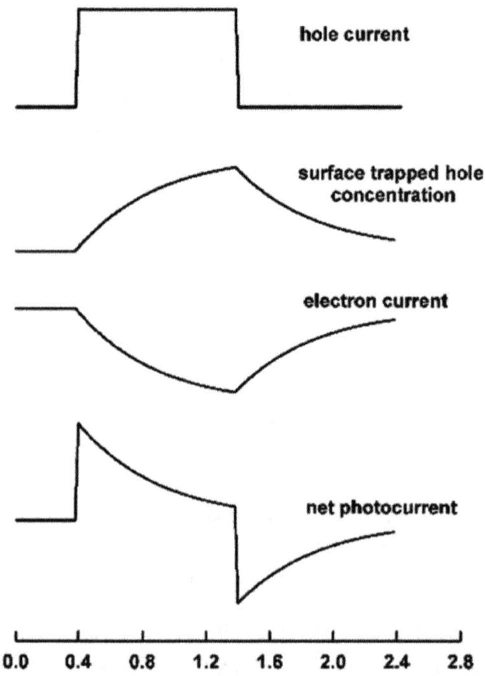

Figure 4.2 Origin of the characteristic shape of the photocurrent response to square wave illumination.
Adapted from ref. 57 with permission from The Royal Society Chemistry.

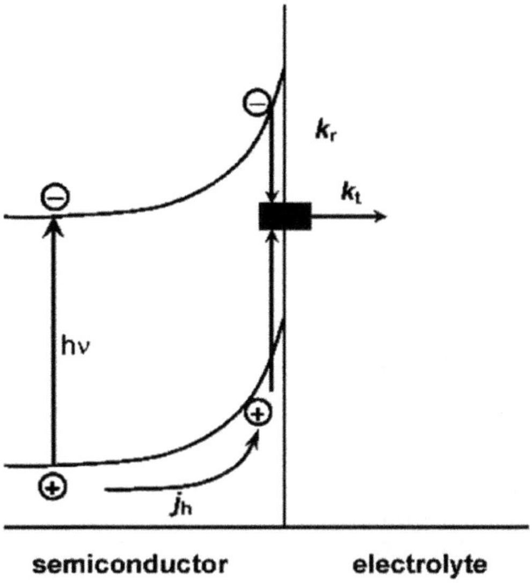

Figure 4.3 Generalised reaction scheme showing the competition between charge transfer and recombination. j_h is the hole flux into the surface and k_t and k_r are pseudo first order rate constants for charge transfer and surface recombination respectively.
Reproduced from ref. 57 with permission from The Royal Society Chemistry.

generation and separation, holes that are stored at the surface with a flux j_h either undergo injection with electrolyte species to generate oxygen with a rate constant k_t, or recombine with conduction band electrons with a rate constant k_r. The k_t and k_r are both pseudo first order rate constants. k_t depends on the relevant electrolyte species concentration (H_2O or OH^-) by $k_t = k_t'[H_2O]$ since $[H_2O]$ is constant. The charge transfer rate is first order with respect to surface hole concentration, p_{surf}. k_r depends on the surface electron concentration by $k_r = k_r' n_{surf}$, where k_r' is determined by the product of the thermal velocity of electrons ν_{th} and the electron capture cross section σ_n of the surface states. Based on the generation of the current transient and its characteristic decay described above, the time constant of the decay of photocurrent is related to $k_t + k_r$, and HCE is given by $k_t/(k_t + k_r)$, which is equivalent to the ratio of the steady state photocurrent to the instantaneous photocurrent observed upon illumination.

Since photocurrent transients indicate the presence of holes trapped at the surface during water oxidation, one can calculate the actual charge passed by integrating the current transient curve. Hamann *et al.* calculated the charge stored at the surface of α-Fe_2O_3 electrodes at different applied potentials.[59] It was observed that the stored charge as a function of potential shows a peak around 1.3 V *vs.* RHE, which agrees with the onset potential for water oxidation (Figure 4.4). This confirms that water oxidation with α-Fe_2O_3 electrodes is initiated by storing holes at the surface state.

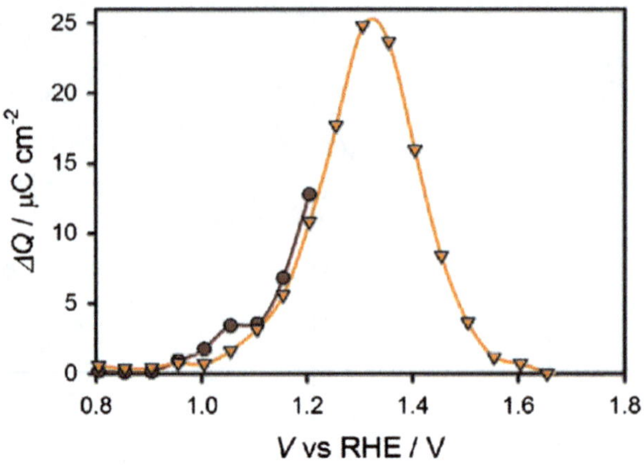

Figure 4.4 Transient anodic (maroon circles) and cathodic (orange triangles) charge passed as a function of potential.
Reproduced from ref. 59 with permission from The Royal Society Chemistry.

4.2.3 Cyclic Voltammetry (CV) Surface Measurements

Since surface trapping of holes means the photooxidation of a chemical species at the surface, and surface recombination is the reduction of this oxidised species, Hamann *et al.* applied cyclic voltammetry (CV) measurements to α-Fe_2O_3 electrodes to observe surface trapping and recombination of holes.[59] A positive bias (~2 V *vs.* RHE) was applied under illumination for 60 s to accumulate holes at the surface (oxidising the surface species), followed by applying a CV scan in the negative direction in the dark, which generates two cathodic peaks that involves the reduction of the oxidised surface species (*i.e.*, surface recombination), as shown in Figure 4.5(a). The successive fast reverse scan in the positive direction is not sufficient to re-oxidise the surface states and no peak is therefore observed in the second CV scan. One of the cathodic peaks is at the same potential as the onset potential of water oxidation of the hematite electrode (~1.25 V *vs.* RHE), and is attributed to the formation of an intermediate species for water oxidation. Further CV measurements in a non-aqueous electrolyte, acetonitrile (MeCN), were performed to help identify the nature of surface state of α-Fe_2O_3.[62] As shown in Figure 4.5(b), no peak was observed in absence of water, which indicates that these surface states are related to a water oxidation intermediate.

Hamann and co-workers also compared *J–V* curves of α-Fe_2O_3 electrodes after annealing at 500 °C and 800 °C to identify the nature of the surface states observed in the CV measurements.[63] Interestingly, only one peak in the CV surface measurement was observed with the 800 °C annealed α-Fe_2O_3, which shows a 400 mV cathodic shift in the onset potential compared with the 500 °C annealed α-Fe_2O_3 (Figure 5.5(c)). It was proposed that for the 500 °C annealed hematite, the surface state which shows a peak around water oxidation onset is essentially the intermediate for water oxidation, while the other one is attributed to a defect state at the surface. This defect state contributes to a late onset and Fermi level pinning of the hematite electrode during water oxidation. After annealing at 800 °C, only one surface state that acts as the intermediate is present and the overpotential required to overcome surface recombination is reduced.

4.2.4 Electrochemical Impedance Spectroscopy (EIS)

Wijayantha *et al.* performed electrochemical impedance spectroscopy (EIS) measurements to study the kinetics of water oxidation at hematite electrodes.[41] Their EIS results were analysed using a phenomenological approach assuming that the oxygen evolution reaction occurs *via* surface states. The rate constant for surface recombination was found to be almost constant in the onset potential region, which indicates Fermi level pinning in this region resulting from accumulation of holes during water oxidation. The phenomenological rate constant for charge transfer increases with applied potential, which suggests that the potential drop in the Helmholtz layer influences the activation energy, and thus the rate-limiting steps, for the oxygen evolution process.

Hamann *et al.* also employed EIS under different conditions including varying applied potential, light intensity, electrolyte and pH to investigate the steps involved in water oxidation.[59,64] An equivalent circuit model as shown in Figure 4.6(a) was used to interpret the EIS data. This equivalent circuit is built based on the model depicted in Figure 4.3, which assumes charge accumulation and charge transfer through the surface state at the electrode–electrolyte interface. It was found that the surface state capacitance, C_{ss}, which represents the charging of the surface state, follows a Gaussian distribution, with its peak potential coinciding with the photocurrent onset potential, as shown in Figure 4.6(b). Meanwhile, the charge transfer resistance from the surface state, $R_{ct,ss}$, reaches its minimum at the water oxidation onset potential. In addition, the coincidence of the photocurrent onset potential, peak position of the charge accumulation in photocurrent transient measurements, peak potential of the cathodic surface state peak in CV measurements and the maximum and minimum potential of C_{ss} and $R_{ct,ss}$, respectively, confirms that the interpretation of EIS data is reasonable, and water oxidation with α-Fe_2O_3 electrodes occurs through the surface state. This phenomenon of water oxidation proceeding by holes trapped in surface states can be interpreted as follows; surface accumulation of holes consists of the oxidation of the electrode surface to generate species with distinct electronic states that correspond to intermediate species in the complex water oxidation mechanism.[59]

According to the Mott–Schottky plot measured in H_2O and $[Fe(CN)_6]^{3-/4-}$ electrolytes (Figure 4.6(c)), the plot measured for water oxidation under illumination shows a horizontal shift compared to the dark plot, which is evidence for Fermi level pinning. The Mott-Schottky plots measured with $[Fe(CN)_6]^{3-/4-}$ are essentially identical both under dark and illumination conditions, and no Fermi level pinning is observed. This further confirms that the surface states are only active under photoelectrochemical water oxidation conditions. It was proposed that water oxidation with α-Fe_2O_3 is initiated by oxidative steps that form high valent Fe centers, which eventually follows a bimolecular decomposition step to form oxygen (eqn (4.1)).[59]

$$2\,Fe(\text{IV}) = O \longrightarrow 2\,Fe(\text{III}) + O_2 \uparrow \tag{4.1}$$

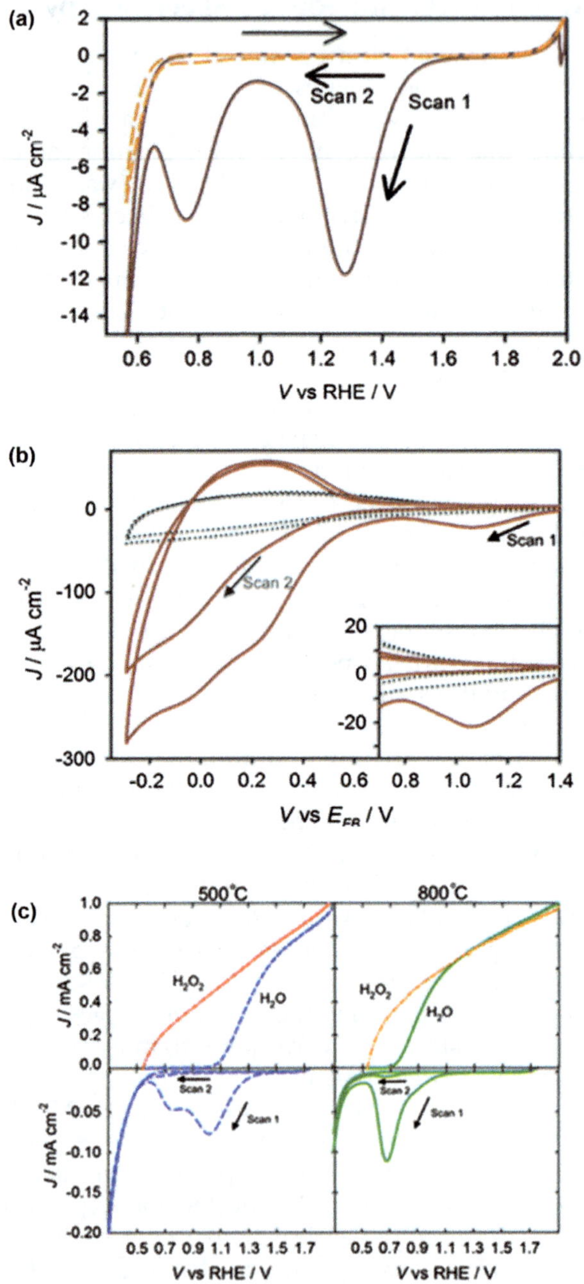

A similar oxygen evolution mechanism has been proposed before, albeit for Co-based oxygen evolution catalysts instead of α-Fe_2O_3 electrodes.[65,66] For hematite-based electrodes it was observed that the surface behaviour in EIS measurements as well as in photocurrent transients is potential-dependent. At lower potentials, there is an absence of maintained steady state photocurrent after the spike in current transients, which coincides with a relatively large low-frequency semicircle in EIS Nyquist plot which represents C_{ss}. As was noted earlier in this chapter, the amount of charge attributed to the surface oxidised Fe species reaches a peak that coincides with the onset of photocurrent. This indicates that a certain concentration of the surface intermediate species is necessary for the turn-on of water oxidation. At very positive potentials (*e.g.* > 1.45 V *vs.* RHE), the current transients are negligible, and the low-frequency semicircle in EIS disappears. A concomitant increase in steady state photocurrent is observed, with an essentially unity charge (hole) collection efficiency. Therefore, at high potentials, once a sufficient concentration of surface intermediates is built up, water oxidation at surface is no longer the rate limiting step for α-Fe_2O_3 electrode.

4.2.5 Intensity Modulated Photocurrent Spectroscopy (IMPS)

Peter *et al.* used intensity modulated photocurrent spectroscopy (IMPS) to investigate the kinetics of water oxidation with α-Fe_2O_3 electrodes.[57] It was supposed that the four-electron water oxidation process at the α-Fe_2O_3 surface involves the formation of higher-valent Fe states (possibly Fe(IV), Fe(V) and Fe(VI)) at the surface sites which proceeds through a bimolecular step similar to the one discussed above (eqn (4.2)):

$$2\ \mathrm{Fe(V)} + 4\ \mathrm{OH^-} \longrightarrow 2\ \mathrm{Fe(III)} + 2\ \mathrm{H_2O} + \mathrm{O_2}\uparrow \tag{4.2}$$

Here the Fe(V) sites are assumed to bind to OH ligands. Similar as the mechanism discussed above for interpreting the EIS results, the surface stored holes can also be thought of as 'surface trapped holes', which could

Figure 4.5 (a) Cyclic voltammetry of α-Fe_2O_3 in H_2O scanned at 200 mV s^{-1} after holding the electrode at 2 V *vs.* RHE for 60 s under one sun illumination. The first cycle is a solid red line and the second cycle is a dashed orange line. (b) CV response of α-Fe_2O_3 electrode measured in contact with MeCN (black dotted line) and after addition of 10 μL H_2O (0.2%) to the electrolyte (red solid line). CVs were recorded at 500 mV s^{-1} in the dark after holding the electrode at 1.5 V *vs.* Ag/AgCl under one sun illumination for 60 s. (c) *J–V* curves under H_2O and H_2O_2 oxidation conditions for α-Fe_2O_3 electrodes annealed at 500 °C and 800 °C. CV curves scanned at 1 V s^{-1} in the dark of electrodes annealed at 500 °C and 800 °C.
(a) was adapted from ref. 59 with permission from The Royal Society Chemistry. (b) was reprinted with permission from B. Klahr, S. Gimenez, O. Zandi, F. Fabregat-Santiago and T. Hamann, *ACS Appl. Mater. Interfaces*, 2015, 7, 7653–7660.[62] Copyright (2015) American Chemical Society. (c) was reprinted with permission from O. Zandi and T. W. Hamann, *J. Phys. Chem. Lett.*, 2014, 5, 1522–1526.[63] Copyright (2014) American Chemical Society.

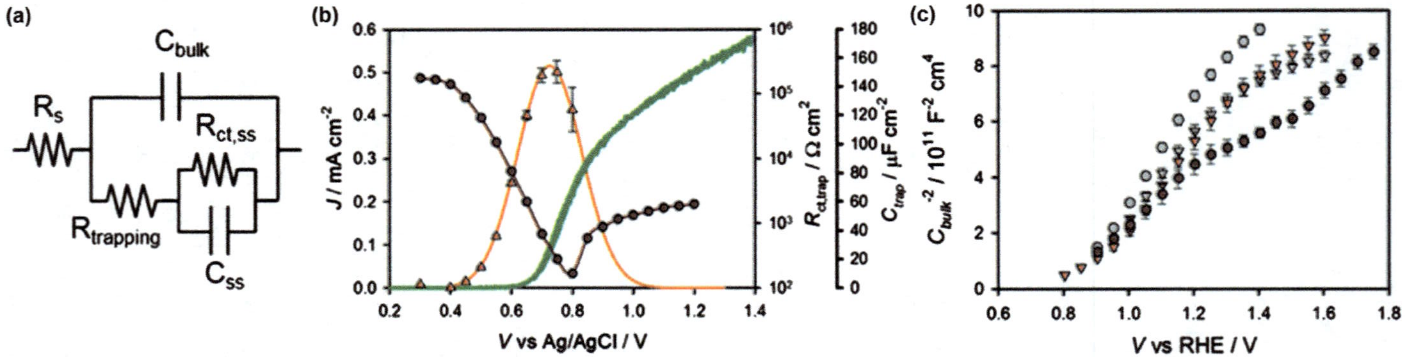

Figure 4.6 (a) Equivalent circuit for fitting EIS data for water oxidation under illumination conditions when two capacitive features are observed. (b) *J–V* curve (green solid line), C_{ss} (orange triangles) and $R_{ct,ss}$ (red circles) values obtained for a thin film α-Fe_2O_3 electrode under 1 sun illumination and pH 6.9. (c) Mott–Schottky plots calculated from space charge capacitance, C_{bulk}, determined from fitting EIS for H_2O (circles) and $[Fe(CN)_6]^{3-/4-}$ (triangles) electrolytes in the dark (grey) and under one sun illumination (coloured).

(a) and (b) were adapted with permission from B. Klahr, S. Gimenez, F. Fabregat-Santiago, T. Hamann and J. Bisquert, *J. Am. Chem. Soc.*, 2012, 134, 4294–4302.[64] Copyright (2012) American Chemical Society. (c) was adapted from ref. 59 with permission from The Royal Society Chemistry.

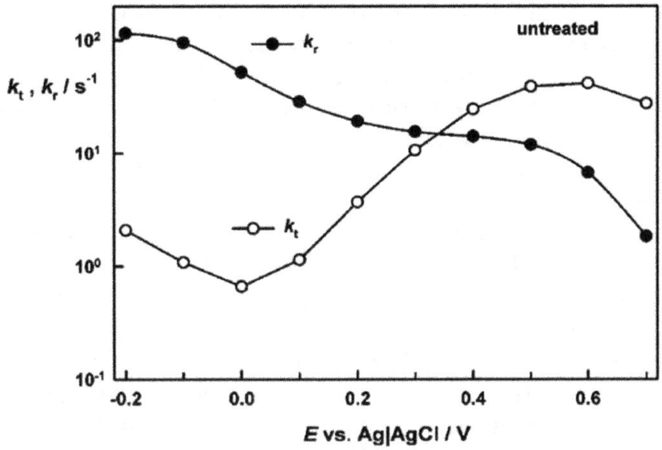

Figure 4.7 Potential dependence of the rate constants k_t and k_r for α-Fe_2O_3 electrode. Illumination 455 nm, dc photon flux 1.1×10^{17} cm^{-2} s^{-1}. Adapted from ref. 57 with permission from The Royal Society Chemistry.

recombine with electrons from the conduction band. These processes follow the same model depicted in Figure 4.3. The rate constants for injection and recombination, k_t and k_r, can be extracted from the frequency-dependent IMPS response, as shown in Figure 4.7. Results reveal that the charge transfer process leading to oxygen evolution is very slow on α-Fe_2O_3, and is competing with fast recombination *via* surface states. The slower decrease in k_r with increasing applied potential compared with ideal semiconductor/electrolyte junction (a factor of 10 for each 59 mV of additional reverse bias) indicates the presence of a high density of surface trap states that effectively cause Fermi level pinning at the surface. These results are consistent with the observations from EIS measurements, and explain the late onset of water oxidation with α-Fe_2O_3 electrodes compared with its flat band potential.

4.3 Mechanism of Water Oxidation

4.3.1 PEC Water Oxidation on Hematite Photoanode

The water oxidation is a four-hole reaction and thus is considered as a series of four elementary steps, each of which includes one hole transfer step (eqn (4.3) to (4.7)). As described above, upon light absorption and subsequent charge separation within an *n*-type semiconductor, the photogenerated holes, residing in the valence band (VB), are collected at the surface of the photoanode and drive the water oxidation reaction. The VB of hematite is composed of hybridised O 2p and Fe 3d orbitals.[67] Thus, the photogenerated holes arriving at the surface may be partially localised on O as well as the Fe ions. Further, the relative formation energies of the water oxidation intermediates depend on the nature as well as the coordination environments of the atoms on the surface. As a result, the rate-determining step (RDS) and the reaction pathway are strongly influenced by the surface structure and surface terminal groups. For example, Rothschild *et al.*[45] recently studied the effect of the crystal orientation on the bulk and surface properties of Sn-doped hematite with different crystal orientations. They observed that under identical PEC water oxidation conditions, the photocurrent onset potentials for (110) and (100) oriented hematite were located ~170 mV more negative than the (001) oriented hematite electrode. In the presence of a fast hole scavenger such as H_2O_2, however, the photocurrent onset potential and the photocurrent density at any given potential were found to be independent of the crystal orientation. Subsequent analysis *via* EIS measurements revealed relatively similar bulk properties (dopant density and flat-band potentials) for these series of hematite electrodes. Based on these observations, it was concluded that the crystal orientation primarily influences the kinetics of charge transfer at the surface of the electrode. In a separate study, Hu *et al.*[68] studied the surface chemistry and the electrical properties of hematite as a function of the concentration of the oxygen vacancies *via* X-ray photoelectron spectroscopy (XPS), resonant photoemission spectroscopy (RPES), and EIS. They showed that upon filling the oxygen vacancies, the density of surface states decreases and as a result the photocurrent density was diminished. However, the Fe-rich hematite surface (oxygen vacant surface) produced superior activity toward water oxidation. Recently, Hellman and Pala[69] theoretically evaluated the thermodynamics of the PEC water oxidation mechanism as a function of surface terminal groups of hematite; $(HO)_3$–Fe–H_3O_3–R, $(HO)_3$–Fe–O_3–R, O_3–Fe–O_3–R, $(HO)_3$–Fe–R and O_3–R surface terminations (R represents the bulk hematite) were evaluated. The stability under PEC conditions and the reactivity of each surface termination toward water oxidation were analysed. It was shown that under relevant conditions of PEC water oxidation at pH 14, only O_3–Fe–O_3–R and O_3–R terminations are stable. To study the reactivity of each surface, the mechanism outlined in eqn (4.3) to (4.7) was proposed and the change in free energy of each elementary step as a function of terminal group were calculated. It was predicted that water oxidation is thermodynamically allowed only on $(HO)_3$–Fe–H_3O_3–R, $(HO)_3$–Fe–O_3–R, and O_3–Fe–O_3–R surfaces.

$$\text{Fe—OH} \xrightarrow{h^+(\text{VB})} \text{Fe—O} + \text{H}^+ \tag{4.3}$$

$$\text{Fe—O} + \text{H}_2\text{O} \xrightarrow{h^+(\text{VB})} \text{Fe—OOH} + \text{H}^+ \tag{4.4}$$

$$\text{Fe—OOH} \xrightarrow{h^+(\text{VB})} \text{Fe} + \text{O}_2 + \text{H}^+ \tag{4.5}$$

$$\text{Fe} + \text{H}_2\text{O} \xrightarrow{h^+(\text{VB})} \text{Fe—OH} + \text{H}^+ \tag{4.6}$$

$$2\ \text{H}_2\text{O} \xrightarrow{4h^+(\text{VB})} \text{O}_2 + 4\ \text{H}^+ \tag{4.7}$$

The three examples provided above collectively emphasise the critical role of the surface structure and the surface terminal groups with regard to the efficiency of light-driven water oxidation on hematite electrodes.

The central point of the reaction scheme outlined above is the formation of the Fe–O and Fe–OOH intermediates during the water oxidation reaction. These groups are consistent with the formation of high valent iron intermediate species proposed from experimental results.[41,48,70–73] For example, Cummings *et al.*[70] studied the *in-situ* formation of the surface intermediate on the mesoporous hematite electrode during the oxygen evolution reaction *via* light-modulated absorbance spectroscopy (LMAS), potential-modulated absorption spectroscopy (PMAS), and impedance spectroscopy. LMAS uses a lock-in amplifier to measure transmission changes over the range 450–1100 nm in response to modulated 365 nm illumination. PMAS also measures transmission changes in the same wavelength range, but in response to a modulated potential. Examples of the PMAS and LMAS spectra for mesoporous hematite as a function of applied potential are shown in Figure 4.8. LMAS measurements are performed at potentials around the photocurrent onset, whereas much more positive potentials are required to acquire PMAS spectra, consistent with the onset of water oxidation current in the dark. At first glance, it can be noted that both measurements produce a very similar absorption profile, however, the intensity of the LMAS response is lower than that of the PMAS spectra. The similarity of the spectra measured in response to light and potential (dark) modulation implies that the water oxidation in light and dark proceeds through the same intermediates. The lower intensity of the LMAS spectra compared to PMAS may be due to increased recombination since it is measured at much more negative potentials (electron Fermi level) which results in a larger surface electron concentration.

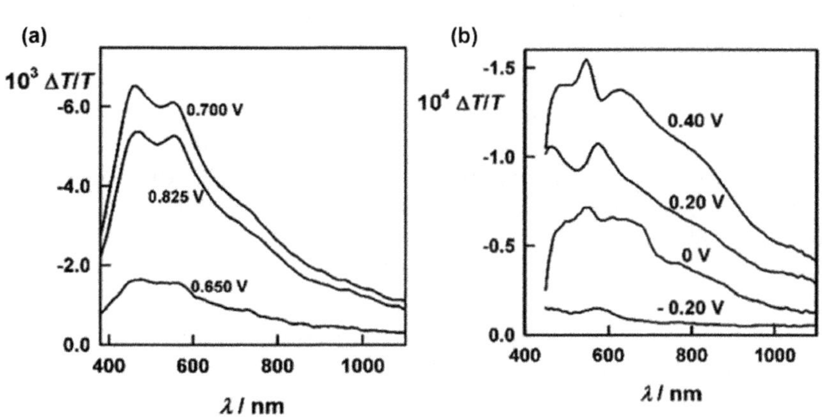

Figure 4.8 (a) PMAS and, (b) LMAS spectra for mesoporous α-Fe_2O_3 as a function of applied potential.
Reprinted with permission from C. Y. Cummings, F. Marken, L. M. Peter, K. G. U. Wijayantha and A. A. Tahir, *J. Am. Chem. Soc.*, 2012, **134**, 1228–1234.[70] Copyright (2012) American Chemical Society.

From Figure 4.8(a), it can be seen that at sufficiently positive potentials where the water oxidation occurs, two prominent peaks located at 450, 560 nm and a shoulder at 770 nm emerge. These peaks were attributed to the oxidation of the surface states and the formation of the water oxidation intermediates at the surface of hematite.[68] In a separate work, Klahr *et al.*[71] studied PEC water oxidation on hematite photoelectrodes through EIS, CV, and *in-situ* spectroelectrochemistry. The change in the absorption spectra as a function of applied potential during PEC water oxidation were measured. Strikingly, very similar absorption spectra to the LMAS and PMAS spectra reported by Cummings *et al.* were measured. The optical absorption peaks were tentatively attributed to the oxidation of low valent Fe^{III} to a high valent $Fe^{IV}{=}O$ at the surface under water oxidation conditions.[69] Moreover, it was shown that the change in the absorption spectrum of hematite strongly correlates with the charging of the surface states (detected by CV and EIS). These observations agree well with the mechanism of hole accumulation at the surface of the electrode prior to water oxidation.

4.3.2 Photochemical Water Oxidation on Iron-based Homogeneous Catalysts

In addition to the heterogeneous electrocatalytic and photoelectrochemical water oxidation on semiconducting electrodes, homogeneous molecular transition metal complexes were extensively studied for chemical and photochemical water oxidation.[74–78] Chemical water oxidation with Fe^{III}-complexes with a turn-over frequency (TOF) of 1.3 s^{-1} was initially demonstrated by Collins *et al.* in 2010.[75,79] Since then, a large number of interesting mechanistic studies with the aim to improve the efficiency of these systems have been published.[74,80–83] Despite the differences in the proposed mechanisms, the common feature is the formation of high valent iron sites, *i.e.* Fe^{IV} or Fe^{V}, as the active intermediates responsible for the formation of the O–O bond.[79,81–84] For example, Panda *et al.*[81] studied the mechanism of chemical and photochemical water oxidation on the Biuret-modified Fe^{III}-TAML complex by means of HRMS, UV-vis, and EPR spectroscopy. The proposed catalytic cycle (Figure 4.9(a)) is initiated by the oxidation of Fe^{III}–OH_2 to Fe^{IV}–OH, which is in a fast equilibrium with a μ-oxo-Fe^{IV} dimer followed by a second hole transfer which forms a reactive $Fe^{V}{=}O$ intermediate. This intermediate then undergoes a nucleophilic attack by the second water molecule resulting in the formation of the O–O bond. The reactive $Fe^{V}{=}O$ intermediate was inferred *via* comparison of the deconvoluted UV-vis spectra of the reaction mixture and the UV-vis spectra of the chemically synthesised $Fe^{V}{=}O$ group (Figure 4.9(b)). Two absorption peaks centered at 445 and 613 nm and the peak located at 950 nm were attributed to $Fe^{V}{=}O$ and $Fe^{IV}(O)$, respectively. Remarkably, these absorption features are very well consistent with the peaks observed by Peter and Klahr (Figure 4.8) during PEC water oxidation on hematite photoanodes.

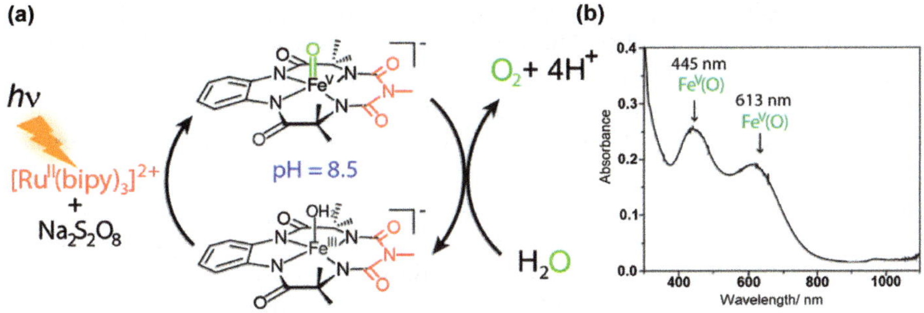

Figure 4.9 (a) Proposed mechanism of catalytic photochemical water oxidation by Biuret-modified Fe^{III}-TAML, (b) The deconvoluted UV-vis spectra of the reaction mixture.
Reprinted with permission from C. Panda, J. Debgupta, D. Díaz Díaz, K. K. Singh, S. Sen Gupta and B. B. Dhar, *J. Am. Chem. Soc.*, 2014, **136**, 12273–12282[81] Copyright (2014) American Chemical Society.

4.3.3 Determination of Water Oxidation Intermediates *via* Operando Infrared Spectroscopy

While the experiments described above strongly indicate that the photogenerated holes accumulate at the surface of hematite as high valent iron intermediates, they do not directly confirm the chemical identity of these intermediates. Most of the surface sensitive techniques are established for gas–solid interfaces, while the reaction of interest here occurs at the interface of two condensed phases, *i.e.* solid–liquid interface. This imposes serious challenges for *operando* mechanistic studies of PEC water oxidation at the surface of the photoanode. Attenuated total reflectance infrared (ATR-IR) spectroscopy is an attractive technique which can be coupled with photochemical and photoelectrochemical cells for *operando* mechanistic studies. It has been successfully utilised to detect intermediate species formed during chemical reactions at solid–liquid interfaces, such as photochemical water oxidation on the surface of TiO_2.[85–89] Nakamura and co-workers, for example, utilised a combination of *in-situ* IR absorption and photoluminescence (PL) to detect the primary intermediate species that evolve during photochemical water oxidation on TiO_2 (rutile).[89,90] Under UV irradiation and in the presence of Fe^{3+} (electron scavenger) a group of new peaks emerged. Through isotope labelling and the effect of pH and hole scavenger, these peaks were attributed to the TiOOH and TiOOTi surface groups. Based on these results, they proposed that photo-evolution of oxygen on the TiO_2 surface proceeds through the nucleophilic attack of H_2O to the photogenerated hole residing on radical oxygen.

In an impressive example, Zandi *et al.*[86] recently utilised ATR-IR to detect the intermediate of the water oxidation reaction on the surface of hematite under PEC *operando* conditions. This study is unique as it is the first example of identification of a water oxidation intermediate under photoelectrochemical water oxidation conditions. The schematic of the *operando* set-up including a three-electrode system, light source, and ATR-IR probe is shown in Figure 4.10(a). The *J*–*V* response measured under illumination in this *operando* set-up (Figure 4.10(b)) showed a modest performance for a hematite electrode under a monochromatic 395 nm LED light source at 10 mW cm^{-2}. The IR spectra of the hematite as a function of applied potential in the dark and under illumination are shown in Figure 4.10(c) and (d), respectively. A prominent peak at 898 cm^{-1} was observed at potentials positive of the (photo)current onset potential (both in light and dark) and was accordingly ascribed to the intermediate of water oxidation on hematite surface. Following further mechanistic study *via* isotope labelling, this peak was assigned to the vibration frequency of the $Fe^{IV}{=}O$ species and a reaction cycle similar to eqn (4.3) to (4.6) was proposed. As can be seen from Figure 4.10(c) and (d), interestingly, the IR spectra in light and dark are very similar. This observation is further in agreement with the previous conclusion drawn by Cummings *et al.*,[70] suggesting that water oxidation on hematite surface in light and dark proceeds through the same mechanism and involves the same intermediates species.

The two examples provided above highlight the versatility of ATR-IR as a powerful technique to be coupled with other characterisation techniques to identify the intermediate species formed *in-situ* during the reaction at the

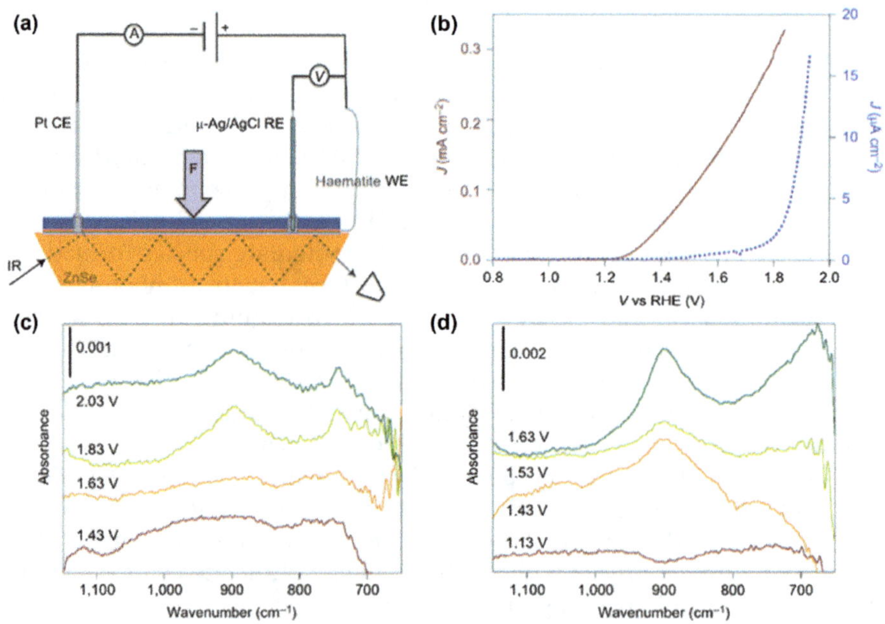

Figure 4.10 (a) The schematic of the experimental set-up for *in-situ* infrared spectroscopy measurements during PEC water oxidation measurements. (b) *J*–*V* curves of a hematite electrode in the operando set-up in the dark (blue) and under illumination (red). Infrared spectra of hematite as a function of the applied potential in (c) dark, and (d) light.[86]

interface of two condensed phases. Furthermore, ATR-IR can be coupled with time-resolved methods to follow the formation and disappearance of chemical intermediates after charge injection.[85,88,91] These techniques would allow further in-depth studies on the kinetics of the individual steps. In addition, the reactivity of different surface sites toward water oxidation can be analysed. It has been shown that hematite possesses two types of surface states, where one of them can be selectively removed by annealing at 800 °C.[63] Therefore, rapid-scan ATR-FTIR is beneficial as it can be utilised to identify the chemical nature of these surface sites. With the time-resolved analysis the activity of each surface site toward water oxidation can be further characterised.

In an interesting study for example, Cuk and co-workers[85] utilised a combination of ultrafast infrared spectroscopy and theoretical modelling to identify the oxyl radical intermediate at the single crystal *n*-$SrTiO_3$/aqueous interface. A polarised IR probe was used to detect the photogenerated intermediates. They observed that the oxyl radical peak appeared exclusively with a *p*-polarised IR probe beam (Figure 4.11(a)) which suggests that this peak is due to the vibrational mode of a surface radical hydroxyl groups with a dipole perpendicular to the sample. This photogenerated absorption peak was attributed to a long-lived Ti-O$^\bullet$ radical group. A life time measurement with and without methanol as a hole scavenger further indicated that the photogenerated oxyl radical species are the surface intermediate formed upon illumination (Figure 4.11(b)). They further compared the magnitude of the absorbance change to the photocurrent in a three-electrode setup and closed-circuit conditions as a function of illumination intensity. As depicted in Figure 4.11(c), the change in photocurrent closely tracks the absorbance change which further indicates that the surface Ti-O$^\bullet$ species are the active intermediate of water oxidation at the surface of $SrTiO_3$.

Notably in an interesting study, Frei and co-workers[88] utilised the rapid-scan ATR-FTIR method to detect the surface intermediates on cobalt oxide nanoparticles formed by hole injection from optically excited $[Ru(bpy)_3]^{2+}$. Remarkably, under photochemical water oxidation conditions two independent intermediates were detected. Isotope labelling and analysis of evolved oxygen gas *via* mass spectroscopy revealed that these surface intermediates are due to the formation of cobalt superoxide (three-electron oxidation) and $Co^{IV}{=}O$ (one electron oxidation) intermediate species. The time-resolved analysis indicated that these two intermediates are independent and they originate from two separate catalytic sites with distinct kinetic behaviours. It was shown that the mechanism that involves the formation of the superoxide has a substantially higher activity than the mechanism for the formation of the oxo-cobalt intermediate. The difference in activity of these sites were ascribed to the presence/absence of neighboring Co^{III}–OH sites coupled *via* bridging oxygen groups. Accordingly, two independent mechanisms involving binuclear and mononuclear cobalt sites were proposed (shown in Figure 4.12).

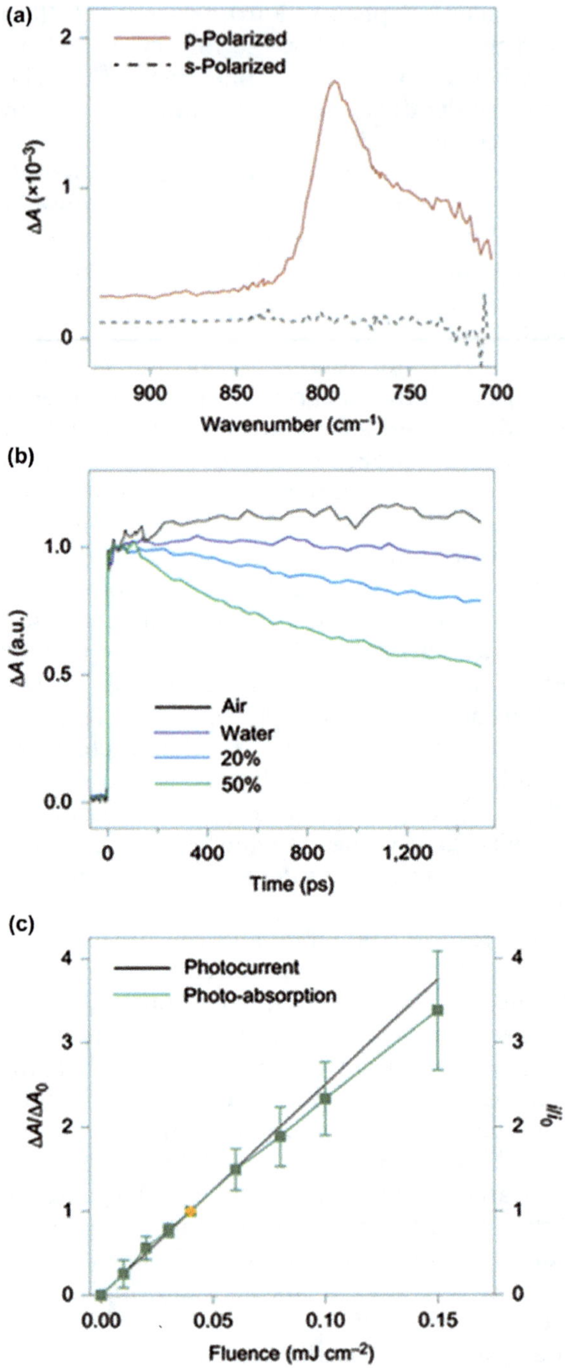
(a)
p-Polarized
s-Polarized
ΔA (×10⁻³)
Wavenumber (cm⁻¹)
(b)
ΔA (a.u.)
Air
Water
20%
50%
Time (ps)
(c)
Photocurrent
Photo-absorption
ΔA/ΔA₀
I/I₀
Fluence (mJ cm⁻²)

4.4 Ternary Metal Oxides for PEC Water Oxidation

As described previously, the PEC water oxidation on hematite strongly correlates to the surface structure and terminal groups. Unlike hematite, ternary metal oxides such as $BiVO_4$ and $CuWO_4$, are comprised of three elements. Therefore, for ternary metal oxides the composition of the surface varies for different crystal orientations. This is crucially important as the dynamics of charge transfer are linked to the catalytic activity of the surface, which is determined by its composition. Therefore, a thorough knowledge of the composition of the outermost surface layer as well as their activities toward water oxidation is exceedingly beneficial. The surface composition of ternary metal oxides further becomes critically important when catalytic modification of the surface is desired. In the next paragraphs, we will discuss these aspects for the ternary photoanode materials $CuWO_4$ and $BiVO_4$.

4.4.1 $CuWO_4$

Similar electrochemical and photoelectrochemical measurements as discussed above for hematite were also performed on other photoanodes for water oxidation in order to understand the general rule behind various metal oxides. For example, the Hamann group recently measured the photocurrent of water oxidation and oxidation of a suitable hole scavenger, Na_2SO_3, as well as IMPS, to quantify the hole collection efficiency of $CuWO_4$.[10] It was found that the hole collection efficiency of water oxidation with $CuWO_4$ is non-optimal at low applied potential due to surface recombination, but is essentially quantitative once it passes 1.23 V *vs.* RHE. EIS, photocurrent transient and fast CV measurements under various illumination conditions and in different electrolytes were done in a follow-up study to elucidate the role of surface states of $CuWO_4$.[92] This study revealed that the surface state of $CuWO_4$ is only available when water oxidation occurs. It was concluded that charging the surface state of $CuWO_4$ is related to forming the water oxidation intermediate species at the photoelectrode surface, similar to what has been observed for α-Fe_2O_3. It will be interesting to see if these observations hold up and support the central idea of water oxidation intermediate species building up on photoelectrode surfaces which can act as recombination centres and hinder the performance.

Figure 4.11 (a) Photoinduced absorption of the oxyl radical as a function of polarisation of infrared probe beam; (b) normalised kinetics of the transient response at 795 cm^{-1} for the 0.1% Nb *n*-doped sample in contact with air (black line), water (blue line), 20% (cyan line) and 50% (green line) methanol in water; (c) the normalised absorbance change at 795 cm^{-1} (green square) and the normalised steady-state photocurrent (black solid line) measured in three electrode setup at 0.0 V *vs.* Ag/AgCl at pH 14 as a function of laser fluence.
These graphs were reprinted by permission from Macmillan Publishers Ltd: *Nature Chemistry* (ref. 85), copyright (2016).

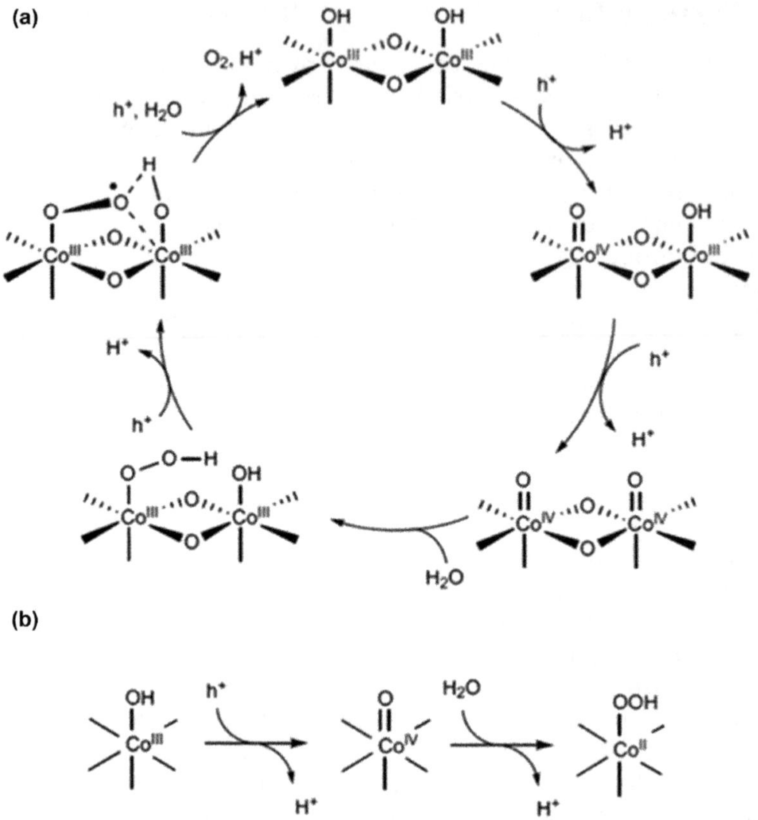

Figure 4.12 Proposed photocatalytic mechanism. Water oxidation mechanism *via* (a) superoxide intermediate and (b) oxo intermediate on the surface of Co_3O_4.
Reprinted by permission from Macmillan Publishers Ltd: *Nature Chemistry* (ref. 88), copyright 2014.

4.4.2 $BiVO_4$

A large body of the literature in the area of solar fuel production and PEC water oxidation has been dedicated to the potential and the opportunities as well as the strategies to improve the PEC efficiency of water oxidation on $BiVO_4$. The interested reader is encouraged to visit theses references.[13,93–95] $BiVO_4$, with a band gap of 2.4 eV and suitable VB band position for water oxidation, is another promising candidate photoanode for PEC water oxidation.[13,96,97] Poor charge separation and sluggish water oxidation kinetics, are the bottle-neck processes suppressing the efficiency of bismuth vanadate.[98–100] Similar to hematite, surface modifications with co-catalysts have shown to be an effective strategy to improve the PEC water oxidation efficiency of $BiVO_4$.[98,101–104] One of the best examples of PEC water oxidation performance of $BiVO_4$ with co-catalyst has been introduced by Choi and co-workers,[93] where they circumvent the poor charge separation efficiency and slow kinetic of water oxidations by nanostructuring and a combination of both hole-storage and catalyst layers, respectively. Various electrochemical and spectroscopic techniques, such as IMPS,[105,106] photoinduced absorption spectroscopy, transient absorption spectroscopy and transient photocurrent measurements[104,107,108] (the last three techniques are discussed in more detail in Chapter 5) have been adopted to uncover the mechanism of water oxidation and the dynamic of charge transfer/recombination processes at the surface of $BiVO_4$. The van de Krol group used IMPS to examine the limitation of charge transfer at the $BiVO_4$ surface during water oxidation.[109] They extracted the charge transfer and recombination rate constants of water oxidation at $BiVO_4$ surface with and without Co-Pi catalyst, and found that with Co-Pi, the recombination rate constant was reduced by 10–20 fold, with minimal changing in the charge transfer rate constants. Control experiments with RuO_x electrocatalyst for oxygen evolution reaction showed no decrease in recombination rate constant and a decrease in photocurrent. Therefore, it was concluded that the charge transfer of $BiVO_4$ for water oxidation is limited by surface recombination rather than surface catalysis. It has also been shown that similar to hematite, the water oxidation at the surface of $BiVO_4$ proceeds through the charge accumulation at the surface (Chapter 5, Section 5.1.2). The formation of these surface states under PEC conditions were deduced from transient absorption spectroscopy. The chemical identity of these states and the mechanism of water oxidation on $BiVO_4$, however, are an important area yet to be explored.

4.5 Outlook

As discussed above, water oxidation on the hematite surface proceeds through the formation of Fe=O intermediate species. Although operando ATR-IR experiments have been successful in identifying this intermediate, the oxidation state of iron at the surface of hematite is another important question that remains to be definitively answered. Mössbauer spectroscopy is a technique with a high sensitivity to iron sites.[110,111] Recently, Stahl and co-workers[112] utilised operando Mössbauer spectroscopy to investigate the oxidation state as well as the role of the iron in NiFeOOH and FeOOH electrocatalysts for water oxidation. Their data provides direct evidence for the formation of Fe^{IV} in NiFe catalyst under electrochemical water oxidation conditions. The *in-situ* Mössbauer is certainly an informative experiment as it has been successfully shown that it can be coupled with the electrochemical cell. Thus, operando- Mössbauer measurements for hematite under PEC conditions may help to assign the oxidation state of the iron during the PEC water oxidation reaction. Although the detection of Fe=O groups at the surface of hematite was a significant advancement in assigning the first step(s) of the catalytic cycle of water oxidation at the hematite photoanode surface, more research is required to identify the subsequent steps and rate constants. Recent important advances towards achieving this are discussed in the following chapter.

4.6 Summary

In this chapter, recent advances made on uncovering the mechanism of water oxidation on hematite photoanodes were discussed. It was shown that water oxidation on the hematite surface proceeds through the accumulation of minority carriers (photogenerated holes) at the surface of the photoanodes. These accumulated charges at the surface can be detected *via* a variety of electrochemical and spectroscopic methods such as CV, EIS, photocurrent transient, photoinduced absorption, and transient absorption spectroscopy. The accumulation of charge was shown to consist of the oxidation of iron sites at the surface to form high valent Fe=O species, which are intermediates in the overall water oxidation mechanism. The chemical identity of these intermediates was confirmed *via in-situ* methods such as spectroelectrochemical and operando ATR-IR measurements. Analogies between the hematite surface and iron-based homogeneous catalysts were drawn, which supports to assignment of Fe=O as a relevant intermediate species for water oxidation. Finally, results of other promising semiconductor materials were introduced, which exhibit similar behavior of surface hole accumulation prior to water oxidation. Thus, we are approaching a general understanding of water oxidation on metal oxide semiconductor surfaces and the effect of these mechanisms on the photoelectrochemical performance.

Acknowledgements

We gratefully thank the National Science Foundation (CHE-1664823) for their continued financial support.

References

1. N. S. Lewis and D. G. Nocera, *Proc. Natl. Acad. Sci.*, 2006, **103**, 15729–15735.
2. D. Gielen, F. Boshell and D. Saygin, *Nat. Mater.*, 2016, **15**, 117–120.
3. M. G. Walter, E. L. Warren, J. R. McKone, S. W. Boettcher, Q. Mi, E. A. Santori and N. S. Lewis, *Chem. Rev.*, 2010, **110**, 6446–6473.
4. L. Zhou, *Renewable Sustainable Energy Rev.*, 2005, **9**, 395–408.
5. G. P. Dinga, *J. Chem. Educ.*, 1988, **65**, 688–691.
6. R. Schlögl, *ChemSusChem*, 2010, **3**, 209–222.
7. M. Grätzel, *Nature*, 2001, **414**, 338–344.
8. J. M. Fraile, G. Lafuente, J. A. Mayoral and A. Pallarés, *Tetrahedron*, 2011, **67**, 8639–8647.
9. Y. Gao, O. Zandi and T. W. Hamann, *J. Mater. Chem. A*, 2016, **4**, 2826–2830.

10. Y. Gao and T. W. Hamann, *Chem. Commun.*, 2017, **53**, 1285–1288.
11. C. R. Lhermitte and B. M. Bartlett, *Acc. Chem. Res.*, 2016, **49**, 1121–1129.
12. A. Fujishima and K. Honda, *Nature*, 1972, **238**, 37–38.
13. Y. Park, K. J. McDonald and K.-S. Choi, *Chem. Soc. Rev.*, 2013, **42**, 2321–2337.
14. J. A. Seabold and K.-S. Choi, *J. Am. Chem. Soc.*, 2012, **134**, 2186–2192.
15. Y. Qiu, W. Liu, W. Chen, W. Chen, G. Zhou, P.-C. Hsu, R. Zhang, Z. Liang, S. Fan, Y. Zhang and Y. Cui, *Sci. Adv.*, 2016, **2**, e1501764–e1501764.
16. O. Zandi, A. R. Schon, H. Hajibabaei and T. W. Hamann, *Chem. Mater.*, 2016, **28**, 765–771.
17. J.-W. Jang, C. Du, Y. Ye, Y. Lin, X. Yao, J. Thorne, E. Liu, G. McMahon, J. Zhu, A. Javey, J. Guo and D. Wang, *Nat. Commun.*, 2015, **6**, 7447.
18. J. Y. Kim, G. Magesh, D. H. Youn, J.-W. Jang, J. Kubota, K. Domen and J. S. Lee, *Sci. Rep.*, 2013, **3**, 2681.
19. S. C. Warren, K. Voïtchovsky, H. Dotan, C. M. Leroy, M. Cornuz, F. Stellacci, C. Hébert, A. Rothschild and M. Grätzel, *Nat. Mater.*, 2013, **12**, 842–849.
20. G. Liu, S. Ye, P. Yan, F. F.-Q. Xiong, P. Fu, Z. Wang, Z. Chen, J. Shi and C. Li, *Energy Environ. Sci.*, 2016, **9**, 1327–1334.
21. H. Hajibabaei, O. Zandi and T. W. Hamann, *Chem. Sci.*, 2016, **7**, 6760–6767.
22. Y. Li, L. Zhang, A. Torres-Pardo, J. M. González-Calbet, Y. Ma, P. Oleynikov, O. Terasaki, S. Asahina, M. Shima, D. Cha, L. Zhao, K. Takanabe, J. Kubota and K. Domen, *Nat. Commun.*, 2013, **4**, 2566.
23. H. Hajibabaei and T. W. Hamann, *Langmuir*, 2017. DOI: 10.1021/acs.langmuir.7b02414.
24. J. Hou, H. Cheng, O. Takeda and H. Zhu, *Energy Environ. Sci.*, 2015, **8**, 1348–1357.
25. The U.S. Geological Survey, https://pubs.usgs.gov/fs/2002/fs087-02/, (accessed September 2017).
26. K. L. Hardee and A. J. Bard, *J. Electrochem. Soc.*, 1976, **123**, 1024–1026.
27. O. Zandi and T. W. Hamann, *Phys. Chem. Chem. Phys.*, 2015, **17**, 22485–22503.
28. K. M. H. Young, B. M. Klahr, O. Zandi and T. W. Hamann, *Catal. Sci. Technol.*, 2013, **3**, 1660–1671.
29. Y. Lin, G. Yuan, S. Sheehan, S. Zhou, D. Wang, J. H. Lee, D. Wang, J. R. Durrant and T. E. Mallouk, *Energy Environ. Sci.*, 2011, **4**, 4862–4869.
30. K. Sivula, F. Le Formal and M. Grätzel, *ChemSusChem*, 2011, **4**, 432–449.
31. G. Horowitz, *J. Electroanal. Chem. Interfacial Electrochem.*, 1983, **159**, 421–436.
32. J. H. Kennedy, *J. Electrochem. Soc.*, 1978, **125**, 709–714.
33. M. P. Dare-Edwards, J. B. Goodenough, A. Hamnett and P. R. Trevellick, *J. Chem. Soc., Faraday Trans. 1*, 1983, **79**, 2027.

34. C. Sanchez, K. D. Sieber and G. A. Somorjai, *J. Electroanal. Chem. Interfacial Electrochem.*, 1988, **252**, 269–290.
35. K. Itoh and J. O. Bockris, *J. Appl. Phys.*, 1984, **56**, 874–876.
36. K. Itoh, *J. Electrochem. Soc.*, 1984, **131**, 1266–1271.
37. S. C. Riha, B. M. Klahr, E. C. Tyo, S. S. Seifert, S. Vajda, M. J. Pellin, T. W. Hamann and A. B. F. Martinson, *ACS Nano*, 2013, **7**, 2396–2405.
38. K. Sivula, R. Zboril, F. Le Formal, R. Robert, A. Weidenkaff, J. Tucek, J. Frydrych and M. Grätzel, *J. Am. Chem. Soc.*, 2010, **132**, 7436–7444.
39. A. Kay, I. Cesar and M. Grätzel, *J. Am. Chem. Soc.*, 2006, **128**, 15714–15721.
40. Y. Lin, S. Zhou, S. W. Sheehan and D. Wang, *J. Am. Chem. Soc.*, 2011, **133**, 2398–2401.
41. K. G. U. Wijayantha, S. Saremi-Yarahmadi and L. M. Peter, *Phys. Chem. Chem. Phys.*, 2011, **13**, 5264.
42. B. M. Klahr, A. B. F. Martinson and T. W. Hamann, *Langmuir*, 2011, **27**, 461–468.
43. B. M. Klahr and T. W. Hamann, *J. Phys. Chem. C*, 2011, **115**, 8393–8399.
44. B. M. Klahr and T. W. Hamann, *Appl. Phys. Lett.*, 2011, **99**, 63508.
45. D. A. Grave, D. Klotz, A. Kay, H. Dotan, B. Gupta, I. Visoly-Fisher and A. Rothschild, *J. Phys. Chem. C*, 2016, **120**, 28961–28970.
46. L. M. Peter and K. G. U. Wijayantha, *ChemPhysChem*, 2014, **15**, 1983–1995.
47. J. E. Thorne, J.-W. Jang, E. Y. Liu, D. Wang, E. Article, J. E. Thorne, J.-W. Jang, E. Y. Liu and D. Wang, *Chem. Sci.*, 2016, **7**, 3347–3354.
48. M. Barroso, C. A. Mesa, S. R. Pendlebury, A. J. Cowan, T. Hisatomi, K. Sivula, M. Gratzel, D. R. Klug and J. R. Durrant, *Proc. Natl. Acad. Sci.*, 2012, **109**, 15640–15645.
49. T. Hisatomi, F. Le Formal, M. Cornuz, J. Brillet, N. Tétreault, K. Sivula and M. Grätzel, *Energy Environ. Sci.*, 2011, **4**, 2512–2515.
50. B. Klahr, S. Gimenez, F. Fabregat-Santiago, J. Bisquert and T. W. Hamann, *J. Am. Chem. Soc.*, 2012, **134**, 16693–16700.
51. H. Hajibabaei, A. R. Schon and T. W. Hamann, *Chem. Mater.*, 2017, **29**, 6674–6683.
52. J. Qiu, H. Hajibabaei, M. R. Nellist, F. A. L. Laskowski, T. W. Hamann and S. W. Boettcher, *ACS Cent. Sci.*, 2017, **3**, 1015–1025.
53. S. R. Pendlebury, A. J. Cowan, M. Barroso, K. Sivula, J. Ye, M. Grätzel, D. R. Klug, J. Tang and J. R. Durrant, *Energy Environ. Sci.*, 2012, **5**, 6304–6312.
54. S. R. Pendlebury, M. Barroso, A. J. Cowan, K. Sivula, J. Tang, M. Grätzel, D. Klug and J. R. Durrant, *Chem. Commun.*, 2011, **47**, 716–718.
55. H. Dotan, K. Sivula, M. Gr, S. C. Warren, M. Grätzel, A. Rothschild, S. C. Warren, M. Gr, S. C. Warren, M. Grätzel, A. Rothschild and S. C. Warren, *Energy Environ. Sci.*, 2011, **4**, 958–964.
56. A. J. Cowan, C. J. Barnett, S. R. Pendlebury, M. Barroso, K. Sivula, M. Grätzel, J. R. Durrant and D. R. Klug, *J. Am. Chem. Soc.*, 2011, **133**, 10134–10140.

57. L. M. Peter, K. G. U. Wijayantha and A. A. Tahir, *Faraday Discuss.*, 2012, **155**, 309–322.
58. A. Duret and M. Grätzel, *J. Phys. Chem. B*, 2005, **109**, 17184–17191.
59. B. Klahr, S. Gimenez, F. Fabregat-Santiago, J. Bisquert and T. W. Hamann, *Energy Environ. Sci.*, 2012, **5**, 7626.
60. P. Iwanski, *J. Electrochem. Soc.*, 1981, **128**, 2128–2133.
61. L. M. M. Abrantes and L. M. M. Peter, *J. Electroanal. Chem.*, 1983, **150**, 593–601.
62. B. Klahr, S. Gimenez, O. Zandi, F. Fabregat-Santiago and T. Hamann, *ACS Appl. Mater. Interfaces*, 2015, **7**, 7653–7660.
63. O. Zandi and T. W. Hamann, *J. Phys. Chem. Lett.*, 2014, **5**, 1522–1526.
64. B. Klahr, S. Gimenez, F. Fabregat-santiago, T. Hamann and J. Bisquert, *J. Am. Chem. Soc.*, 2012, **134**, 4294–4302.
65. T. R. Cook, D. K. Dogutan, S. Y. Reece, Y. Surendranath, T. S. Teets and D. G. Nocera, *Chem. Rev.*, 2010, **110**, 6474–6502.
66. H. Dau, C. Limberg, T. Reier, M. Risch, S. Roggan and P. Strasser, *ChemCatChem*, 2010, **2**, 724–761.
67. M. N. Huda, A. Walsh, Y. Yan, S.-H. Wei and M. M. Al-Jassim, *J. Appl. Phys.*, 2010, **107**, 123712.
68. Y. Hu, F. Boudoire, I. Hermann-Geppert, P. Bogdanoff, G. Tsekouras, B. S. Mun, G. Fortunato, M. Graetzel and A. Braun, *J. Phys. Chem. C*, 2016, **120**, 3250–3258.
69. A. Hellman and R. G. S. Pala, *J. Phys. Chem. C*, 2011, **115**, 12901–12907.
70. C. Y. Cummings, F. Marken, L. M. Peter, K. G. U. Wijayantha and A. A. Tahir, *J. Am. Chem. Soc.*, 2012, **134**, 1228–1234.
71. B. Klahr and T. Hamann, *J. Phys. Chem. C*, 2014, **118**, 10393–10399.
72. T. Takashima, K. Ishikawa and H. Irie, *J. Phys. Chem. C*, 2016, **120**, 24827–24834.
73. C. Y. Cummings, F. Marken, L. M. Peter, A. A. Tahir and K. G. U. Wijayantha, *Chem. Commun.*, 2012, **48**, 2027.
74. J. D. Blakemore, R. H. Crabtree and G. W. Brudvig, *Chem. Rev.*, 2015, **115**, 12974–13005.
75. W. C. Ellis, N. D. McDaniel, S. Bernhard and T. J. Collins, *J. Am. Chem. Soc.*, 2010, **132**, 10990–10991.
76. N. D. McDaniel, F. J. Coughlin, L. L. Tinker and S. Bernhard, *J. Am. Chem. Soc.*, 2008, **130**, 210–217.
77. S. W. Gersten, G. J. Samuels and T. J. Meyer, *J. Am. Chem. Soc.*, 1982, **104**, 4029–4030.
78. J. Limburg, G. W. Brudvig and R. H. Crabtree, *J. Am. Chem. Soc.*, 1997, **119**, 2761–2762.
79. D.-L. Popescu, M. Vrabel, A. Brausam, P. Madsen, G. Lente, I. Fabian, A. D. Ryabov, R. van Eldik and T. J. Collins, *Inorg. Chem.*, 2010, **49**, 11439–11448.
80. A. Singh and L. Spiccia, *Coord. Chem. Rev.*, 2013, **257**, 2607–2622.
81. C. Panda, J. Debgupta, D. Díaz Díaz, K. K. Singh, S. Sen Gupta and B. B. Dhar, *J. Am. Chem. Soc.*, 2014, **136**, 12273–12282.

82. L. D. Wickramasinghe, R. Zhou, R. Zong, P. Vo, K. J. Gagnon and R. P. Thummel, *J. Am. Chem. Soc.*, 2015, **137**, 13260–13263.
83. F. T. de Oliveira, A. Chanda, D. Banerjee, X. P. Shan, S. Mondal, L. Que, E. L. Bominaar, E. Munck and T. J. Collins, *Science*, 2007, **315**, 835–838.
84. J.-U. Rohde, L. Q. J. J. U. Rohde, J. H. In, M. H. Lim, W. W. Brennessel, M. R. Bukowski, A. Stubna, E. Munck and W. Nam, *Science*, 2003, **299**, 1037–1039.
85. D. M. Herlihy, M. M. Waegele, X. Chen, C. D. Pemmaraju, D. Prendergast and T. Cuk, *Nat. Chem.*, 2016, **8**, 549–555.
86. O. Zandi and T. W. Hamann, *Nat. Chem.*, 2016, **8**, 778–783.
87. R. Nakamura, T. Tanaka and Y. Nakato, *J. Phys. Chem. B*, 2005, **109**, 8920–8927.
88. M. Zhang, M. de Respinis and H. Frei, *Nat. Chem.*, 2014, **6**, 362–367.
89. R. Nakamura and Y. Nakato, *J. Am. Chem. Soc.*, 2004, **126**, 1290–1298.
90. A. Imanishi, T. Okamura, N. Ohashi, R. Nakamura and Y. Nakato, *J. Am. Chem. Soc.*, 2007, **129**, 11569–11578.
91. X. Chen, S. N. Choing, D. J. Aschaffenburg, C. Das Pemmaraju, D. Prendergast and T. Cuk, *J. Am. Chem. Soc.*, 2017, **139**, 1830–1841.
92. Y. Gao and T. W. Hamann, *J. Phys. Chem. Lett.*, 2017, **8**, 7–11.
93. T. W. Kim and K.-S. Choi, *Science*, 2014, **343**, 990–994.
94. K.-S. Choi, *J. Phys. Chem. Lett.*, 2010, **1**, 2244–2250.
95. M. Zhou, S. Zhang, Y. Sun, C. Wu, M. Wang and Y. Xie, *Chem. – Asian J.*, 2010, **5**, 2515–2523.
96. B. Xie, H. Zhang, P. Cai, R. Qiu and Y. Xiong, *Chemosphere*, 2006, **63**, 956–963.
97. S. Tokunaga, H. Kato and A. Kudo, *Chem. Mater.*, 2001, **13**, 4624–4628.
98. F. F. Abdi and R. van de Krol, *J. Phys. Chem. C*, 2012, **116**, 9398–9404.
99. S. P. Berglund, A. J. E. Rettie, S. Hoang and C. B. Mullins, *Phys. Chem. Chem. Phys.*, 2012, **14**, 7065.
100. J. Su, L. Guo, N. Bao and C. A. Grimes, *Nano Lett.*, 2011, **11**, 1928–1933.
101. X. Chang, T. Wang, P. Zhang, J. Zhang, A. Li and J. Gong, *J. Am. Chem. Soc.*, 2015, **137**, 8356–8359.
102. M. Zhong, T. Hisatomi, Y. Kuang, J. Zhao, M. Liu, A. Iwase, Q. Jia, H. Nishiyama, T. Minegishi, M. Nakabayashi, N. Shibata, R. Niishiro, C. Katayama, H. Shibano, M. Katayama, A. Kudo, T. Yamada and K. Domen, *J. Am. Chem. Soc.*, 2015, **137**, 5053–5060.
103. D. K. Zhong, S. Choi and D. R. Gamelin, *J. Am. Chem. Soc.*, 2011, **133**, 18370–18377.
104. Y. Ma, F. Le Formal, A. Kafizas, S. R. Pendlebury and J. R. Durrant, *J. Mater. Chem. A*, 2015, **3**, 20649–20657.
105. C. Zachäus, F. F. Abdi, L. M. Peter and R. van de Krol, *Chem. Sci.*, 2017, **8**, 3712–3719.

106. F. Yu, F. Li, T. Yao, J. Du, Y. Liang, Y. Wang, H. Han and L. Sun, *ACS Catal.*, 2017, **7**, 1868–1874.
107. Y. Ma, S. R. Pendlebury, A. Reynal, F. Le Formal and J. R. Durrant, *Chem. Sci.*, 2014, **5**, 2964–2973.
108. Y. Ma, C. A. Mesa, E. Pastor, A. Kafizas, L. Francàs, F. Le Formal, S. R. Pendlebury and J. R. Durrant, *ACS Energy Lett.*, 2016, **1**, 618–623.
109. C. Zachäus, F. F. Abdi, L. M. Peter and R. van de Krol, *Chem. Sci.*, 2017, **8**, 3712–3719.
110. L. Guerlou-Demourgues, L. Fournès and C. Delmas, *J. Electrochem. Soc.*, 1996, **143**, 3083–3088.
111. D. A. Corrigan, R. S. Conell, C. A. Fierro and D. A. Scherson, *J. Phys. Chem.*, 1987, **91**, 5009–5011.
112. J. Y. C. Chen, L. Dang, H. Liang, W. Bi, J. B. Gerken, S. Jin, E. E. Alp and S. S. Stahl, *J. Am. Chem. Soc.*, 2015, **137**, 15090–15093.

CHAPTER 5

Rate Law Analysis of Water Splitting Photoelectrodes

LAIA FRANCÀS,[a] CAMILO A. MESA,[a] ERNEST PASTOR,[a] FLORIAN LE FORMAL[b] AND JAMES R. DURRANT*[a]

[a] Department of Chemistry, Imperial College London, South Kensington Campus, London SW7 2AZ, UK; [b] Institut des Sciences et Ingénierie Chimiques, Ecole Polytechnique Fédérale de Lausanne, Station 6, CH-1015 Lausanne, Switzerland
*Email: j.durrant@imperial.ac.uk

5.1 Introduction

5.1.1 Rate Law Analysis for Solar Fuels Production

Hydrogen has emerged as a promising clean and potentially inexpensive solar fuel when produced by photoelectrochemical (PEC) water splitting. Despite advances in the synthesis of new catalysts and efficient systems, the mechanism of action of these materials still remains not yet well understood. Catalytic processes are key in solar fuels production. These processes are extremely challenging, especially those light-driven. In these systems, the light absorption and charge separation events are coupled to the catalytic process, thus increasing the complexity of the overall reaction. As a result, often, these systems suffer from poor efficiencies and stabilities. To some extent, this is originated from a kinetic mismatch between the light-initiated events (*i.e.*: charge separation and charge transfer) and the ultimate catalytic reaction (*i.e.*: chemical bond rearrangement), which favours loss paths and

Energy and Environment Series No. 20
Advances in Photoelectrochemical Water Splitting: Theory, Experiment and Systems Analysis
Edited by S. David Tilley, Stephan Lany and Roel van de Krol

Published by the Royal Society of Chemistry, www.rsc.org

photodegradation processes. In order to guide the design of efficient and robust photocatalytic systems we need a solid understanding of the kinetics and reaction mechanisms.

Traditionally mechanistic information has been obtained by means of rate law analyses.[1] One well-known strategy to approach the rate law of any reaction is to keep all the reactants in excess except one. The concentration of this reactant under study is systematically changed to determine the order of reaction against the other species. This process can be repeated for each reactant, providing information about the rate constant and the order of the reaction until the complete rate law has been determined.

At this point it is important to distinguish between order of reaction and molecularity: the order of reaction is experimentally determined based on the rate law and can refer to a specific reagent or to all reagents (overall reaction order); on the other hand, the molecularity is a theoretical concept that refers to the number of molecules or atoms that need to collide to drive the reaction. Molecularity and order of reaction can coincide for elementary step reactions. This book chapter deals with the experimental determination of the order of reaction. We note that a rate law, alongside complementary experiments can, in some cases, be related with the molecularity of some of the elementary steps of the catalytic cycle.

The order of reaction determination is relatively easy and has been extensively employed to study dark homogenous systems, where the initial concentrations can be perfectly controlled. In the literature there are several examples using this approach to obtain insights into the mechanism of the reactions involved in solar fuels generation.[2–5] However, the study of heterogeneous systems is more challenging. Several studies have focused on studying the rate law with respect to one of the initial reagents such as CO_2 or H^+, but not the catalyst *per se*.[6–9] However, directly monitoring catalytic species is necessary to extract additional relevant information about the catalytic mechanism, such as the rate limiting step or the amount of extracted electrons necessary to reach the most stable intermediate, therefore the steady state of the catalysis.

Traditionally for dark electrocatalytic systems rate law analyses have been done evaluating the current voltage (J–V) curve characteristics or the Tafel plots.[10] For ideal, well-behaved systems, these plots give relevant mechanistic information. In these analyses the Tafel slope is related with a specific rate limiting step.[11] However, most of the widely used systems for solar fuels generation do not involve ideal metallic electrodes, but nanostructured materials with multi-redox chemistry.[12] The interpretation of Tafel plots for these types of systems is non-trivial and is hampered by the presence of different surface and/or intraband states.[13] Furthermore, in photocatalytic devices, these electrocatalysts are typically coupled with a light harvesting unit forming photoelectrodes. This arrangement presents an extra challenge for the interpretation of the Tafel plots because the J–V data strongly depend on the electron–hole recombination, which modulates the flux of charge carriers to the surface as a function of applied potential. In this chapter, we describe a method to experimentally assess the order of reaction and determine the rate law of heterogeneous systems, which perform both electrocatalytic and photoelectrocatalytic reactions, necessary to build a mechanistic model.

This method, recently reported by our group, allows the study of these non-ideal systems.[14] Our approach is based on accessing the species accumulated during the steady state using optical and simultaneously electrochemical measurements. The concentration (population) of these states can be modified by modulating the applied potential and/or the irradiated light intensity. As it will be further discussed below this allows us to determine the rate law of the studied catalytic reaction as a function of surface accumulated charges (electrons or holes depending on the studied reaction). This new approach opens the possibility to gain mechanistic insights of non-ideal systems and to compare photoactivated and non-photoactivated systems.

The optical measurement is based on photo-induced absorption (PIA) spectroscopy. This optical pump–probe technique has been extensively used to study the excited state of photoactive materials in timescales ranging from milliseconds to seconds. As opposed to transient absorption spectroscopy (TAS), which employs ultrafast excitation pulses, in PIA experiments relatively long pulses of light are used. Such pulse characteristics enable the generation and monitoring of steady-state populations. PIA spectroscopy—without simultaneous TPC measurement—has been used to study mid band gap states in Si and Ge semiconductors using a 6 ms argon laser as pump.[15–17] More recently, Boschloo and co-workers have introduced the concept of PIA analysis and the use of longer LED pulses to study recombination kinetics in dye-sensitised solar cells (DSSC) and perovskite solar cells.[18–21] Additionally, PIA measurements have been used alongside spectroelectrochemistry analysis to study charge transfer kinetics in DSSCs between the interfaces TiO_2/dye/electrolyte.[22–24]

However, to the best of our knowledge, the use of PIA concomitant with TPC measurements was first reported by our group[14] to study the charge transfer of photogenerated holes at the semiconductor/electrolyte interface of hematite photoanodes. Our group used this technique to gain a deeper understanding into the kinetics of the photogenerated hole transfer from the metal oxide semiconductors to the electrolyte as well as the mechanism of multi-electron redox reactions in non-ideal photoelectrochemical systems.[14,25–29]

5.1.2 Kinetic Model

We consider a simple kinetic model to describe the behaviour of minority charge carriers, which accumulate on a PEC electrode surface. The equation and model will be derived for an *n*-type semiconductor in contact with an electrolyte and under depletion. In this case, an electric field is built to repulse majority carriers (electrons) from the semiconductor–liquid interface and attract minority carriers (holes or electron–holes) to the surface, where they can react with water. We note here that analogous development can be performed for *p*-type semiconductors, where bulk holes and surface electrons reacting with electrolyte species should be considered.

For an *n*-type semiconductor, the density of surface-accumulated holes, p_s, can be expressed as in eqn (5.1) where its variation corresponds to the sum of an incoming hole flux (J_{holes}) and two consuming processes: hole transfer to the electrolyte (water oxidation for example) and surface hole recombination. The hole current is induced by light absorption followed by charge separation in the space charge layer, therefore being a function of both light intensity and applied voltage. Water oxidation and surface recombination are both expressed in a general way, with apparent rate constants, k_{WO} and k_{SR} and a power law dependence on the surface hole density with β and γ as the respective rate orders. It is also important to note that we also consider a semiconducting material that is stable under experimental conditions, meaning holes are neither consumed in the material (*i.e.* self-oxidation), nor in other side processes such as competitive chemical reactions (*i.e.*, Faradaic efficiency of unity for water oxidation).

$$\frac{dp_s}{dt} = J_{\text{holes}} - k_{\text{WO}} \cdot p_s^{\beta} - k_{\text{SR}}\, p_s^{\gamma} \tag{5.1}$$

The multiple parameters that govern the dynamic population of interfacial charge carriers in eqn (5.1) complicate the easy extraction of useful kinetic data such as the order of reaction and the rate constant. Under large anodic bias conditions (for example, 1.7 V_{RHE} and 1.5 V_{RHE} for $BiVO_4$ and α-Fe_2O_3, respectively), the model that has been employed can be further simplified by neglecting hole consumption from back electron recombination.[25,30–32] The only pathway for surface accumulated holes to be extracted is thus to react with the electrolyte. At high current densities, the hole transfer in hematite has been shown to result only in oxygen evolution;[33] we note that in some cases, such as at lower current densities, alternative oxidation products including hydrogen peroxide may also be important. The faradaic efficiency for $BiVO_4$ and TiO_2 has also been characterised as equal or close to unity at high applied voltage or high current density.[34,35] The simplified model is then written in eqn (5.2) and depicted graphically in Figure 5.1.

$$\frac{dp_s}{dt} = J_{\text{holes}} - k_{\text{WO}} \cdot p_s^{\beta} \tag{5.2}$$

We used optical measurements (PIA, see details in Section 5.1.3) to quantify directly the density of accumulated minority charge carriers in semiconductors. The presence of photogenerated carriers induces changes in the absorption of the sample. At certain defined wavelengths, these absorption changes have been assigned to the absorption of light by

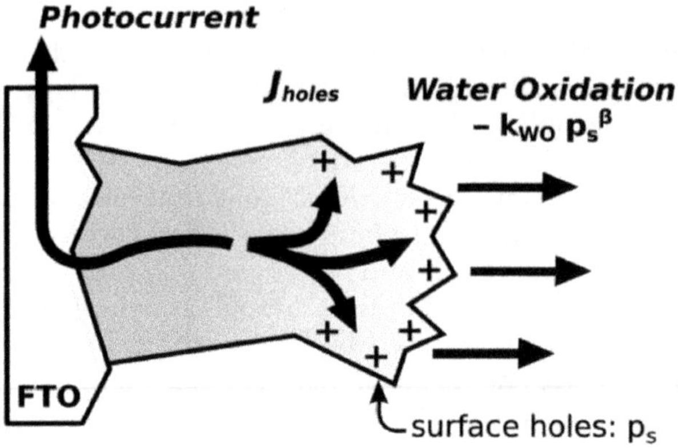

Figure 5.1 Schematic representation of the simplified kinetic model employed to describe the hole density accumulated on the surface of an *n*-type semiconductor during water oxidation.
Adapted with permission from Le Formal *et al.*, *J. Am. Chem. Soc.*, 2015, **137**, 6629–6637.[14] Copyright (2015) American Chemical Society.

minority charge carriers: a broad optical feature peaking around 600 nm assigned to holes or higher valent iron species (Fe(IV)=O)[36] in α-Fe_2O_3,[37–40] an analogous broad absorption peaking at 550 nm for $BiVO_4$ holes[41] and at 500 nm for TiO_2 holes.[42] In addition to the direct characterisation of minority charges, photocurrent density recording provides useful information on the flow of majority charge carriers (electrons for *n*-type) in the external circuit.

This model can be further analysed when considering two particular conditions: at steady state and immediately after illumination is interrupted. These two conditions correspond both to the detection of the electronic charge a long time after their photogeneration. It consequently suggests that only minority charge carriers accumulated at the surface are optically probed due to their very small lifetime in the bulk and to the electric field drawing them to the interface with the electrolyte.

First, steady state is attained when both the photocurrent and the PIA signal reach a constant level with time. This is approximately achieved 5–6 s after illumination is switched on for α-Fe_2O_3, $BiVO_4$, TiO_2 and the $[Cu_2O]/RuO_x$ photocathode.[14,25–27,29] Under these conditions, the density of holes is not changing with time and the photocurrent, *i.e.* the electron flow recorded in the external circuit during operation, equals the flux of holes towards the surface, as detailed in eqn (5.3) (flux's absolute values). We recall here that we consider stable materials, hence with no reductive or oxidative corrosion processes and high faradaic efficiencies towards the studied reaction (OER in this example).

$$\frac{dp_s}{dt} = 0 \Rightarrow \text{Photocurrent} = J_{\text{holes}} = k_{\text{WO}} \cdot p_s^{\beta} \tag{5.3}$$

The reaction kinetic parameters can therefore be obtained with logarithmic plot of the photocurrent shown against the hole density obtained by optical measurements (see example on Figure 5.10).

From the same optical data, a second method can be used to corroborate the insight gained in reaction kinetics, when scrutinising the decay of the PIA signal immediately after illumination is stopped. Under these conditions, the flux of holes towards the surface is null as no more minority carriers are photogenerated within the material. Moreover, due to the high voltage applied (first simplification used from eqn (5.1) to eqn (5.2)), the high electric field in the space charge region prevents the establishment of an electron current from the electric circuit to recombine with the holes accumulated on the surface. Therefore, the only process causing the

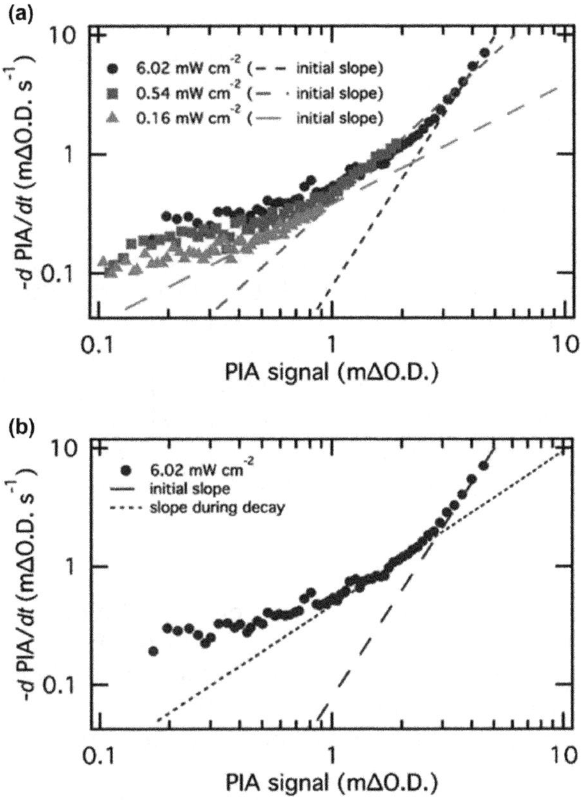

Figure 5.2 Additional data treatment methods used to extract kinetic information about the OER on a hematite electrode surface where the derivative of the PIA signal is plotted against the PIA signal, (a) probing the initial slope at different light intensities and (b) the slope during the decay at one light intensity.
Adapted with permission from Le Formal *et al.*, *J. Am. Chem. Soc.*, 2015, **137**, 6629–6637.[14] Copyright (2015) American Chemical Society.

extinction of surface holes is the reaction with electrolyte species (OER in our example); eqn (5.2) can therefore be simplified in eqn (5.4).

$$J_{\text{holes}} = 0 \Rightarrow \frac{dp_s}{dt} = -\, k_{\text{WO}} \cdot p_s^{\beta} \tag{5.4}$$

Figure 5.2 shows two experiments that utilise this simplification to extract kinetic information from the OER on a hematite surface. The initial slope of the PIA signal has been recorded after equilibration under different light intensities, which leads to different surface hole concentration at steady state before the light is switched off (see different first point on Figure 5.2(a), first point corresponding to the highest PIA signal). In addition, the slope of the hole population derivative can be followed during the decay of the hole population as shown on Figure 5.2(b).

It is important to note here that the kinetic parameters obtained from these different experiments or data treatment method (Photocurrent *vs.* PIA at steady state and different decay of the PIA signal after illumination) have shown excellent agreement for materials investigated for the OER (α-Fe_2O_3,[14] $BiVO_4$[25] and TiO_2[27]).

5.1.3 Experimental Set-up

As described above, the rate law of a photocatalytic material can be investigated by measuring simultaneously the PIA and the TPC response of the material under continuous illumination conditions. Therefore, this technique can be described as two independent types of measurements: one optical and one electrochemical (see Figure 5.3). In this section, these two types of measurements are discussed focusing on the simultaneity of the analysis. This technique allows us to study independently but concurrently minority (optically) and majority (electrochemically) charge carriers.

5.1.3.1 Optical System

The optical system is based on a pump-probe technique, here both components will be discussed (Figure 5.3, optical system).

The *pump* is composed by one or two UV (365 nm) light emitting diodes (LEDs), which deliver a photon flux of usually between 1 and 40 mW cm^{-2} depending on the current passing through the LEDs (usually between 0.02 and 0.6 A). The length and symmetry of the pump are controlled by a function generator, using a MOSFET as interface between the function generator and the LEDs for a fast switch response.

The *probe* consists of a constant monochromatic beam from a 100 W tungsten lamp. It is focused and filtered by a series of lenses, long-pass and band-pass filters that block the scatter from the LEDs pump light. A silicon photodiode is used to monitor the absorbing species in the material in a time-resolved basis. This signal is recorded and averaged without

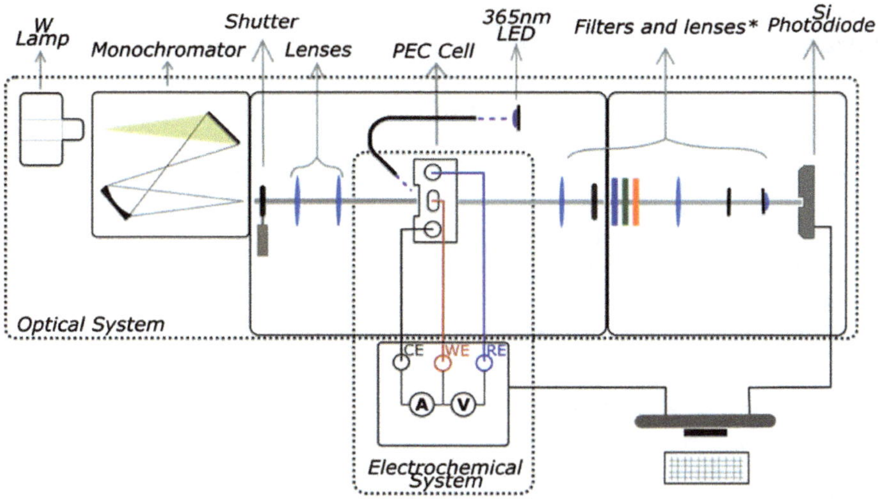

Figure 5.3 Schematic representation of the simultaneous PIA (optical system) and TPC (electrochemical system) measurements. The set of filters and lenses marked with * can be replaced with a second monochromator.

amplification by a data acquisition (DAQ) card and a LabView based controlling software, which is triggered simultaneously with the pump.

5.1.3.2 Electrochemical System

The *electrochemical* system uses a standard set-up of a photoelectrochemical (PEC) measurement. A three-electrode system was used to monitor the TPC while the material is being illuminated by the LEDs. In addition, the current is also monitored during darkness. For the examples shown in this chapter, all the measurements have been done holding the photoelectrode at a defined applied potential between the reference, a KCl saturated Ag/AgCl electrode, and the working (photoelectrode) electrodes. The current passing between the working and the counter, a Pt mesh, electrodes was measured by a potentiostat and recorded by an oscilloscope to average the photocurrent signals. This system was triggered simultaneously with the optical system by the same square wave from the function generator.

5.2 Case Studies

In this section, we will show the potential of using photo-induced absorption spectroscopy alongside transient photocurrent measurements through several practical examples.

The first part will be focused on photoanodes. Although rapid progress has been made improving the performance of metal oxide photoanodes, such as α-Fe_2O_3, $BiVO_4$, TiO_2 and WO_3 among others,[43] the OER mechanism and its limiting factors remain elusive. In this section, we will discuss the analysis of the rate law on two well-known photoanodes for water oxidation: α-Fe_2O_3 and $BiVO_4$. In addition to this, we will also discuss the oxidation of $K_4[Fe(CN)_6]$ and the role of CoPi catalyst overlayers on $BiVO_4$.

The second part of this section is focused on reductive reactions. The studied photocathode consists of a multilayer structure of Cu_2O/Al : ZnO/TiO_2.[44] The Al : ZnO (AZO) and TiO_2 overlayers serve a double role: firstly, by forming a buried junction which allows a more efficient charge extraction and secondly, by protecting the Cu_2O against photocorrosion (reduction of Cu(I) centres to Cu(0) by photogenerated electrons. However, as a secondary effect, the passivation layer also blocks the catalytic activity towards proton reduction. To bypass this limitation, a catalytic layer of a porous RuO_x is deposited on top of the protection layer.[45]

5.2.1 Oxidation Reactions

5.2.1.1 α-Fe_2O_3

In this section, we discuss the spectroscopy of a state-of-the-art α-Fe_2O_3 photoanode, from the hole signal assignation to the implications of the rate law results, upon illumination and applied potential.

The α-Fe_2O_3 films studied herein were grown by atmospheric pressure chemical vapour deposition (APCVD) following a procedure developed by Grätzel and co-workers.[46,47] Figure 5.4 shows a typical current–potential (*J*–*V*) response in alkaline conditions, under chopped (1 sun) and electrode–electrolyte interface (EE) illumination. Under these conditions, a photocurrent onset at ~1.0 V *vs.* the reversible hydrogen electrode (RHE) and photocurrent of ~2.7 mA cm^{-2} at 1.5 V_{RHE} are measured.

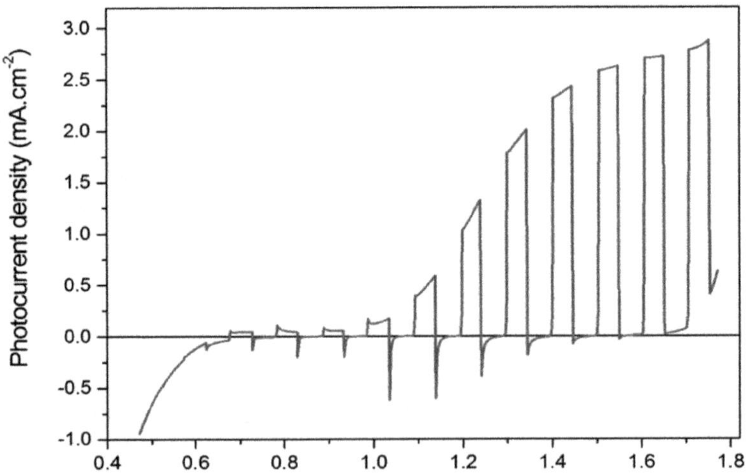

Figure 5.4 Current/potential response of an APCVD α-Fe_2O_3 photoanode under simulated 1 sun chopped EE illumination in 0.1 M NaOH aqueous solution.

We start by studying the species involved in the oxidation of water by means of transient absorption spectroscopy (TAS) from microsecond to seconds. This time scale allows the study of the recombination as well as the charge transfer of these species to the electrolyte upon light excitation at different conditions of applied potential. Figure 5.5(a) shows the TA spectrum obtained upon UV excitation at open circuit potential (OCP) (~0.8 V_{RHE}) and 1.5 V_{RHE}. At OCP we observe a ground state bleaching signal at 580 nm and a weak positive absorption peaking around 650–700 nm (Figure 5.5(a), grey). This positive absorption is enhanced (Figure 5.5(a), black) when the applied potential increases to strong oxidative conditions (1.5 V_{RHE}), where photogenerated holes are less prompt to recombine with electrons resulting in a higher population of holes in the hematite. These photogenerated holes are shown to live long enough (~100 ms) to drive the OER. Based on this, we assign the positive absorption peaking in ~650–700 nm to photogenerated holes. Further supporting this assignment we observe that the ~650–700 nm signal is quenched in the presence of a well-known chemical hole scavenger, methanol, (Figure 5.5(b), light grey) due to the faster extraction of holes from the surface of the hematite photoanode.[37]

Once the optical signal corresponding to photogenerated holes has been assigned to the TA ~650–700 nm, we turn to discuss the effect of increasing the oxidative potential to achieve useful photocurrents, shown in the *J–V* curve (Figure 5.4), which has been previously studied in our group.[38,48,49] Figure 5.6 shows the TA decay dynamics of the photogenerated holes in the nanostructured hematite photoanode as a function of applied potential. At 0.8 V_{RHE}, ~OCP conditions, where no significant photocurrent is observed, the optical signal follows a power law decay, where dissipative, non-geminate electron–hole recombination takes place. As the oxidative potential is

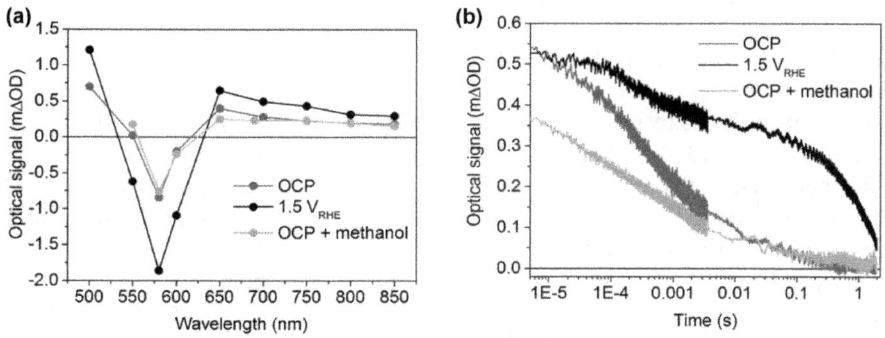

Figure 5.5 Transient absorption response of a hematite photoanode, (a) spectral response at 100 ms in OCP (grey), OCP + methanol (light grey) and 1.5 V_{RHE} (black) conditions, and (b) optical signal decays from 1 μs to 1 s after 0.2 mJ cm^{-2}, 355 nm EE excitation, probed at 650 nm, at OCP (~0.8 V_{RHE}) (grey), 1.5 V_{RHE} (black) and at OCP in presence of methanol as hole scavenger (light grey). OCP = open circuit potential.

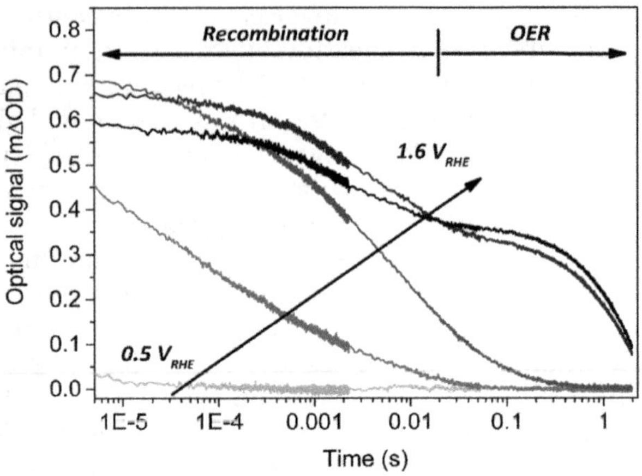

Figure 5.6 Transient absorption decays for an APCVD α-Fe_2O_3 photoanode as a function of applied potential under 355 nm, 0.20 mJ cm^{-2} EE excitation, probed at 650 nm, where photogenerated holes have been assigned to absorb.
Adapted from ref. 38 with permission from The Royal Society of Chemistry.

increased until it is close to the onset of photocatalysis (~1.0 V_{RHE}) the fast phase (μs to ms), increases in amplitude and lifetime. We associate this to the suppression of the aforementioned recombination process as the space charge layer widens. We note here that bulk recombination, which takes place in the ps timescale, is substantially retarded upon the application of a high oxidative potential.[50] A second phase (slow phase) appears in the ms to s timescale, which increases in amplitude when the applied potential is within ~1.0 to 1.4 V_{RHE} (from the photocurrent onset to the beginning of the plateau photocurrent). We assign this process to the oxidation of water.[38] This slow phase signal (~100 ms) exhibits a direct correlation with the photocurrent showing the need of long-lived holes, which avoid recombination, to oxidise water. Further supporting this, the plateau photocurrent is reached at high oxidative potentials when back electron–hole recombination is suppressed and long-lived holes are generated.[32] However, the kinetics and mechanism of the water oxidation reaction on a hematite surface, which are the parameters controlling the reaction once the photogenerated holes have overcome the recombination, is not yet well understood. Therefore, it is necessary, to study the kinetics of accumulation and reaction of the photogenerated holes in steady-state conditions by means of PIA spectroscopy measured simultaneously with the TPC, as described in the experimental set-up section of this chapter.

As previously described, the 650–700 nm optical absorption measured at 100 ms in TAS has been assigned to photogenerated holes in α-Fe_2O_3, due to an increase in the signal amplitude when applying a positive potential and the signal quenching when a hole scavenger is added. Figure 5.7 shows the

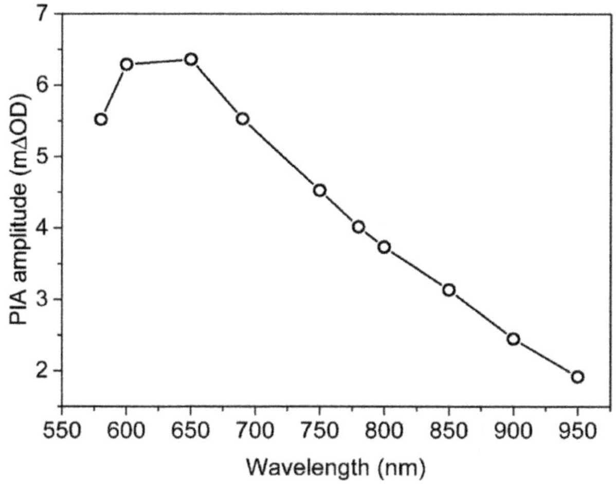

Figure 5.7 PIA spectra of the α-Fe_2O_3 photoanode under 16 mW cm^{-2}, 365 nm illumination, measured at 1.5 V_{RHE}, where water oxidation occurs and recombination has stopped.

spectral response of a hematite photoanode after 4 s of LED 365 nm excitation at 1.5 V_{RHE}, conditions where photogenerated holes live long enough to drive water oxidation. This spectral fingerprint peaks at 650–700 nm matching the signal assigned to photogenerated holes by means of TA measurements

We turn now to establish the proportionality between the optical absorption at 650 nm and the photogenerated surface hole density. As can observed in the *J–V* curve shown in Figure 5.4, at applied potentials where no water oxidation occurs (0.9 V_{RHE}), anodic and cathodic transient spikes are present due to back electron–hole recombination. When the light is switched on, varying the light intensity (PIA signal, Figure 5.8(a)), a transient positive current spike is observed (Figure 5.8(b)) without the establishment of a steady-state water oxidation photocurrent. The integration of this current with time result in the number of positive charges accumulated at the interface. When light is switched off, a cathodic current spike is observed corresponding to electrons driven to the surface to recombine with the accumulated holes. The integration of the negative current is also a measurement of the accumulated hole density. This correlation, between the plateau PIA signal and the amount of charges extracted is obtained by extrapolating the linear correlation up to 6 mΔOD, (see Figure 5.8(c)).[14,32]

As discussed in the introduction, we can evaluate the rate law of chemical reaction by measuring the reaction output as a function of the concentration of each of the reactants. Our PIA and TPC data as a function of light intensity (Figure 5.9) provides us such information. Whilst the optical signal provides us with an assay of the concentration of reactive minority carriers (holes) accumulated at the surface, p_s, the photocurrent data correlates with the flux of these carriers towards the surface, J_{holes}, in eqn (5.3).

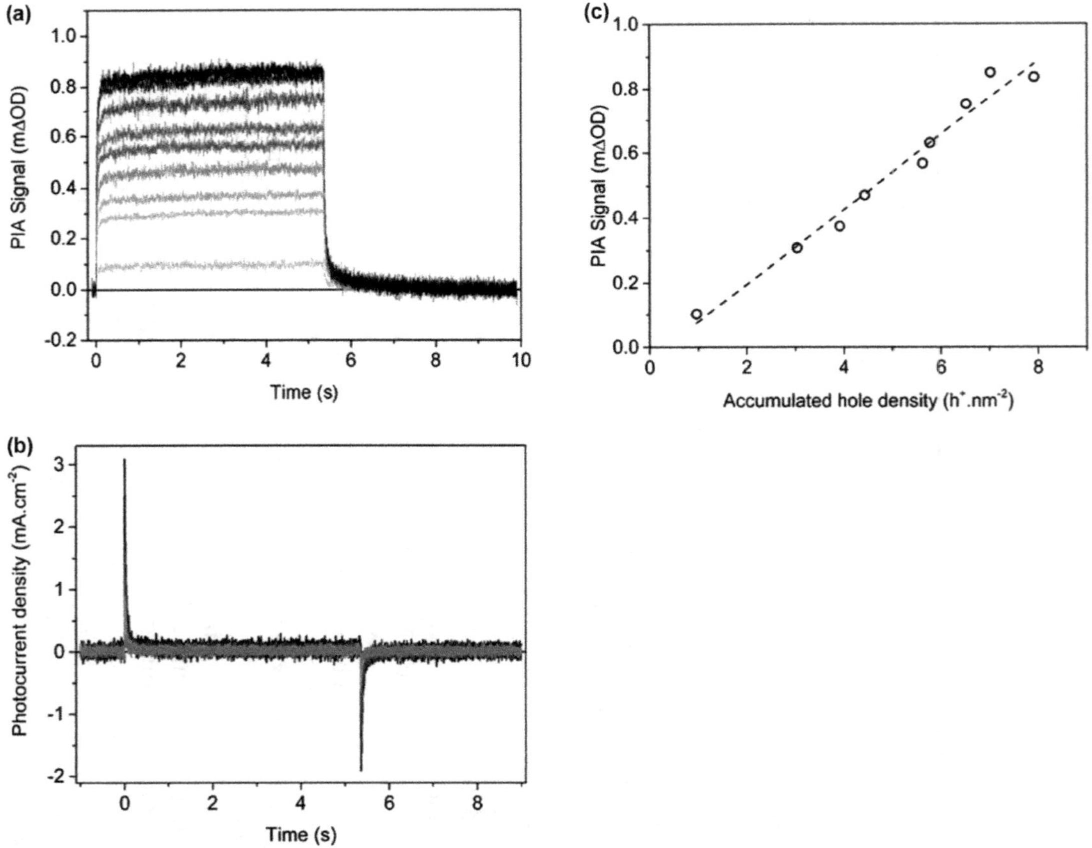

Figure 5.8 (a) PIA signals of the reacting species, monitored at 650 nm of the APCVD α-Fe_2O_3 photoanode during 5 s ON/5 s OFF pulsed 365 nm illumination, measured at 0.9 V_{RHE}, (b) transient photocurrent density produced by the extraction and subsequent back electron–hole recombination as a function of light intensity and (c) correlation between the integrated extracted and subsequently recombined charges with the steady-state PIA signal, offering a conversion factor of 1 mΔOD = 0.4459 h^+ nm^{-2} real area (with a roughness factor of ∼21).
Adapted with permission from Le Formal *et al.*, *J. Am. Chem. Soc.*, 2015, **137**, 6629–6637.[14] Copyright (2015) American Chemical Society.

Figure 5.9(a) shows the time resolved PIA measurements of the 650 nm signal at different illumination intensities. As expected, the signal amplitude increases with increasing illumination intensity, indicating a greater accumulation of photogenerated holes. When the light is turned on, we observe a relatively slow rise in the PIA signal due to the generation and surface accumulation of photogenerated holes that react with the electrolyte. At approximately 1 s, the PIA signal plateaus once the accumulation of photogenerated holes at the surface of the photoelectrode reaches a kinetic equilibrium with the holes reacting with the electrolyte, as detailed in the kinetic model section. When light is turned off and charge carrier generation

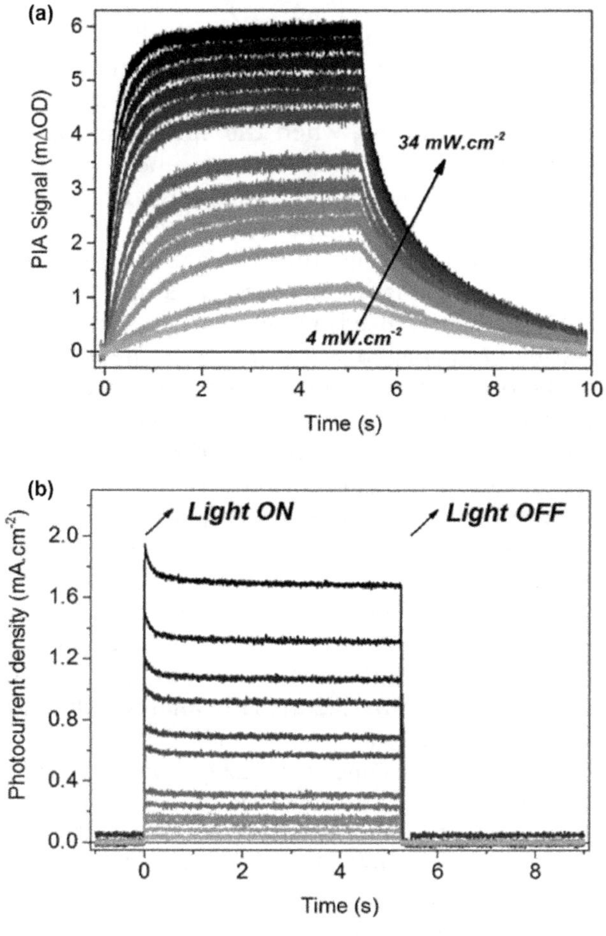

Figure 5.9 (a) PIA signals of the reacting species, monitored at 650 nm of the APCVD α-Fe_2O_3 photoanode during 5 s ON/5 s OFF pulsed 365 nm illumination, measured at 1.5 V_{RHE}, and (b) photocurrent density produced by the water oxidation reaction as a function of light intensity.
Adapted with permission from Le Formal *et al.*, *J. Am. Chem. Soc.*, 2015, **137**, 6629–6637.[14] Copyright (2015) American Chemical Society.

stops, the PIA signal decays as the holes accumulated at the surface of the hematite react or transfer to the electrolyte.[14]

Unlike the PIA signal, transient photocurrent (TPC) densities raise sharply when the LED illumination is turned ON and therefore the steady-state is achieved earlier for the TPC measurements (see Figure 5.9(b)). This is due to the faster majority carrier mobility in hematite compared to the hole mobility as detailed in the theoretical background section. The steady-state photocurrent (related directly to water oxidation, as indicates its almost unity faradaic efficiency) drops rapidly to zero when the light is turned OFF, with the absence of cathodic current spikes, which correspond to back electron–hole or surface recombination.[14,32,51,52]

We now turn to the analysis of the rate law of water oxidation with respect to the density of accumulated surface holes. As described in the kinetic model section, we consider two conditions for this analysis: at steady-state and comprising the initial rates when the light is turned off.

Under steady-state conditions, when the species accumulated to overcome the rate limiting step of the reaction are under observation, there is not net change in the PIA signal, *i.e.*: the change in the density of accumulated holes at the surface with the time is zero ($dp_s/dt = 0$, eqn (5.3)). This PIA signal can be converted into surface hole density (p_s) by the conversion factor shown in Figure 5.8(c). Such conversion can be done for the range of light intensities used. Figure 5.10 shows the water oxidation photocurrent as log $[J_{\text{holes}}]$ against the log $[p_s]$. The gradient of this plot corresponds to the order of reaction (β) with respect to the surface holes and the intercept of the fitted linear function refers to the apparent water oxidation rate constant (k_{WO}). This figure shows two distinctive regions where the reaction order transitions from 1st to 3rd as the surface hole density increases. Notably such transition from first to third order behavior occurs when there is an accumulation of $\sim$1 h^+ nm^{-2}, when the 'turn over frequency' per hole is $\sim$0.4 s^{-1}.

This behaviour was corroborated using alternative kinetic analysis (Figure 5.10(b)), including a TAS kinetic analysis. The initial slope analysis (Figure 5.10(b), light grey squares) that relies only in the optical measurements was also shown in Figure 5.2(a). This analysis shows an analogous logarithmic plot to Figure 5.10(a), corresponding to the initial rate law analysis of the accumulated holes once the flux of holes towards the surface of the hematite is zero. Therefore, the rate law depends on the initial rate of reaction and the density of accumulated holes at the surface (eqn (5.4)). The initial rate of reaction is obtained by calculating the slope of the optical decay when the light is turned OFF. Similarly, to the steady-state rate law, the gradient of this plot relates to the order of reaction and therefore to the number of holes that need to be accumulated to overcome the rate limiting step of the reaction. This transitioning behavior from 1st to 3rd order of reaction at $\sim$1 h^+ nm^{-2} is also found from these entirely optical analyses in agreement with the rate law calculated from the optical and electrochemical measurements.

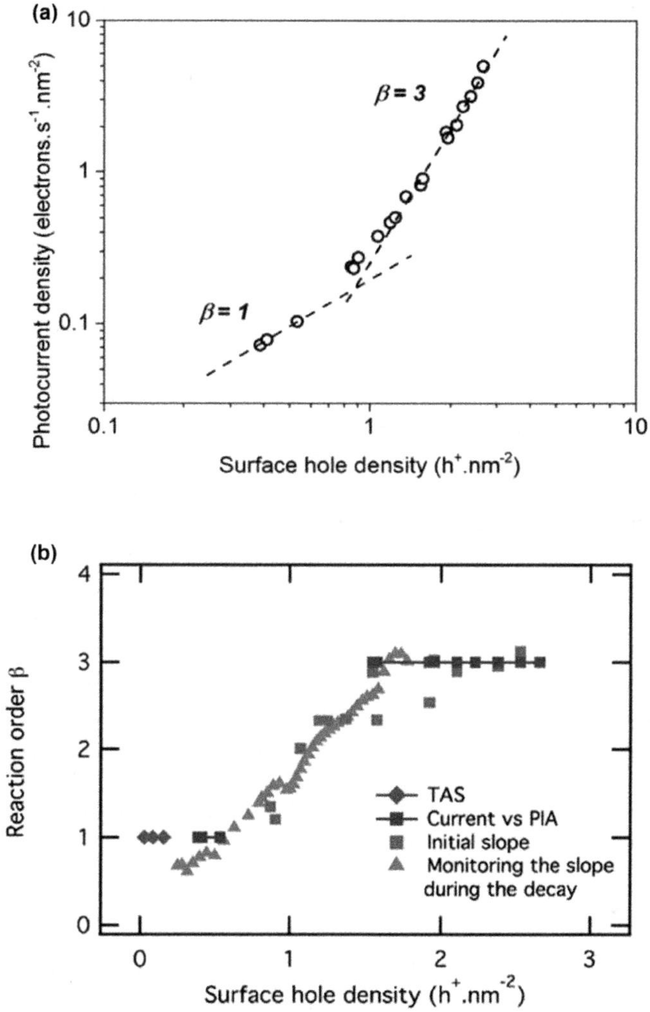

Figure 5.10 (a) Steady-state rate law analysis, photocurrent density, J_{holes}, *vs.* surface hole density, p_s, for the water oxidation on the APCVD α-Fe_2O_3 photoanode measured at 1.5 V_{RHE}, the dashed lines represent the fitted functions for the first and third order regimes. (b) Change in reaction order with increased surface hole density. The reaction order determined by different techniques (see legend) is shown against the surface hole density accumulated during water oxidation.
Adapted with permission from Le Formal *et al.*, *J. Am. Chem. Soc.*, 2015, **137**, 6629–6637.[14] Copyright (2015) American Chemical Society.

The particular transitioning behavior from 1st to 3rd order of reaction implies a change in the rate limiting step of the catalytic cycle, potentially related to a change in the mechanism upon increasing the surface hole density. This order of reaction is directly related to the number of absorbing reacting species (Fe(IV)=O)[36] accumulated in the rate limiting step;

therefore, information about the number of surface holes that need to be accumulated to overcome the rate limiting step of the reaction can be extracted from the gradient in Figure 5.10. Based on the aforementioned premises, this particular transitioning behaviour suggests that the rate limiting step of the reaction is conditioned to the ability of the material to accumulate holes at the surface.

Water is known to require four oxidative equivalents to evolve oxygen; however, it is striking to find that only three photogenerated holes need to be accumulated at the surface of the hematite photoanode to overcome the rate limiting step under operating conditions. At low light intensities, where the gradient of the rate law curve is one, the rate limiting step of the reaction depends on the transfer of a hole arriving to the surface with a characteristic apparent rate constant of $0.27\ s^{-1}$. We note that the transfer of one hole to the electrolyte might produce ·OH radicals which would produce O_2 through the formation of H_2O_2. At high surface hole accumulation, the OER undergoes kinetics of $0.38\ s^{-1}\ hole^{-2}\ nm^{-4}$, and requires the accumulation of three photogenerated holes to overcome the rate limiting step of the reaction. A reaction mechanism has not been clearly depicted in the literature, with two main proposals still in debate: a water nucleophilic attack over a metallic center with a subsequent formation of a MO–OH species and the reaction between two metal oxide units generating the species MO–OM as intermediate,[39,40,53–56] with the formation of the O–O bond considered as the rate limiting step of the OER in both cases. In the studied case this last step would require three surface holes to be formed in accordance with the presented order of reaction. These analyses suggest that the first to third order behavior transition is due to the formation of a different transition state and hence a different mechanism of reaction.

5.2.1.2 *$BiVO_4$*

Another promising photoanode, $BiVO_4$, has also been studied using this technique. As for α-Fe_2O_3, the first step is to determine the optical fingerprint of photogenerated holes. As can be observed in Figure 5.11 the PIA signal under steady state conditions (recorded after *ca.* 5 s illumination, Figure 5.11(a))[57] is similar to that obtained by means of TA spectroscopy at strong applied potentials (Figure 5.11(b)) with a signal maxima within 500–600 nm assigned to holes.[41,58] This assignment is further supported by studies in the presence of scavengers with the signal being quenched in the presence of the hole scavenger, Na_2SO_3 (Figure 5.11(c)). In analogy to α-Fe_2O_3 we assign the steady state signal to photogenerated surface holes involved in the water oxidation catalysis. Furthermore, we can associate these surface holes to the ones accumulated during the steady state, and thus the ones needed to overcome the rate limiting step of the catalytic process.

Figure 5.12(a) and (b) show the PIA and TPC signals at 550 nm and 1.7 V_{RHE} applied potential as a function of light intensity. As expected an increase in the light intensity results in larger PIA signals and larger

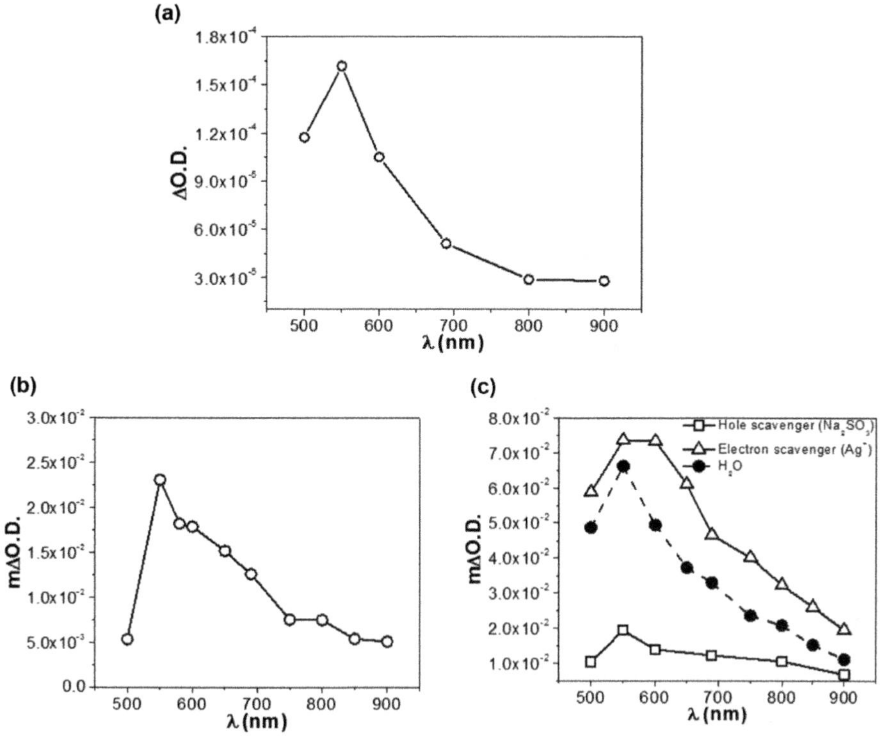

Figure 5.11 (a) PIA spectrum in pH 7 phosphate buffer at 1.7 V_{RHE} recorded under 5 s LED illumination; (b) transient absorption spectrum in pH 7 phosphate buffer at 1.2 V_{RHE}, at 100 ms after laser excitation (355 nm, 100 μJ cm^{-2}, 0.33 Hz) and (c) transient absorption spectra in aqueous $AgNO_3$ (triangles), de-ionised water (circles) and aqueous Na_2SO_3 (squares), recorded 50 ms after laser excitation (355 nm, 100 μJ cm^{-2}, 0.33 Hz).
(a) Adapted from Y. Ma, C. A. Mesa, E. Pastor, A. Kafizas, L. Francàs, F. Le Formal, R. R. Pendlebury and J. R. Durrant, *ACS Energy Lett.*, 2016, **1**, 618.[25] Copyright (2016) American Chemical Society. (b) Reproduced from ref. 41 with permission from The Royal Society of Chemistry. (c) Reproduced from ref. 58 with permission from The Royal Society of Chemistry.

photocurrent. Analogously to the analysis with α-Fe_2O_3 we can determine the order of reaction for the water oxidation on $BiVO_4$. The rate law analysis plot correlating the photocurrent with hole density is shown in Figure 5.12(c). Similarly, to what we measured with α-Fe_2O_3, we observe a change in reaction order from 1 at low surface hole density to a reaction order of 3 when the surface density reaches 1 h^+ nm^{-2}.

Following eqn (5.3), the rate law plot (Figure 5.12(c)) also gives information about the rate constant of the catalytic reaction. In this case we can estimate two different apparent rate constants: one for the first order regime, $K_{WO} = 0.5\ s^{-1}$; and $K_{WO} = 0.8\ s^{-1}\,hole^{-2}\,nm^{-4}$ for the third order reaction. These rate constants suggest that the reaction increases its rate at high hole

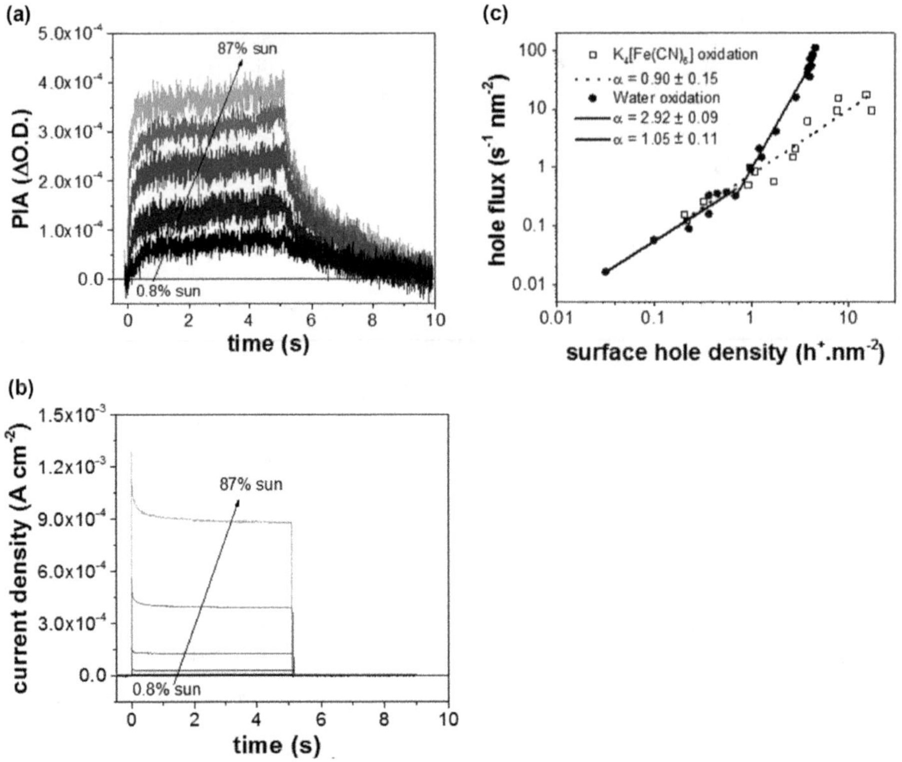

Figure 5.12 (a) PIA signals of the reacting species, monitored at 550 nm of the $BiVO_4$ photoanode during 5 s ON/5 s OFF pulsed 365 nm illumination, measured at 1.7 V_{RHE}; (b) photocurrent density produced by the water oxidation reaction as a function of light intensity and (c) relationship of PIA amplitude and quasi-steady-state photocurrent density for water oxidation (black circles) and $K_4[Fe(CN)_6]$ oxidation (empty squares), the lines represent the fitted functions for water oxidation (solid lines for 1st and 3rd order) and $K_4[Fe(CN)_6]$ oxidation (dotted line). Adapted with permission from Y. Ma, C. A. Mesa, E. Pastor, A. Kafizas, L. Francàs, F. Le Formal, R. R. Pendlebury and J. R. Durrant, *ACS Energy Lett.*, 2016, **1**, 618.[25] Copyright 2017 American Chemical Society.

surface densities. Furthermore, a comparison of these apparent rate constants with those found for α-Fe_2O_3 ($K_{WO} = 0.27\ s^{-1}$ and $K_{WO} = 0.38\ s^{-1}\ hole^{-2}\,nm^{-4}$), reveals a faster reaction kinetics in $BiVO_4$ in both regimes (1st and 3rd order). This different kinetics could be associated with a deeper position of the $BiVO_4$ (2.6 eV) valence band than hematite (2.4 eV). This difference could result in a larger driving force for the reaction to occur when the photogenerated hole is on $BiVO_4$ than on α-Fe_2O_3.

Analogously to hematite, different analyses have been carried out and compared in order to confirm these results: differentiation of the gradient of one complete PIA decay after light-off; an initial slopes analysis of several PIA decays under different excitation intensities; and kinetic analysis of TAS data

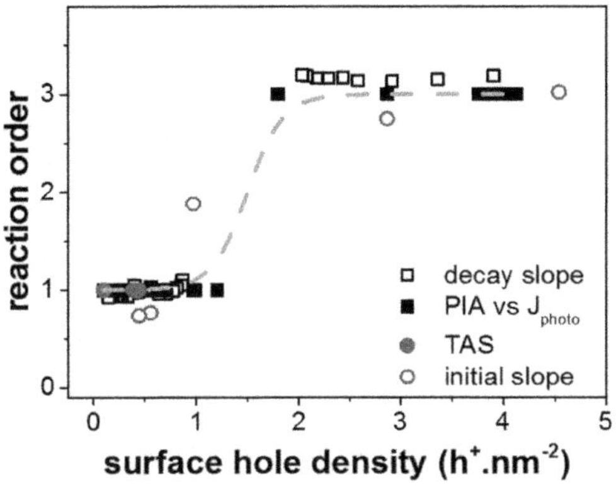

Figure 5.13 Change in reaction order as a function of $BiVO_4$ surface hole density, determined from four independent methods of analysis. The dotted line is a guide for the eye (not fit to a model).
Adapted with permission from Y. Ma, C. A. Mesa, E. Pastor, A. Kafizas, L. Francàs, F. Le Formal, R. R. Pendlebury and J. R. Durrant, *ACS Energy Lett.*, 2016, **1**, 618.[25] Copyright 2017 American Chemical Society.

previously recorded, showing a monoexponential decay of the surface hole as a function of time for water oxidation. As can be observed in Figure 5.13 all the results are in agreement confirming the previously described results.

Now we turn to study the applicability of this technique to other reactions. To this aim, the well-known $K_4[Fe(CN)_6]$ hole scavenger has been chosen as it undergoes a one electron oxidation:

$$[Fe(CN)_6]^{4-} + 1h^+ \rightarrow [Fe(CN)_6]^{3-} \tag{5.5}$$

The simplicity of this reaction is in sharp contrast with the previously studied water oxidation, which will help us to validate our technique. The first order reaction in accumulated surface holes on $BiVO_4$ was estimated from the rate law plot (Figure 5.12(c) empty squares), with a rate constant of 0.8 s^{-1}. This rate constant is higher than that obtained for the 1st order regime of water oxidation. This is in agreement to a lower required potential reaction (0.36 V_{RHE}) than to oxidise water (1.23 V_{RHE}).

This experiment validates our technique to study solar driven catalytic mechanisms and emphasises how our experimental approach can be used to study both relatively simple reactions as well as more challenging reactions such as water oxidation.

5.2.1.3 Role of Catalyst Layers: $BiVO_4$/CoPi

One of the limiting factors for solar fuels generation has been attributed to the slow kinetics of the photocatalysts to oxidise water.[52,59] In the literature this has been overcome by the addition of electrocatalysts layers on top of the metal oxide photocatalysts.[60,61] This strategy has led to an improvement in both photocurrent and onset of the photocatalysis. However, it is not clear if this improvement is due to a hole transfer to the catalyst layer, and so an improved water oxidation kinetics[62–64] or due to a retardation of surface hole recombination.[32,48,49,52,65–67] In this section we will focus on study the effect of the CoPi layer on top of a dense $BiVO_4$ using our PIA and TPC technique.[57]

Previous TA spectroscopy studies on this system have shown that the additional CoPi layer can retard recombination losses in the photoanode.[58] However by using the PIA spectroscopy alongside TPC measurements described herein, we can access the species accumulated under catalytic quasi-steady state conditions, and determine the kinetics in which they oxidise water. In order to obtain a global picture of the composite photoanode the PIAS and TPC studies are complemented by spectroelectrochemical experiments on bare CoPi electrodes. Figure 5.14 presents the spectroelectrochemistry of CoPi electrocatalyst; such data give us an assay of the population of the species responsible for the catalysis, in addition to the pre-catalytic ones. As can be observed the main increase in absorbance is due to a pre-catalytic species, since its appearance is not directly correlated with an increase of current, compare black circles (optical signal) and white squares (current density) in Figure 5.14(b). In the literature, this pre-catalytic species has been associated with Co(III) species.[56,68,69] The high absorbance of this pre-catalytic species (620 ± 60 M^{-1} cm^{-1} at 550 nm) hampers the detection of the real catalytic ones, which are associated with Co(IV) species.[56,68,69]

The strong absorption of the pre-catalytic species mentioned above prevents us from determining a rate law for the catalytic species. However, our technique allows us to assess the amount of Co(III) accumulated on the CoPi/$BiVO_4$ photoanodes under pseudo-steady state conditions. As is shown in Figure 5.15 under strong oxidative conditions the photoanode containing CoPi presents higher currents and higher ΔO.D. The current is doubled in the presence of the catalyst; however, the optical signal is increased 100 times.

Now we turn to analyse the spectra of both $BiVO_4$ and CoPi/$BiVO_4$ (Figure 5.16). The presence of CoPi on top of $BiVO_4$ modifies the spectral response of the film, which evidences similarities to the spectrum of a bare CoPi film (please refer to Figure 5.14), indicating that under pseudo steady state conditions the CoPi catalyst is oxidised to Co(III). Under these conditions the population of Co(III) can be estimated to be 3.2×10^6 cm^{-2}, which would produce a negligible current, around 100 μA cm^{-2} in the dark (see Figure 5.14(b)). This strongly suggests that in the modified photoanode the population of Co(III) is not enough to drive the high-water oxidation currents that are observed, which would require the accumulation of at least three times more Co(III) (1×10^{17} Co(III) cm^{-2}, Figure 5.14(b)). Thus, from this analysis we can estimate that 5% of the photocurrent is due to water oxidation from the CoPi layer, and the remainder of the photocurrent derives directly from the valence band of the $BiVO_4$.

From all the studies mentioned above and other studies in the literature, a schematic representation of the different time scales for each process could

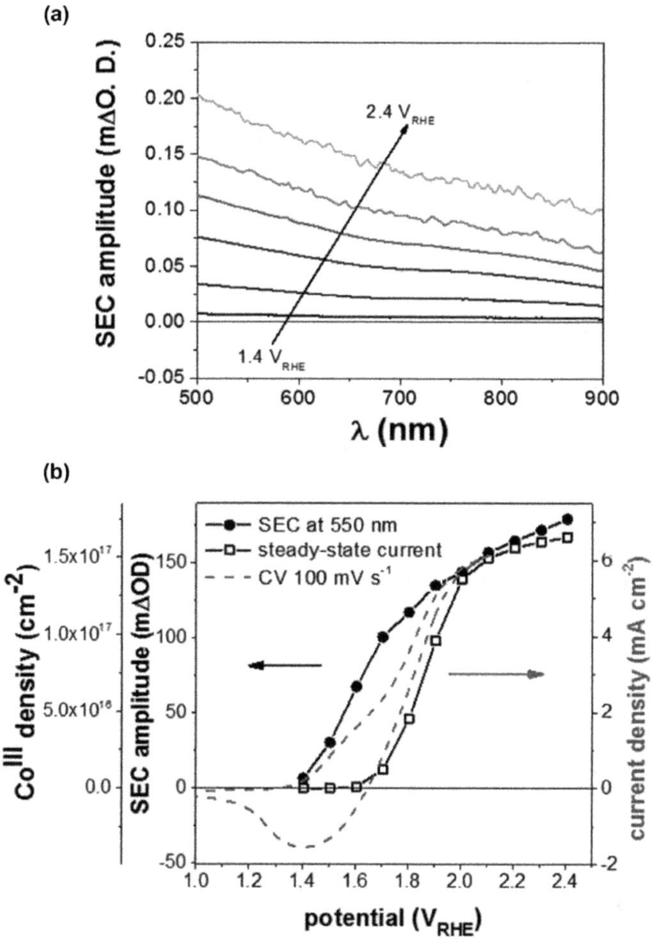

Figure 5.14 (a) Spectroelectrochemical (SEC) difference spectra of CoPi/FTO measured as a function of applied potential from 1.4 to 2.4 V_{RHE}. All difference spectra are generated by subtracting the spectrum measured under applied potential by the spectrum measured at open-circuit (without applied bias); (b) the amplitude of the CoPi/FTO spectra at 550 nm and calculated Co(III) density (cm^{-2}) as a function of applied potential (solid black line and circles); the steady-state current recorded during SEC measurements of CoPi/FTO (solid black line and squares); CV measurement of CoPi/FTO as a function of applied potential with a scan rate of 100 $mV s^{-1}$ (dashed grey lines).
Adapted from Ma *et al.*, Photoinduced Absorption Spectroscopy of CoPi on BiVO4: The Function of CoPi during Water Oxidation, *Adv. Funct. Mater.*, 2016, **26.**[57]

be drawn (Figure 5.17). Here it is worth mentioning that the scheme is built excluding bulk recombination and back electron–hole recombination, so it is valid only for strong positive applied potential conditions.

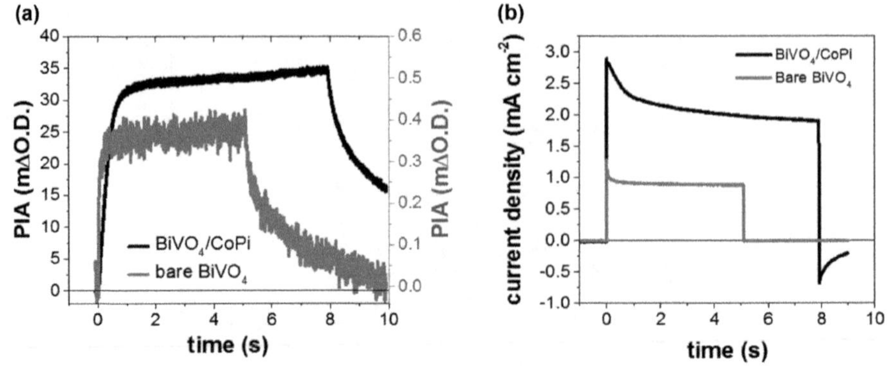

Figure 5.15 (a) PIA response of the unmodified (grey, under 1.7 V_{RHE}) and CoPi-modified (black, under 1.4 V_{RHE}) $BiVO_4$ photoanode. The probe wavelength is 550 nm; the excitation wavelength is 365 nm (excitation intensity: 87% of AM 1.5). Length of excitation pulse: 5 s for unmodified $BiVO_4$ and 8 s for CoPi-modified $BiVO_4$ and (b) photocurrent response measured at the same time as PIA measurements.
Adapted from Ma *et al.*, Photoinduced Absorption Spectroscopy of CoPi on BiVO4: The Function of CoPi during Water Oxidation, *Adv. Funct. Mater.*, 2016, **26.**[57] © 2016 WILEY-VCH Verlag GmbH & Co. KGaA, Weinheim.

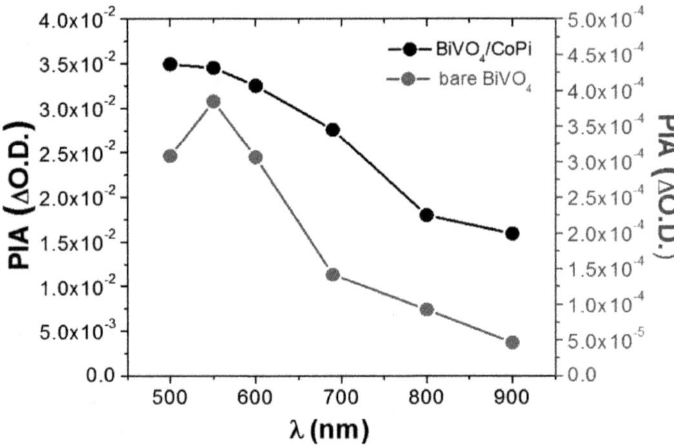

Figure 5.16 Spectra of the unmodified and CoPi-modified $BiVO_4$.
Adapted from Ma *et al.*, Photoinduced Absorption Spectroscopy of CoPi on BiVO4: The Function of CoPi during Water Oxidation, *Adv. Funct. Mater.*, 2016, **26.**[57] © 2016 WILEY-VCH Verlag GmbH & Co. KGaA, Weinheim.

As shown in Figure 5.17 we can assume that the slow hole transfer to the CoPi catalyst layer and the slow water oxidation kinetics of the CoPi catalyst are the main causes for the catalysis taking place directly from

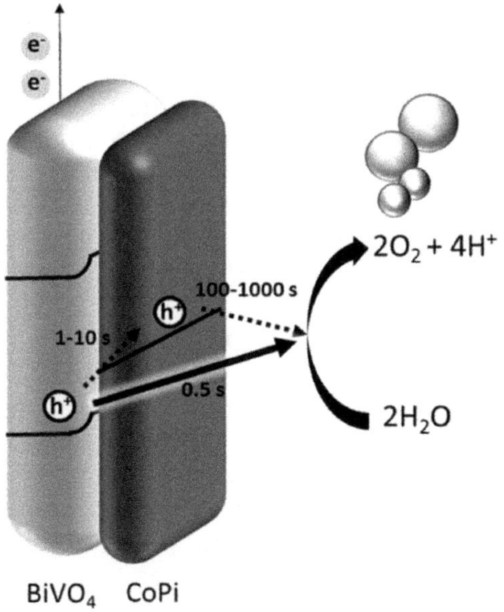

Figure 5.17 Schematic representation of the kinetics of hole transfer at the $BiVO_4$ surface, based on the results discussed herein. Time scale for CoPi water oxidation has been defined according to the literature TOF values $2 \times 10^{-3}\ s^{-1}$ [69] and 1×10^{-2} and 1×10^{-3}.[70] Note that the difference in value is due to a different measurement of active sites. The timescale for water oxidation from $BiVO_4$ was determined by previous studies.[41,57,58] Finally, the timescale of hole transfer to the CoPi layer from $BiVO_4$ was determined by an estimation of the time needed to accumulate the detected population taking into account a life-time of 100 s. Moreover, this is approximately the same time scale found in impedance spectroscopy.[63,71]

the valence band of the $BiVO_4$ photoanode. This can be related with the deep valance band position of $BiVO_4$, which results with a large overpotential to drive water oxidation, consistent with the fast water oxidation kinetics compared with the CoPi electrocatalyst. This clear example of a slow hole transfer from the photoanode to the catalyst overlayer suggests that a good junction between the photoanode and the catalyst plays an essential role in the performance of the composite material. This is not obvious from a thermodynamic point of view since the process should be downhill, denoting a kinetic problem in this process. From the scheme, the competition between the hole injection to the catalyst and to oxidise water is clear, favouring the water oxidation from $BiVO_4$ directly. The improvement of performance of the CoPi/$BiVO_4$ is the result of the retardation of the recombination due to the formation of a Schottky-type junction between the $BiVO_4$ and the CoPi. The latter enhances the band bending in the $BiVO_4$ slowing down the recombination.

5.2.2 Reduction Reactions: Proton Reduction on [Cu_2O]/RuO_x

In this section, we turn to study the catalytic processes taking place on photocathodes. As for the composite CoPi/$BiVO_4$ photoanode we compare electro- and the photoelectrocatalytic systems. Here we focus on a well-established electrocatalyst for the hydrogen evolution reaction (HER) based on RuO_x.[72] We study the function of a nanostructured, amorphous and highly porous RuO_x when employed as a dark electrocatalyst deposited on FTO and when used as part of a multi-layer photocathode based on Cu_2O (Figure 5.18). This Cu_2O/AZO/TiO_2/RuO_x photocathode (herein referred as [Cu_2O]/RuO_x) has shown promising efficiencies,[45] and can be considered as an example of a buried junction system in which the generation and separation of charges takes place in the Cu_2O/AZO/TiO_2 layers and the catalysis occurs on the RuO_x deposited at the surface. In this section, we aim to compare the catalytic mechanism in the dark and under light conditions. The key challenge to undertake these studies is the determination of the density of active species in each catalytic system. As previously discussed in dark electrochemical systems such mechanistic information is often obtained from the analysis of current–voltage (*J*–*V*) characteristics or Tafel plots.[11] However, for non-ideal semiconducting or metallic electrodes with multiple redox states as well as surface or intraband states, the interpretation of Tafel slopes is not trivial. For photochemically driven systems, the interpretation of *J*–*V* plots is even more challenging due to the voltage dependence on electron-hole recombination.[32] Herein we show that it is possible to use the combined electrochemical and optical rate law methodology presented in this chapter to bypass these problems and obtain insightful mechanistic information of both electrocatalyst and photoelectrocatalysts.

Typically, the first step towards evaluating a (photo)electrocatalytic mechanism is the analysis of the current–voltage (*J*–*V*) characteristics.

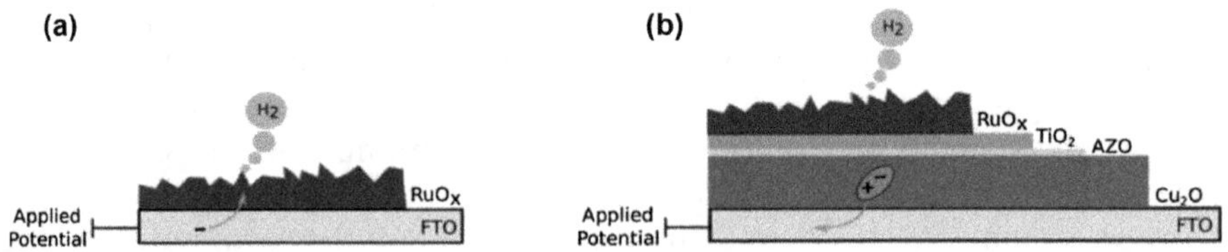

Figure 5.18 Schematic representation of the HER dark and light electrocatalytic systems. (a) RuO_x/FTO cathode and (b) FTO/Cu_2O/AZO/TiO_2/RuO_x (herein [Cu_2O]/RuO_x) photocathode. The diagram is not drawn to scale. Adapted from E. Pastor, F. Le Formal, M. T. Mayer, S. D. Tilley, L. Francàs, C. A. Mesa, M. Grätzel and J. R. Durrant, *Nat. Commun.* 2017, **8**, 14280,[26] http://dx.doi.org/10.1038/ncomms14280. Copyright © 2017, The Authors. Published under the terms of the CC BY 4.0 licence, https://creativecommons.org/licenses/by/4.0/.

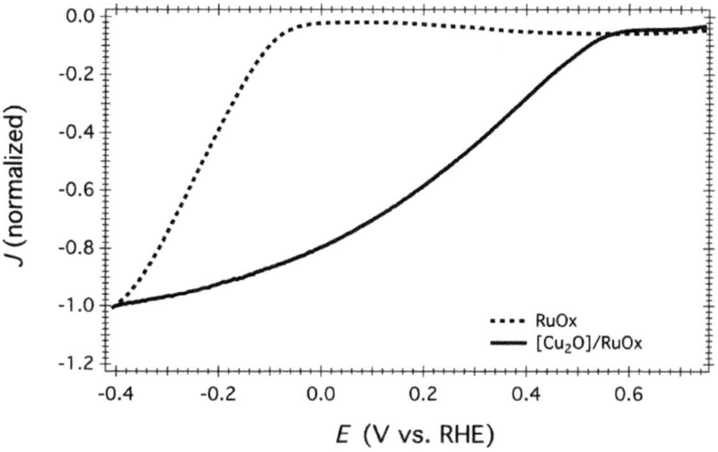

Figure 5.19 Current–Voltage characteristics of the dark electrocatalyst (dashed line) and the illuminated photocathode (solid line). For clarity, the photocurrent response has been normalised at an applied potential where HER occurs in both systems. The samples were measured in a pH = 5 phosphate-sulphate electrolyte.
Adapted from E. Pastor, F. Le Formal, M. T. Mayer, S. D. Tilley, L. Francàs, C. A. Mesa, M. Grätzel and J. R. Durrant, *Nat. Commun.* 2017, **8**, 14280,[26] http://dx.doi.org/10.1038/ncomms14280.

Figure 5.19 shows the *J–V* of the RuO_x in the dark and the $[Cu_2O]/RuO_x$ under illumination. The negative currents are assigned to the reduction of protons to hydrogen. Comparison of the *J–V* curves shows that there is an approximately 0.5 V difference in the onset of current as expected from the photovoltage provided by the Cu_2O/AZO buried junction.[73] In both cases we observe a small negative current at potentials before the onset, E_{onset} (or E^{ph}_{onset}), which has been associated with the reduction of RuO_2 to, most likely, $Ru(OH)_3$ (herein referred to as $RuO_x(1-)$).[74,75] Beyond these similarities it is apparent that the shape of the *J–V* response between the dark and light system are qualitatively different despite tracking the same catalytic reaction (HER on RuO_x). In particular, the current onset is sharper for the dark, electrochemical driven reaction. We associate this difference to the impact of the applied potential on the charge separation and recombination of photogenerated carriers in the light driven system.[32,67] Such dependence introduces a further modulation in the flux of electrons to the surface and hampers the direct comparison of the *J–V* characteristics for dark and light-driven systems.[26]

One approach to study the reaction mechanisms is the use of the combined electrochemical and optical method described in this chapter. We start by comparing the spectroelectrochemical characteristics of the RuO_x employed as a dark electrode and as part of the $[Cu_2O]$ photocathode.

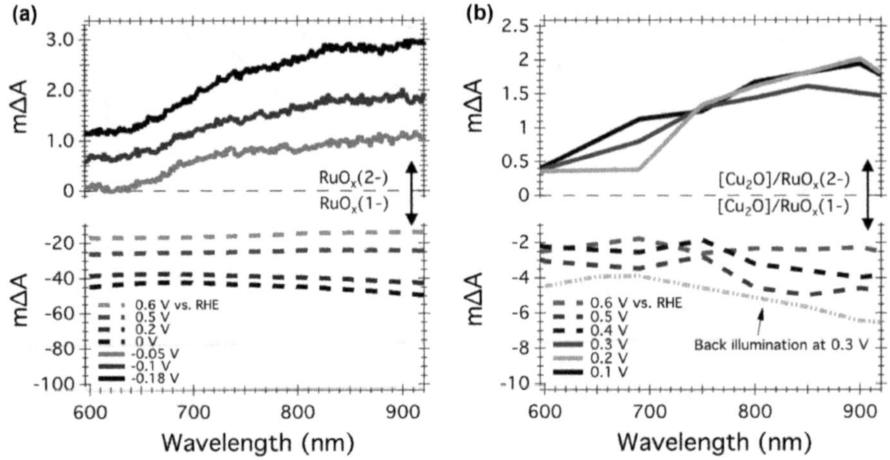

Figure 5.20 Changes in absorbance when RuO_x acts as: (a) a dark electrocatalyst and (b) as part of the [Cu_2O] photocathode under illumination with 365 nm light. In (a) the lower box is plotted relative to the absorbance at OCP (0.78 V_{RHE}) and the upper box the spectra are shown relative to the absorbance at E_{onset} of HER. In (b) the absorbance changes are plotted as a difference between light on and light off. For both device types, the negative ΔA signals observed for $E > E_{onset}$ are assigned to RuO_x reduction to RuO_x(1−), and the positive ΔA signals observed for $E < E_{onset}$ are assigned to the further reduction to RuO_x(2−) species.
Adapted from E. Pastor, F. Le Formal, M. T. Mayer, S. D. Tilley, L. Francàs, C. A. Mesa, M. Grätzel and J. R. Durrant, *Nat. Commun.* 2017, **8**, 14280,[26] http://dx.doi.org/10.1038/ncomms14280.

Figure 5.20(a) shows the spectroelectrochemical data of RuO_x plotted as the absorbance difference against wavelength at different applied potentials. At applied potentials negative of the open circuit potential (OCP), a broad negative change in absorbance is monitored (dashed lines), corresponding to the bleaching of the ground state absorbance at OCP. This loss in absorbance increases with increasing reducing potential until 0 V_{RHE}, the onset of electrocatalytic HER, E_{onset}. Further increase in the applied potential results in the appearance of a positive change in absorbance with respect to the absorbance at E_{onset} (solid lines). We assign the negative signal observed at $E > E_{onset}$ to the reduction of RuO_x to a singly reduced species, RuO_x(1−) and the positive signal to the formation of a further reduced catalytic species, RuO_x(2−). Comparison of the amplitude of the optical signal with the catalytic current assigned to HER reveals a correlation between the generation of the reduced RuO_x(2−) and the electrocatalysed HER current (Figure 5.21). Importantly, the optical data indicates that RuO_x(2−) is only formed once the formation of RuO_x(1−) is completed. The reduction of RuO_x to RuO_x(1−) is most probably associated with the reduction of Ru(IV) to Ru(III) leading to the generation of hydroxo species and a lattice

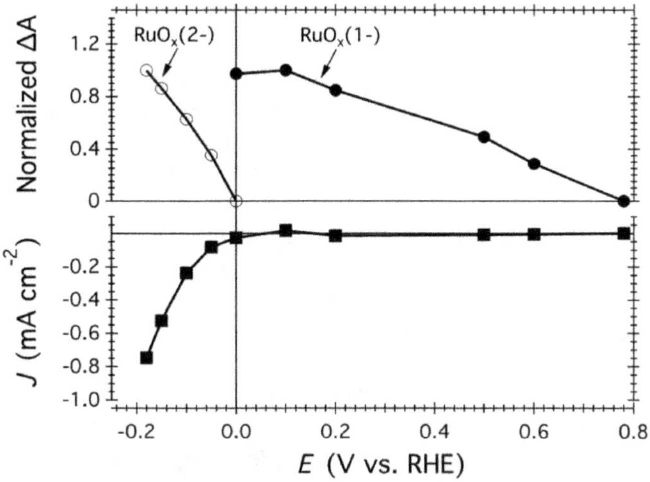

Figure 5.21 Correlation between the changes in absorbance and the HER current. The ΔA was calculated as the average of the optical signal between 850 and 900 nm from the raw data shown in Figure 5.20 and represented in a normalised scale for clarity. The steady state current density was calculated as the average current measured within 300–350 s. Adapted from E. Pastor, F. Le Formal, M. T. Mayer, S. D. Tilley, L. Francàs, C. A. Mesa, M. Grätzel and J. R. Durrant, *Nat. Commun.* 2017, **8**, 14280,[26] http://dx.doi.org/10.1038/ncomms14280.

expansion.[74–76] The second reduction to RuO_x(2−) is most likely associated with the formation of hydride-like species (see discussion on the mechanism at the end of this section).

We next turn to the spectroelectrochemical analysis of the [Cu_2O]/RuO_x photocathode. For this analysis, we employ the PIA alongside TPC technique described in the previous Sections (5.1.2 and 5.1.3). In this system the current, hence the reduction of RuO_x, is dependent on the light and the voltage. We monitor the change in absorbance of [Cu_2O]/RuO_x induced by UV illumination at different applied potentials (Figure 5.20) with control data showing negligible changes in the absence of Cu_2O.[26] The spectra changes show qualitative similarities to those obtained for the dark spectroelectrochemistry of RuO_x. At applied potentials positive of the onset of photocurrent (E^{ph}_{onset}) we observe a broad negative change in absorbance. On the other hand, a broad positive change in absorbance dominates at applied potentials negative of the E^{ph}_{onset}. The remarkable similarity between the spectroelectrochemistry of RuO_x and [Cu_2O]/RuO_x suggests that in both cases we are monitoring the same species: RuO_x(1−) and RuO_x(2−). Further evidence that RuO_x dominates the change in absorbance is found upon back illumination of the photocathode under moderate reducing applied potential. Under these conditions in which negligible photocurrent is drawn only a negative change in absorbance assigned to the non-catalytically active $RuO_x(1-)$ is observed.

Next we consider the correlation between the density of catalytically active RuO_x species ($RuO_x(2-)$) and the HER current (eqn (5.3), rate law analysis). For the dark electrocatalyst this information is provided by the optical/current data as a function of applied potential (Figure 5.21). For the photocathode, the analysis has been done by monitoring the change in absorbance as a function of illumination intensity at a fixed applied potential such as for α-Fe_2O_3 and $BiVO_4$. Figure 5.22 shows the change in absorbance of $[Cu_2O]/RuO_x$ at 900 nm (associated with $RuO_x(2-)$ species) and the associated photocurrent at different illumination conditions. As expected an increase in the light intensity results in an increase of the absorbance change signal and the steady state photocurrent indicating the generation of more active species. When the light is turned off, both the optical signal and photocurrent decay. We assign this decay to primarily proton reduction taking place in the 0.1–1 s timescale. Such relatively slow timescale is similar to the one we have monitored for α-Fe_2O_3 photoanodes.[38]

Figure 5.23 summarises the different experimental assays of the catalytic function of RuO_x in the dark (grey squares) and under illumination (white and black circles) plotted as the logarithm of the current (or photocurrent) against the change in absorbance measured at ~900 nm (*i.e.*, $RuO_x(2-)$). For the RuO_x dark electrocatalyst this analysis shows a linear dependence with a gradient of 2.1. Similarly, the RuO_x as part as the photocathode also

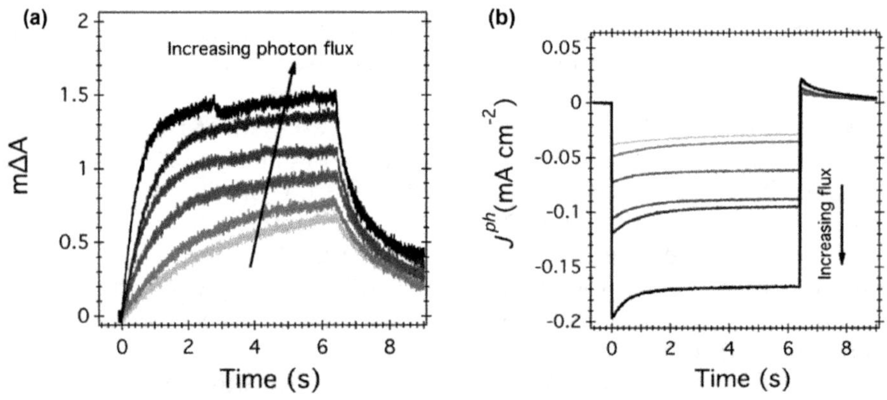

Figure 5.22 (a) Photo-induced absorbance changes of the photocathode upon 365 nm light illumination at different photon fluxes (0.5–1.5 mW cm^{-2}) at a fixed applied potential of 0.1 V_{RHE} and (b) the photocurrent measured simultaneously.
Adapted from E. Pastor, F. Le Formal, M. T. Mayer, S. D. Tilley, L. Francàs, C. A. Mesa, M. Grätzel and J. R. Durrant, *Nat. Commun.* 2017, **8**, 14280,[26] http://dx.doi.org/10.1038/ncomms14280.

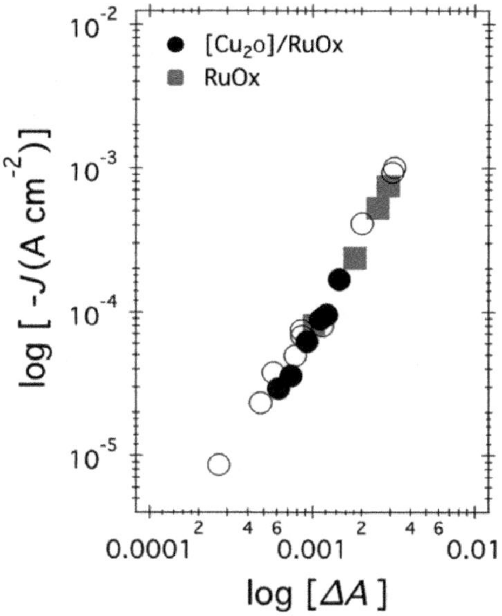

Figure 5.23 Relationship between (photo)current and absorbance changes. For RuO_x (squares), the current-absorbance characteristics are represented as the log [*J*] *vs.* the absorbance of the catalytic RuO_x species at different applied potentials (−0.05 to −0.2 V_{RHE}). For the [Cu_2O]/RuO_x photocathode (circles) the photocurrent-absorbance characteristics are represented as the log [J^{ph}] *vs.* the photo-induced absorbance of the catalytic RuO_x species generated at 0.1 V_{RHE} upon 365 nm illumination at different photon fluxes. For the photocatalyst two sets of data are shown, collected with (empty circles) and without (solid circles) the addition of 0.01 M surfactant (Tritron X-100) used to promote facile bubble release. The independent data sets have been normalised to be plotted in the same graph.
Adapted from E. Pastor, F. Le Formal, M. T. Mayer, S. D. Tilley, L. Francàs, C. A. Mesa, M. Grätzel and J. R. Durrant, *Nat. Commun.* 2017, **8**, 14280,[26] http://dx.doi.org/10.1038/ncomms14280.

shows a linear dependence with a slope of 1.9. This data provides evidence that in both systems the current is proportional to approximately the square of the density of $RuO_x(2-)$. In analogy to the rate law of photoanodes, we equate this behavior with a second order in electron density.

The agreement in our data for RuO_x and [Cu_2O]/RuO_x is striking given the different experimental methods employed: analysis of the dark current as a function of applied potential and of the photocurrent as a function of light intensity. In both cases we observe an approximately second order dependence of the HER current density upon $RuO_x(2-)$. Such observation provides evidence that in the systems studied the catalytic function is independent of (i) the substrate (*i.e.*, FTO or buried Cu_2O/AZO/TiO_2 junction); (ii) the mechanism triggering the reaction (*i.e.*, electrochemical or photoelectrochemical) and (iii) the means by which the density of $RuO_x(2-)$ is modulated (*i.e.*, *via* the applied potential or light intensity). Furthermore, the data suggests that, at least in catalytic terms, the $[Cu_2O]/RuO_x$ photocathode can be considered as a buried junction device in which the function of the catalyst is independent of the underlying layers. The similarities between the dark electrocatalysis and light driven photocatalytic systems are difficult to elucidate *a priori* from an inspection of the *J*–*V* characteristics shown in Figure 5.19 and highlight the potential of the methodology discussed herein to assay and compare the catalytic function of photo- and electrocatalyst.

From the rate law analysis we can gain insight into the catalytic mechanism of HER on RuO_x. The reduction of protons is often described to occur *via* a (i) homolytic mechanism (Volmer–Tafel), in which the H_2 evolution step involves the interaction of two hydride and the reductive elimination of H_2 (eqn (5.6)); or *via* a (ii) heterolytic mechanism (Volmer–Heyrovsky), in which the hydride intermediate undergoes further reduction and protonation to generate H_2 (eqn (5.7)).[72,77–79]

$$2[M^{n+1} - H] \rightarrow 2M^{n} + H_2 \tag{5.6}$$

$$[M^{n+1} - H] + e^{-} + H^{+} \rightarrow M^{n} + H_2 \tag{5.7}$$

Previous studies have proposed a heterolytic mechanism for the HER on RuO_x on the basis of Tafel analyses.[80,81] The same type of analysis based on the data shown in Figure 5.19, yields a Tafel slop of 140 mV dec^{-1} for our RuO_x. This value differs from those measured for metallic RuO_2 which are typically in the 40–60 mV dec^{-1} range.[72] We associate this discrepancy to the amorphous, non-stoichiometric nature of our catalyst. These surface properties significantly complicate the elucidation of surface coverage and the interpretation of the Tafel slope.[82] In contrast our combined electrochemical and optical rate law analysis points towards a second order reaction on the concentration of $RuO_x(2-)$. This behaviour is consistent with a bimolecular mechanism and thus indicative of a homolytic path involving two $RuO_x(2-)$ species. Based on previous literature we propose the following tentative mechanism: initially before the onset, RuO_x undergoes a one electron reduction to generate $Ru(OH)_3$ ($RuO_x(1-)$) as previously characterised. Following this pre-catalytic step, a further reduction takes place generating the active species herein referred to as $RuO_x(2-)$. Finally, two active species undergo the reductive elimination of H_2 regenerating the pre-catalytic state. The identification of the active species requires more detailed and structural sensitive studies however, based on previous literature we hypothesise this to be a hydride-like species. Although further studies are needed to unequivocally elucidate the reaction mechanism, our methodology allows us to distinguish between the two most common reaction paths thus providing important mechanistic information.

5.3 Conclusions

In this chapter, we have presented a new approach to study the rate law of reactions involved in solar fuels production on (photo)electrodes. All these processes are multi-electron / multi-redox catalytic process, and consequently traditional analysis can be hampered by multi-redox states, the surface properties or by electron-hole recombination when the process is light-driven. The versatility of our methodology towards the study of diverse catalytic reactions have been probed by the different presented examples.

To summarise, the first step of the study is the determination of the optical signal associated with the catalytic species, then the correlation between different population of these species and the generated photocurrent (TPC) allow to obtain the rate law plot. Subsequently, fitting these data, following eqn (5.3), we can extract kinetic and mechanistic information. The second one allows us to relate the amount of catalytic species we need to accumulate in order to overcome the rate limiting step of the catalytic process, which together with complementary experiments sheds light on the mechanism of reaction on (photo)electrodes, allowing a rational design of future generations. In addition, further information on surface recombination could be extracted from solving eqn (5.1) at lower applied potentials where surface recombination and catalytic processes are competing.

Acknowledgements

We would like to thank Dr Yimeng Ma and Dr Stephanie Pendlebury for providing us the data on $BiVO_4$ and α-Fe_2O_3. We gratefully acknowledge financial support from the European Research Council (project Intersolar 291482). C.A.M also thanks COLCIENCIAS for funding, L.F. thanks the EU for a Marie Curie fellowship (658270) and E.P. thanks the EPRSC for a DTP scholarship.

References

1. P. Atkins, *Physical Chemistry*, Oxford University Press, Oxford, 6th edn, 2001.
2. J. M. Smieja, E. E. Benson, B. Kumar, K. A. Grice, C. S. Seu, A. J. M. Miller, J. M. Mayer and C. P. Kubiak, *Proc. Natl. Acad. Sci. U. S. A.*, 2012, **109**(39), 15646.
3. P. Kang, C. Cheng, Z. Chen, C. K. Schauer, T. J. Meyer and M. Brookhart, *J. Am. Chem. Soc.*, 2012, **134**(12), 5500.
4. P. Kang, T. J. Meyer and M. Brookhart, *Chem. Sci.*, 2013, **4**(9), 3497.
5. L. Duan, F. Bozoglian, S. Mandal, B. Stewart, T. Privalov, A. Llobet and L. Sun, *Nat. Chem.*, 2012, **4**, 418.
6. M. Dunwell, Q. Lu, J. M. Heyes, J. Rosen, J. G. Chen, Y. Yan, F. Jiao and B. Xu, *J. Am. Chem. Soc.*, 2017, **139**(10), 3774.
7. Q. Lu and F. Jiao, *Nano Energy*, 2016, **29**, 439.

8. X. Fan, Y. Liu, Z. Peng, Z. Zhang, H. Zhou, X. Zhang, B. I. Yakobson, W. A. Goddard, X. Guo, R. H. Hauge and J. M. Tour, *ACS Nano*, 2017, **11**(1), 384.
9. R. Mao, N. Li, H. Lan, X. Zhao, H. Liu, J. Qu and M. Sun, *Environ. Sci. Technol.*, 2016, **50**(7), 3829.
10. A. Holewinski and S. Linic, *J. Electrochem. Soc.*, 2012, **159**(11), H864.
11. T. Shinagawa, A. T. Garcia-Esparza and K. Takanabe, *Sci. Rep.*, 2015, **5**(1), 13801.
12. L. Hammarström, *Acc. Chem. Res.*, 2015, **48**, 840.
13. M. X. Tan, P. E. Laibinis, S. T. Nguyen, J. M. Kesselman, C. E. Stanton; N. S. Lewis, in *Progress in Inorganic Chemistry*, ed. D. Karlin, John Wiley & Sons, Inc, Hoboken, NJ, USA, 2007, pp. 21–144.
14. F. Le Formal, E. Pastor, S. D. Tilley, C. A. Mesa, S. R. Pendlebury, M. Grätzel and J. R. Durrant, *J. Am. Chem. Soc.*, 2015, **137**(20), 6629.
15. Z. Vardeny, P. O'Connor, S. Ray and J. Tauc, *Phys. Rev. Lett.*, 1980, **44**(19), 1267.
16. P. O'Connor and J. Tauc, *Phys. Rev. B*, 1982, **25**(4), 2748.
17. M. Olszakier, Z. Vardeny and R. Weil, *J. Non-Cryst. Solids*, 1985, **77–78**(1), 563.
18. G. Boschloo and A. Hagfeldt, *Chem. Phys. Lett.*, 2003, **370**(3–4), 381.
19. G. Boschloo and A. Hagfeldt, *Inorg. Chim. Acta*, 2008, **361**, 729.
20. D. Bi, L. Yang, G. Boschloo, A. Hagfeldt and E. Johansson, *J. Phys. Chem. Lett.*, 2013, **4**, 1532.
21. M. Pazoki, U. B. Cappel, E. M. J. Johansson, A. Hagfeldt and G. Boschloo, *Energy Environ. Sci.*, 2017, **672**(10), 672.
22. U. B. Cappel, E. A. Gibson, A. Hagfeldt and G. Boschloo, *J. Phys. Chem. C*, 2009, **113**(15), 6275.
23. U. B. Cappel, S. M. Feldt, J. Schö, A. Hagfeldt and G. Boschloo, *J. Am. Chem. Soc.*, 2010, **132**, 9096.
24. W. Yang, M. Pazoki, A. I. K. Eriksson, Y. Hao and G. Boschloo, *Phys. Chem. Chem. Phys.*, 2015, **17**, 16744.
25. Y. Ma, C. A. Mesa, E. Pastor, A. Kafizas, L. Francás, F. Le Formal, S. R. Pendlebury and J. R. Durrant, *ACS Energy Lett.*, 2016, **1**, 618.
26. E. Pastor, F. Le Formal, M. T. Mayer, S. D. Tilley, L. Francàs, C. A. Mesa, M. Grätzel and J. R. Durrant, *Nat. Commun.*, 2017, **8**, 14280.
27. A. Kafizas, Y. Ma, E. Pastor, S. R. Pendlebury, C. Mesa, L. Francá, F. Le Formal, N. Noor, M. Ling, C. Sotelo-Vazquez, C. J. Carmalt, I. P. Parkin and J. R. Durrant, *ACS Catal.*, 2017, 7, 4896.
28. L. Francàs, R. Matheu, E. Pastor, A. Reynal, S. Berardi, X. Sala, A. Llobet and J. R. Durrant, *ACS Catal.*, 2017, 7(8), 5142.
29. C. A. Mesa, A. Kafizas, L. Francàs, S. R. Pendlebury, E. Pastor, Y. Ma, F. Le Formal, M. Mayer, M. Grätzel and J. R. Durrant, *J. Am. Chem. Soc.*, 2017, just accepted.
30. L. M. Peter, K. G. U. Wijayantha and A. A. Tahir, *Faraday Discuss.*, 2012, **155**(0), 309.

31. B. Klahr, S. Gimenez, F. Fabregat-Santiago, T. Hamann and J. Bisquert, *J. Am. Chem. Soc.*, 2012, **134**(9), 4294.
32. F. Le Formal, S. R. Pendlebury, M. Cornuz, S. D. Tilley, M. Grätzel and J. R. Durrant, *J. Am. Chem. Soc.*, 2014, **136**(6), 2564.
33. B. Iandolo, B. R. Wickman, B. Seger, I. Chorkendorff, I. Zoric and A. Hellman, *Phys. Chem. Chem. Phys.*, 2014, **16**, 1271.
34. B. J. Trześniewski and W. A. Smith, *J. Mater. Chem. A*, 2016, **4**(8), 2919.
35. L. Kavan, B. O'regan, A. Kay and M. Grätzel, *J. Electroanal. Chem.*, 1993, **346**(346), 291.
36. O. Zandi and T. W. Hamann, *Nat. Chem.*, 2016, 8.
37. S. R. Pendlebury, M. Barroso, A. J. Cowan, K. Sivula, J. Tang, M. Grätzel, D. Klug and J. R. Durrant, *Chem. Commun.*, 2011, **47**(2), 716.
38. S. R. Pendlebury, A. J. Cowan, M. Barroso, K. Sivula, J. Ye, M. Grätzel, D. R. Klug, J. Tang and J. R. Durrant, *Energy Environ. Sci.*, 2012, **5**(4), 6304.
39. C. Y. Cummings, F. Marken, L. M. Peter, K. G. U. Wijayantha and A. A. Tahir, *J. Am. Chem. Soc.*, 2012, **134**, 1225.
40. B. Klahr and T. Hamann, *J. Phys. Chem. C*, 2014, **118**(19), 10393.
41. Y. Ma, S. R. Pendlebury, A. Reynal, F. Le Formal and J. R. Durrant, *Chem. Sci.*, 2014, **5**(8), 2964.
42. X. Wang, A. Kafizas, X. Li, S. J. A. Moniz, P. J. T. Reardon, J. Tang, I. P. Parkin and J. R. Durrant, *J. Phys. Chem. C*, 2015, **119**(19), 10439.
43. I. Roger, M. A. Shipman and M. D. Symes, *Nat. Rev. Chem.*, 2017, **1**(1), 3.
44. A. Paracchino, V. Laporte, K. Sivula, M. Grätzel and E. Thimsen, *Nat. Mater.*, 2011, **10**(6), 456.
45. S. D. Tilley, M. Schreier, J. Azevedo, M. Stefik and M. Graetzel, *Adv. Funct. Mater.*, 2014, **24**(3), 303.
46. A. Kay, I. Cesar and M. Grätzel, *J. Am. Chem. Soc.*, 2006, **128**(49), 15714.
47. F. Le Formal, M. Grätzel and K. Sivula, *Adv. Funct. Mater.*, 2010, **20**(7), 1099.
48. M. Barroso, C. A. Mesa, S. R. Pendlebury, A. J. Cowan, T. Hisatomi, K. Sivula, M. Grätzel, D. R. Klug and J. R. Durrant, *Proc. Natl. Acad. Sci.*, 2012, **109**(39), 15640.
49. M. Barroso, S. R. Pendlebury, A. J. Cowan and J. R. Durrant, *Chem. Sci.*, 2013, **4**(7), 2724.
50. S. R. Pendlebury, X. Wang, F. Le Formal, M. Cornuz, A. Kafizas, S. D. Tilley, M. Grätzel and J. R. Durrant, *J. Am. Chem. Soc.*, 2014, **136**(28), 9854.
51. L. M. Peter, *Chem. Rev.*, 1990, **90**(5), 753.
52. L. Peter, *J. Solid State Electrochem.*, 2013, **17**(2), 315.
53. M. A. Henderson, *Surf. Sci. Rep.*, 2011, **66**(6), 185.
54. J. Rossmeisl, A. Logadottir and J. K. Nørskov, *Chem. Phys.*, 2005, **319**(1), 178.
55. R. Nakamura and Y. Nakato, *J. Am. Chem. Soc.*, 2004, **126**(4), 1290.
56. M. Zhang, M. De Respinis and H. Frei, *Nat. Chem.*, 2014, **6**, 362.

57. Y. Ma, A. Kafizas, S. R. Pendlebury, F. Le Formal and J. R. Durrant, *Adv. Funct. Mater.*, 2016, **26**(27), 4951–4960.
58. Y. Ma, F. Le Formal, A. Kafizas, S. R. Pendlebury and J. R. Durrant, *J. Mater. Chem. A*, 2015, **3**(41), 20649–20657.
59. A. J. Cowan and J. R. Durrant, *Chem. Soc. Rev.*, 2013, **42**(6), 2281.
60. X. Li, J. Yu, J. Low, Y. Fang, J. Xiao and X. Chen, *J. Mater. Chem. A*, 2015, **3**, 2485.
61. G. Liu, K. Du, S. Haussener and K. Wang, *ChemSusChem*, 2016, **9**(20), 2878.
62. D. K. Zhong, S. Choi and D. R. Gamelin, *J. Am. Chem. Soc.*, 2011, **133**(45), 18370.
63. B. Klahr, S. Gimenez, F. Fabregat-Santiago, J. Bisquert and T. W. Hamann, *J. Am. Chem. Soc.*, 2012, **134**(40), 16693.
64. D. Wang, R. Li, J. Zhu, J. Shi, J. Han, X. Zong and C. Li, *J. Phys. Chem. C*, 2012, **116**(8), 5082.
65. C. Y. Cummings, F. Marken, L. M. Peter, A. A. Tahir and K. G. U. Wijayantha, *Chem. Commun.*, 2012, **48**(14), 2027.
66. M. Barroso, A. J. Cowan, S. R. Pendlebury, M. Grätzel, D. R. Klug and J. R. Durrant, *J. Am. Chem. Soc.*, 2011, **133**(38), 14868.
67. F. Le Formal, K. Sivula and M. Grätzel, *J. Phys. Chem. C*, 2012, **116**(51), 26707.
68. M. W. Kanan and D. G. Nocera, *Science*, 2008, **321**(5892), 1072.
69. Y. Surendranath, M. W. Kanan and D. G. Nocera, *J. Am. Chem. Soc.*, 2010, **132**(46), 16501.
70. K. Klingan, F. Ringleb, I. Zaharieva, J. Heidkamp, P. Chernev, D. Gonzalez-Flores, M. Risch, A. Fischer and H. Dau, *ChemSusChem*, 2014, **7**(5), 1301.
71. G. M. Carroll and D. R. Gamelin, *J. Mater. Chem. A*, 2016, **4**(8), 2986.
72. H. Over, *Chem. Rev.*, 2012, **112**(6), 3356.
73. A. Paracchino, N. Mathews, T. Hisatomi, M. Stefik, S. D. Tilley and M. Grätzel, *Energy Environ. Sci.*, 2012, **5**(9), 8673.
74. J. Juodkazytė, R. Vilkauskaitė, B. Šebeka and K. Juodkazis, *Int. J. Surf. Eng. Coatings*, 2007, **85**(4), 194.
75. Y. Sugawara, A. P. Yadav, A. Nishikata and T. Tsuru, *J. Electrochem. Soc.*, 2008, **155**(9), B897.
76. R. K. B. Karlsson, A. Cornell and L. G. M. Pettersson, *J. Phys. Chem. C*, 2016, **120**(13), 7094.
77. C. Gimbert-Suriñach, J. Albero, T. Stoll, J. Fortage, M.-N. Collomb, A. Deronzier, E. Palomares and A. Llobet, *J. Am. Chem. Soc.*, 2014, **136**(21), 7655.
78. V. S. Thoi, Y. Sun, J. R. Long and C. J. Chang, *Chem. Soc. Rev.*, 2013, **42**(6), 2388.
79. N. Elgrishi, B. D. McCarthy, E. S. Rountree and J. L. Dempsey, *ACS Catal.*, 2016, **6**(6), 3644.
80. I. M. Kodintsev and S. Trasatti, *Electrochim. Acta*, 1994, **39**(11–12), 1803.
81. L. Chen, D. Guay and A. Lasia, *J. Electrochem. Soc.*, 1996, **143**(11), 3576.
82. J. C. F. Boodts and S. Trasatti, *J. Appl. Electrochem.*, 1989, **19**(2), 255.

CHAPTER 6

Emerging Semiconductor Oxides for Direct Solar Water Splitting

K. SIVULA

Ecole Polytechnique Fédérale de Lausanne (EPFL), Institut des sciences et ingénierie chimiques, Station 6, CH-1015 Lausanne, Switzerland
Email: kevin.sivula@epfl.ch

6.1 Introduction

The simple concept of direct photoelectrochemical (PEC) water splitting—where illuminated semiconductors generate photopotential to drive the water oxidation and reduction reactions *via* direct contact with an aqueous electrolyte—is no-doubt attractive towards the economic viability of PEC devices, but it also puts challenging limitations on the semiconductor materials employed.[1] Ideal semiconductors for PEC application are composed of abundant and inexpensive component atoms, are economical to fabricate into their final form, harvest a significant fraction of the solar spectrum, and, critically, must be sufficiently robust to endure continuous and long-term PEC operation in the presence of water and oxygen. Metal oxide semiconductors generally are attractive from a stability point-of-view, especially as photoanodes, given that the relatively high electronegativity of atomic oxygen results in the formation of oxidatively stable compounds based on O^{2-} anions. The first decades of PEC research identified the classic binary transition metal oxide semiconductors like TiO_2 and WO_3 which are easy to prepare and are remarkably stable under appropriate PEC

Energy and Environment Series No. 20
Advances in Photoelectrochemical Water Splitting: Theory, Experiment and Systems Analysis
Edited by S. David Tilley, Stephan Lany and Roel van de Krol

Published by the Royal Society of Chemistry, www.rsc.org

conditions.[2,3] As a drawback, the band gap energy, E_g, of these classic oxides (3.0–3.2 eV for TiO_2 and 2.7 eV for WO_3) are large due to the incorporation of d^0 cations that lead to empty d-bands and an electronic bang gap structure based on the O 2p → Ti 3d or W 5d transition (see Figure 6.1). This results in poor solar light harvesting as only a small fraction of solar illumination contains photons with energy above 3.0 eV and thus prevents these materials from being the main light absorbers for efficient solar-to-fuel energy conversion.

As an alternative, binary metal oxides containing cations with partially filled d-orbitals exhibit smaller band gap energy due to d-to-d transitions and may still have suitable robustness for PEC application. For example, α-Fe_2O_3 (hematite) has a band gap of 2.0–2.2 eV with a valence band position suitable for water oxidation (see Figure 6.1). However, photogenerated charge transport and separation are severe limitations in this oxide due to the formation of small polarons.[4] While significant effort has been focused on studying and modifying hematite as a photoanode its poor semiconducting properties remain an unresolved challenge.[5,6] In contrast, cuprous oxide, Cu_2O, possesses suitable photogenerated charge transport and separation properties for efficient photon harvesting. With a band gap energy of 1.9–2.2 eV and a suitable conduction band position for water photoreduction, it has been extensively investigated as a p-type photocathode material.[7] Unfortunately, the electrochemical potential to reduce Cu_2O to Cu^0 is more positive than the water reduction potential making photocorrosion essentially unavoidable under PEC operation with this semiconductor in direct contact with the aqueous electrolyte.[8]

Overall, while the binary metal oxides have provided a platform to study photoelectrochemical solar fuel production systems for decades, not one material has been found to satisfy all of the requirements for efficient, robust, and inexpensive photoelectrochemical water splitting. However, the experimentally observed limitations of binary oxides uncovered to date have not detracted from the promising potential of semiconducting oxides in general. In recent years research focus has shifted to ternary and more complex multinary oxides considering the wealth of possibilities. Indeed, counting a possible ~50 candidate metals in the periodic table, there are at least 19 000 conceivable ternary metal oxides, and more than 220 000 quaternary oxides.[9] These large numbers give optimism to the possible existence of an ideal oxide or family of oxides that combines the essential requirements for efficient and robust photoelectrochemical solar fuel production. However, developing unambiguous strategies to identify promising metal combinations is challenging. Modern computational design and screening approaches offer hope for the identification of promising oxide systems (see Chapters 2 and 3 of this book) and experimental approaches using combinatorial chemistry can also help to identify potential complex oxide candidates.[9] However, complexities in materials synthesis such as crystal polymorphs, doping, and defects have an important effect on optoelectronic properties, and cannot always be predicted or controlled in combinatorial

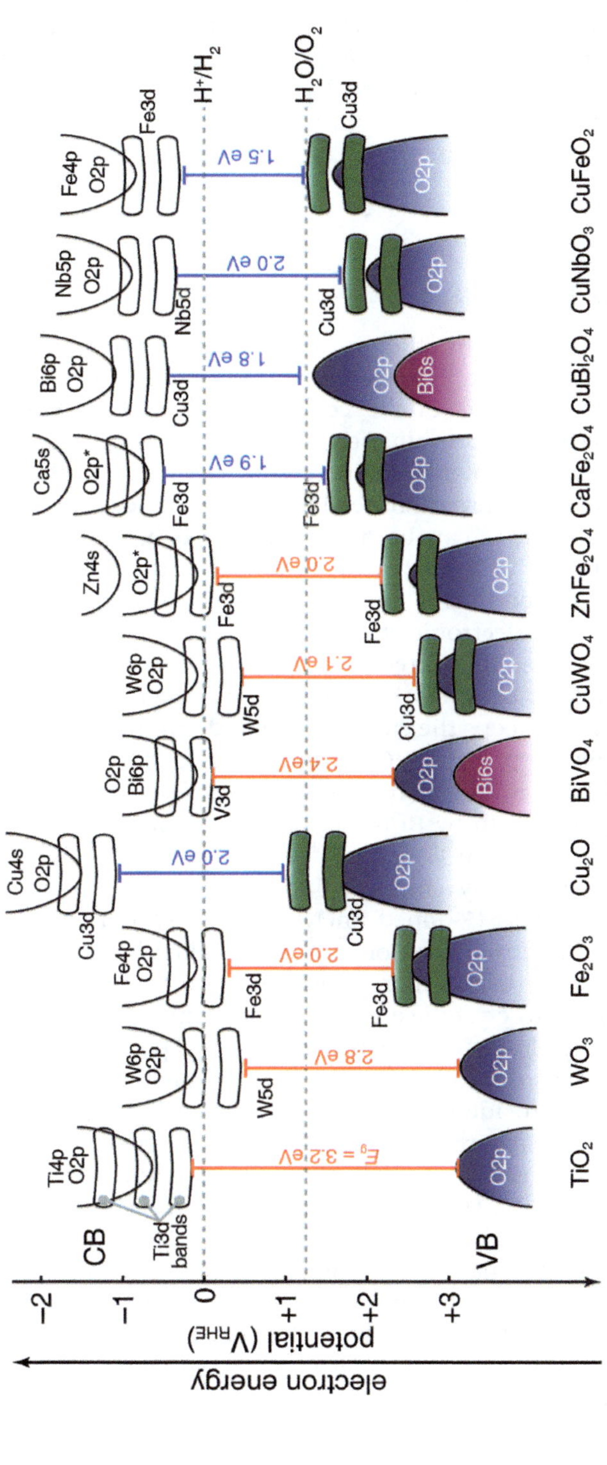

Figure 6.1 Schematic of energy band diagrams for select oxide semiconductors discussed in this work. The conduction band (CB) and valence band (VB) states are indicated with respect to the main contributing orbitals. The band gap energy (E_g) for each material is shown and the energy level positions with respect to the reduction and oxidation potentials for water is given *vs.* the potential from the reversible hydrogen electrode (RHE).

approaches with a high accuracy. Despite these intricacies and the vast number of possible oxides, quite a few potentially promising multinary metal oxides have recently emerged as viable semiconductor materials for both application as photocathodes and photoanodes. In this chapter, a selection of the most prominent multinary semiconductor metal oxide photoelectrode materials are presented together with the rationale behind their development, the current state of the art, and a general outlook for future development.

6.2 Bismuth Vanadate

One method to rationally choose the metal cations for a ternary or multinary oxide semiconductor is to simply assume that the resulting band gap energy and the band edge positions of the complex oxide will be close to the stoichiometrically-weighted combination of the electronic properties of the parent binary oxides.[10,11] Notably, added complexities due to the resulting crystal structure and other deviations lead to significant deviations from this approximation. However, it can be a good place to begin a search. The electronic configuration of a successful ternary metal oxide photoanode, $BiVO_4$, can be rationalised in this way. By mixing the s^2 cation of Bi^{3+} with the d^0 cation of V^{5+} the electronic structure can exhibit a coupling between the *s* and the O 2p states that forces an upward dispersion of the valence band while the coupling between the d bands from the d^0 cation and the p band from the s^2 cation keeps the conduction band minimum low.[12] Indeed the monoclinic polymorph of $BiVO_4$ (clinobisvanite) has a band gap of 2.4–2.5 eV derived from a valence band edge composed of a mixture of un-hybridised O 2p (pushed up by Bi 6s states) and a conduction band edge composed of V3d as shown in Figure 6.1.[13,14]

$BiVO_4$ can be prepared by simple solution processing approaches on standard conducting glass (F-doped SnO_2) substrates. Methods including spray deposition[15,16] electrodeposition,[17,18] chemical bath deposition,[19] and sol–gel coating[20,21] have been reported, where an annealing step in air at 450–500 °C is required to crystallise $BiVO_4$ in its photo active form.

Interestingly, while $BiVO_4$ has been considered a potential photoactive material for the last two decades,[22] only recently has the efforts of many groups come together to identify and overcome the limitations with this material making it arguably the top-performing photoanode for solar water oxidation to date. First, the poor photogenerated charge carrier transport[23] has been addressed *via* the substitutional doping of 6+ cations in place of the V^{5+} to increase the majority carrier density.[24,25] Since doping also has the effect to reduce the space charge width at the semiconductor-liquid junction a clever gradient technique was also reported (see Figure 6.2(a)) that allows an internal electrical field to be present in the 'bulk' of the semiconductor to drive photogenerated carriers to the semiconductor–liquid junction by drift as well as diffusion.[16,26] In this case the equilibration of the fermi-level within the bulk of the semiconductor layer together with

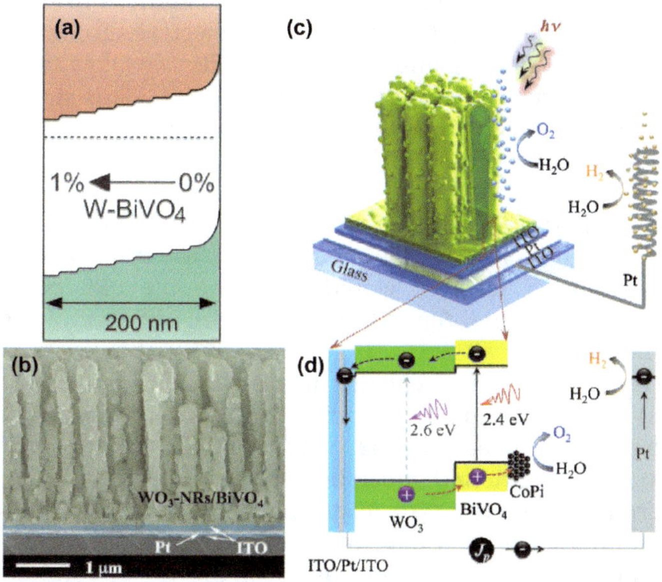

Figure 6.2 Strategies to improve the performance of $BiVO_4$. Panel (a) shows how the gradient doping approach with tungsten from 0% to 1% can affect the band bending in the semiconductor. Reprinted by permission from Macmillan Publishers Ltd: *Nature Communications* (ref. 16), copyright 2013; (b) and (c) show the heterostructure scaffold approach with $WO_3/BiVO_4$; (b) shows a cross-sectional SEM image of the tungsten oxide nanorodes (WO_3-NRs) with the deposited $BiVO_4$ overlayer, which is schematically shown in panel (c); the electronic structure of this heterojunction as a photoanode for water oxidation is shown in (d) where CoPi represents the cobalt catalyst layer and the counter electrode is platinum. Reproduced from Y. Pihosh, I. Turkevych, K. Mawatari, J. Uemura, Y. Kazoe, S. Kosar, K. Makita, T. Sugaya, T. Matsui, D. Fujita, M. Tosa, M. Kondo and T. Kitamori, *Sci. Rep.*, 2015, **5**, 11141,[30] http://dx.doi.org/10.1038/srep11141. Published under the terms of the CC-BY 4.0 licence, http://creativecommons.org/licenses/by/4.0/.

the step-like variation of the doping concentration creates a gradually rising valence band energy and the aforementioned electric field which enhances carrier transport. Another strategy to improve photogenerated charge carrier harvesting is to use a thin layer of a guest semiconductor ($BiVO_4$) on a suitable high surface area host scaffold (*e.g.* nanostructured WO_3)[27–30] This strategy has been particularly successful with bismuth vanadate. For example, recent work from Pihosh *et al.* reported a glancing angle deposition (used to create nanorod arrays of WO_3) which were then coated with $BiVO_4$ by electrodeposition/annealing.[30] Figure 6.2(b) and (c) show, respectively, a cross-sectional scanning electron microscope image of the electrode morphology, and a schematic cartoon of the photoanode structure. The $WO_3/BiVO_4$ junction has the desired energy level configuration for photoanode operation. The alignment of the conduction bands affords the injection of photogenerated electrons from the $BiVO_4$ into the WO_3 while the valence band positions conversely allow photogenerated holes in WO_3 to transfer to the $BiVO_4$ (Figure 6.2(d)). This structure has led to the best performance with $BiVO_4$ as discussed below.

Importantly, doping or nano-hetero-structuring are not sufficient to give good solar water oxidation in $BiVO_4$. Indeed, despite efforts using these approaches to increase the charge transport to the semiconductor–liquid junction, the transfer of these charges across the interface *via* the catalysis of the oxygen evolution reaction would still limit the PEC performance of $BiVO_4$. In particular, initial reports found that the oxygen evolution could be drastically improved by adding a suitable catalyst to drive water oxidation.[24] Many overlayers containing cobalt, iron and nickel based oxyhydroxide catalysts have shown to greatly improve oxygen evolution activity of $BiVO_4$ photoanodes.[31] For example in the $WO_3/BiVO_4$ nanostructured heterojunction photoanode mentioned previously the water oxidation photocurrent increased from 5.4 mA cm^{-2} to 6.7 mA cm^{-2} at 1.23 V *vs.* the Reversible Hydrogen Electrode (RHE) at one sun illumination in potassium phosphate buffer solution ($pH = 7$) after application of the Co–Pi catalyst.[30] This represents the benchmark performance for this ternary oxide.

Overall, the efforts in understanding and overcoming the performance limitations of $BiVO_4$ by many groups have led to impressive progress in the past 10 years as summarised by Abdi and Berglund in a recent review[32] and shown by the plot in Figure 6.3, which gives the measured one sun photocurrent for water oxidation at 1.23 V *vs.* RHE as a function of the year of the report. The theoretical maximum value of the photocurrent (based on the solar AM 1.5 G light absorption limit) has been achieved to about 90% using the $WO_3/BiVO_4$ nanostructured heterojunction approach at this applied potential while photoanodes with only $BiVO_4$ as the light absorber are not far behind at 5.8 mA cm^{-2} out of a maximum possible AM 1.5 G solar photocurrent of 7.5 mA cm^{-2}. Moreover, the relatively good onset potential of $BiVO_4$ (as low as 0.4 V *vs.* RHE) given its high-lying flat band potential of about 0.1 V *vs.* RHE means that it can be employed in tandem cells for overall photoelectrochemical water splitting.[33] In combination with a double junction GaAs/GaInP solar cell, $BiVO_4$ with various doping, structuring, and catalyst overlayers has shown a solar to hydrogen efficiency of over 8%[26] and stability in the 10 h range.[29]

Despite the success of engineering $BiVO_4$ from an initially poor performance of less than 1 mA cm^{-2} in 2006 to the state of the art in 2017 there are critical drawbacks that will prevent it from being incorporated in a viable photoelectrochemical solar fuel device. First and foremost is the relatively large band gap of 2.4 eV that limits the one sun photocurrent density and results in a maximum solar to hydrogen conversion efficiency under 10%. Moreover, $BiVO_4$ exhibits serious stability issues despite being an oxide. Investigations of $BiVO_4$ have mainly been performed under neutral conditions ($pH \sim 7$) because it is chemically unstable and gradually dissolves in

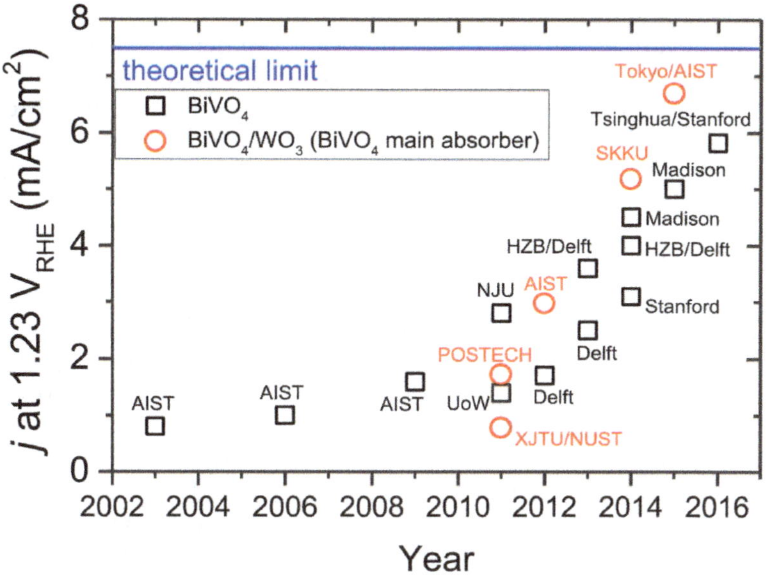

Figure 6.3 Reported photocurrent of $BiVO_4$ (squares) and $BiVO_4/WO_3$ guest–host (circles) photoanodes at 1.23 V *vs.* RHE. Except for the first two points from AIST, the photocurrents were measured under AM 1.5 illumination. The theoretical maximum photocurrent for $BiVO_4$ based on its bandgap of 2.4 eV is shown as the horizontal blue line.
Reprinted from F. F. Abdi and S. P. Berglund, *J. Phys. D: Appl. Phys.*, 2017, **50**, 193002.[32]

strong basic and acidic solutions (where photoelectrochemical cells function best, generally). Indeed, the stability has also shown to strongly depend on the electrolyte and on surface modification by appropriate surface co-catalysts.[20,33–36] The stability can also be improved with semiconducting overlayers.[37] However, it is not clear if $BiVO_4$ would ever be able to endure years of continuous photoelectrochemical operation.[38]

6.3 Multinary Ferrites

The success story of engineering $BiVO_4$ from a poor performance to the nearly perfect quantum yields possible today gives promise to the development of other multinary metal oxides with suitable band gaps, energy positions, and robustness. The ternary ferrites represent a promising class of materials. In a ferrite, an iron(III) containing oxide, the partially filled d-orbital-based states from the Fe^{3+} cation can lead to a higher lying valence band (compared to the O 2p states) while the included alternative cation can hybridise with the conduction band to afford a tuning of the band-gap energy and avoid the localised Fe 3d–3d transitions as in binary Fe_2O_3.

Many ferrites are spinels with the formula AFe_2O_4, where A represents various metal cations in the 2+ oxidation state. Spinel ferrites usually adopt a crystal motif consisting of cubic close-packed (fcc) oxides (based on O^{2-}) with the A cations occupying one eighth of the tetrahedral sites and Fe^{3+} occupying half of the octahedral holes. Alternatively, if one eighth of the tetrahedral sites are filled by iron, then one quarter of the octahedral sites are occupied by the A cation and the other quarter by iron giving the inverse spinel structure. Mixed structure spinel ferrites with varying degrees of inversion are also possible.

Spinel ferrites (with A cations of Ba, Ca, Cu, Co, Mg, Mn, Ni, and Zn) have been examined for photoelectrochemical application[39–43] as photoanodes (*e.g.* $ZnFe_2O_4$)[44] or photocathodes (*e.g.* $CaFe_2O_4$)[45] depending on the position of their conduction and valence bands (see Figure 6.1 for these two examples). These materials can generally be prepared in the thin film electrode format using many solution-based processes to form amorphous or less-crystalline precursors that are then annealed to finally crystallise the spinel crystal structure. For example, a common method involves the chemical bath deposition of a less-crystalline iron(III) oxyhydroxide, followed by the solution deposition of the soluble version of the A cation and an annealing step at high enough temperature (*e.g.* 600 °C) or using microwave conditions to avoid hematite formation.[44] The performance of these materials as the primary light absorber in a photoelectrode has been promising, but overall low compared to the performance of binary Fe_2O_3. For example, $CuFe_2O_4$ has been reported solar photocurrents in the 10 μA cm^{-2} range,[46] and about 200 μA cm^{-2} at 1.23 V *vs.* RHE was observed for $ZnFe_2O_4$ for water oxidation.[47]

In contrast to the few reports of photoelectrodes with only spinel ferrites as the light absorber, heterojunctions using spinel ferrites thin-films deposited onto conventional metal oxide layers have been more extensively reported.[48] A classic example is the use of $ZnFe_2O_4$ on Fe_2O_3 to enhance charge extraction.[49,50] Figure 6.4 shows the band alignment of this system and the improvement found with $Fe_2O_3/ZnFe_2O_4$ composite electrodes prepared *via* a simple surface treatment performed on Fe_2O_3 nanorod films. Similar to the $WO_3/BiVO_4$ system, photogenerated holes in the Fe_2O_3 can be transferred to the $ZnFe_2O_4$. Effectively increasing the photogenerated charge extraction from Fe_2O_3. While a clear performance increase is observed in the current density-voltage curves (Figure 6.4(c)) and the quantum efficiency (Figure 6.4(d)) the active wavelength range does not extend towards the band edge of the $ZnFe_2O_4$ suggesting that the charge extraction performance of the spinel ferrite is more important in this case than the light absorbing properties. Kinetic limitations are still likely an issue in this particular system as the onset potential does not significantly improve as expected when including a charge extraction overlayer.

While the overall performance of the spinel ferrites has been slightly lacklustre to date, the relatively simple preparation of these materials in general and the absence of extensive and in-depth studies on these materials

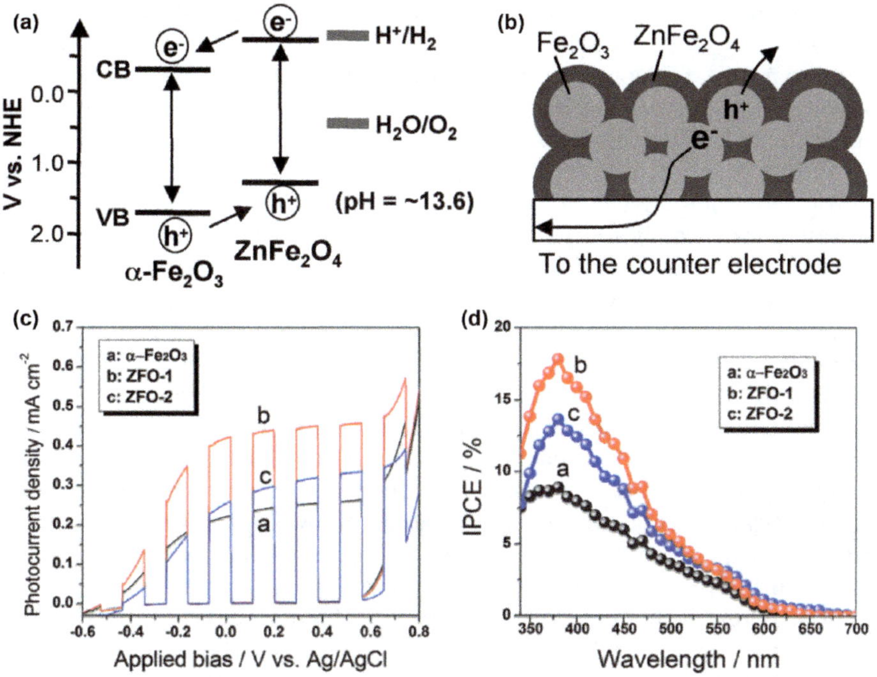

Figure 6.4 Using $ZnFe_2O_4$ as a charge extraction layer. (a) shows the energy level schematic of the $Fe_2O_3/ZnFe_2O_4$ heterojunction *vs.* the normal hydrogen electrode (NHE) and (b) shows a schematic of a cross-section of a typical electrode structure. Reprinted with permission from K. J. McDonald and K.-S. Choi, *Chem. Mater.*, 2011, **23**, 4863–4869.[50] Copyright 2011 American Chemical Society. (c) Photocurrent–potential curves and (d) IPCE of pristine α-Fe_2O_3 and $ZnFe_2O_4$ modified α-Fe_2O_3 (ZFO-1 and ZFO-2) nanorod films as photoanodes at 0.5 V *vs.* Ag/AgCl. ZFO-1 and ZFO-2, respectively, were obtained by spinning coating $ZnFe_2O_4$ precursor solutions with different concentrations (*e.g.*, 9 and 45 mg mL^{-1}). Reproduced from ref. 49 with permission from The Royal Society of Chemistry.

leaves a clear opportunity to better determine if their light absorption/charge transport properties can be tuned to afford improved light harvesting efficiency.

6.4 Copper-based Oxides

Taking advantage of the high-lying copper 3d energy states in the Cu^+ or Cu^{2+} cations while including additional cations to circumvent the instability of binary Cu_2O or CuO has been a recurring theme in the development of multinary oxides for photoelectrochemical application. Indeed, copper bismuthates, niobates, tantalates, tungstates and vanadates have been recently examined as photoelectrodes in photoelectrochemical application.[7] By controlling the other cation and the oxidation state of copper between Cu^+

and Cu^{2+} materials for either active photoanodes or photocathodes can be realised.

For example, for copper(II)-based oxides, $CuWO_4$, is a well-studied photoanode material containing Cu^{2+} and W^{6+} in a triclinic crystal structure.[51–58] The relatively low lying and empty W 5d energy levels (see Figure 6.1) give an indirect band gap energy of approximately 2.1 eV and appropriate energy levels for water oxidation. While the charge transfer at the semiconductor liquid junction can be improved in this material using the standard water oxidation catalyst overlayers,[59] charge transport in the thin film and likely mid-gap states limit the performance of this material.[53] To date, the highest performing $CuWO_4$ photoanode has given a photocurrent density of up to 0.3 mA cm^{-2} (1.23 V *vs.* RHE). Higher values >1 mA cm^{-2} (1.5 V *vs.* RHE) can be realised by incorporating Ag nanowires.[58] However, these values are small compared to theoretical current densities that $CuWO_4$ can generate based on its band gap energy thus limitation of charge separation and transport need to be addressed. Similar rational of the band-gap energy and position used to investigate $CuWO_4$ could in principle be used to motivate the study of copper(II) titianate, $CuTiO_3$, which may in principle have improved transport characteristics. However, this material is relatively unexplored due to its difficultly to be synthesised as a pure phase. This difficulty is noted as one of the drawbacks of engineering ternary or multinary oxides: while the rationale for including certain cations to tune the band-gap energy or other optoelectronic properties may be straightforward, it is not always possible to prepare the intended material.

Despite difficulty to realise any desired phase, other copper(II) oxides have appeared for PEC application. Most notable is $CuBi_2O_4$, which was interestingly identified as a possible photocathode material in 2007 through a combinatorial chemistry study.[60] It has a unique crystal structure of stacks of $[CuO_4]$, between which Bi atoms are connected to six O atoms. This structure plays a significant role in establishing its electronic properties where the O 2p and Cu 3d orbitals determine the valence band energy and conduction band energy positions ($E_g = 1.7$ eV) while Cu atom vacancies introduce states near the valence band edge giving rise to the p-type behaviour needed for photocathode operation.[61] Since the first report then there have been a few recent studies that suggest promise with this material.[62,63] Importantly poor charge carrier transport, primarily for holes, was measured, which limited the obtainable photocurrent density to about 0.5 mA cm^{-2} at 0.4 V *vs.* RHE. However, an optimised $CuBi_2O_4$/Ag-doped $CuBi_2O_4$ photocathode showed achieved a photovoltage higher than 1 V for water reduction, and improved stability.[62] In addition and in contrast to the high temperature conditions needed to prepare ferrites, photoactive $CuBi_2O_4$ can be prepared at low temperature (80 °C) with hydrothermal methods,[64] or by electrodeposition.[65] Further efforts with nanostructuring and doping this material represent an opportunity to improve performance.

Copper(I) multinary oxides have been relatively more studied compared to copper(II) versions and given the higher valence and conduction bands present for copper(I) compounds, they have found application mostly as p-type photocathodes. A recent review by Maggard and co-workers describes results with many phases of copper niobates, tantalates, and vanadates.[7] $CuNbO_3$ in particular has a bandgap of about 2.0 eV (see Figure 6.1) and its relative $CuNb_2O_6$ ($E_g = 1.9$ eV) have been shown to be active for the reduction of water and CO_2,[66,67] however solar photocurrents are generally smaller than 1 mA cm^{-2}. Copper(I) tantalates have shown slightly higher photocurrent density up to *ca.* 2 mA cm^{-2}, but have higher bandgap energies (*ca.* 2.5 eV) which limit their performance as photocathodes for solar application. Overall, the initially demonstrated performance of the copper(I) niobates, tantalates, and vanadates is promising, but most importantly these materials need to be demonstrated as highly robust during extended photocathode operation given the possibility for copper reduction in the film.

In general, for the class of the multinary oxides containing copper(II) or copper(I), there are many relatively new materials for photoelectrochemical application. More study is required to better characterise their charge generation, excited state life time, and charge transport processes as well as the stability in order to motivate their development in a photocathode/photoanode tandem cell.

6.5 Delafossites

Another class of ternary oxides that can be grouped by their crystal structure is the delafossites. The delafossite crystal structure, has a general formula of ABO_2. A is a cation in the +1-oxidation state and typically is Ag, Cu, Pd, or Pt, while B is a cation in the +3-oxidation state and can be an element from the p-block (Al, In, Ga), a transition metal (Co, Fe, Mn, Rh, Y), or a rare earth element (La, Nd, Eu). The structure itself is a layered superposition of A cations in triangular pattern, and edge-sharing BO_6 octahedra (see Figure 6.5). Silver (Ag^+) and copper (Cu^+) based A-site cation delafossites, have been investigated for solar fuel production recently by a few groups. For the silver based systems Ouyang *et al.* first looked at the photocatalytic activity of $AgAlO_2$ and $AgCrO_2$.[68] Interestingly, the authors found that the Cr^{3+} ($E_g = 1.68$ eV) version crystallised in the delafossite phase while the Al^{3+} version ($E_g = 2.95$ eV) adopted an orthorhombic (β) phase. Later the authors prepared a mixed B site cation version with Al^{3+} and Ga^{3+} (*i.e.* β-$AgAl_{1-x}Ga_xO_2$) and theoretically studied electronic structure.[69] The valence bands of the $AgAl_{1-x}Ga_xO_2$ were found to be composed of Ag 4d and O 2p states, while the conduction bands were hybridised by Ag 5s5p, Al 3s3p, and Ga 4s4p states. The level of the conduction band edge can be modulated by varying the molar ratio of Ga/Al while the level of the valence band edge does not shift toward the negative position. Therefore, this material keeps both high visible-light responsive ability (E_g as low as 2.3 eV) and strong oxidisation activity, which has the potential as an application as a viable photoanode. However, the synthesis of silver based ABO_2 oxides is quite difficult due to the inertness of silver. A cation exchange method is typically

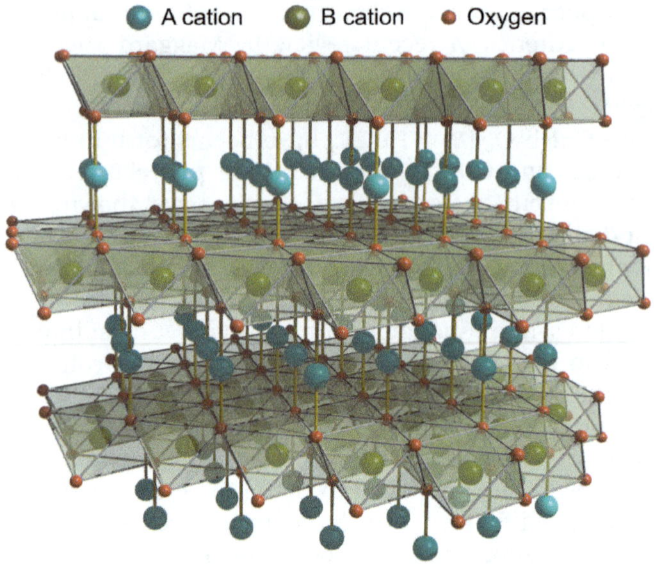

Figure 6.5 The delafossite crystal structure shows the A^+ cations in layers forming a triangular pattern separated by and edge-sharing BO_6 octahedra. Reproduced with permission from Macmillan Publishers Ltd: *Nat. Rev. Mater.* (ref. 92), copyright 2016.

used by first preparing the Na^+ based compound (*e.g.* $NaAlO_2$) followed by treatment with $AgNO_3$, and KNO_3 under an inert atmosphere. KNO_3 is used as a fluxing medium to prevent the reduction of Ag^+ to Ag^0. Despite synthetic difficulties, microparticles with smooth surface were recently prepared and investigated in the photocatalytic degradation of methyl orange under visible light irradiation (wavelength >420 nm). The optimum composition was found to be $AgAl_{0.2}Ga_{0.8}O_2$ due to its small bandgap of 2.3 eV which allowed the most absorption of visible light wavelengths.[70] Interestingly, it is not clear how the crystal structure, whether the delafossite (α) phase or the orthorhombic (β) phase plays a role in the photoactivity, independent of the cations, in this system. This could be a point of further study.

The copper based delafossites can be directly prepared without ion exchange and have been extensively studied as possible p-type transparent conducting oxides. The p-type conductivity of the $CuBO_2$ delafossites is believed to arise from Cu vacancies and interstitial O in the Cu layers of its structure. In particular $CuGaO_2$,[71–73] and $CuAlO_2$,[74–76] have low visible light absorption and reasonable conductivities hence their consideration as viable alternatives to other p-type transparent conductive oxides like NiO.[77,78] Recently, $CuRhO_2$[79] and $CuFeO_2$[80–82] have come forward as a new class of promising photocathode materials, with good stability in aqueous environments. $CuFeO_2$, in particular, is a promising candidate photocathode as it is composed of earth-abundant atoms, possesses favourable properties including an optical band gap, E_g, of 1.5 eV and a large absorption coefficient (up to $\alpha \sim 10^7$ m^{-1}). Furthermore, its flat-band potential, which is positioned ~1 V *vs.* RHE (~0.8 eV below the conduction band edge), suggests the capability to develop a high photovoltage for the reduction of water to H_2, or CO_2 to formic acid or other carbon-based fuels. Notably, the E_g of $CuFeO_2$ is lower than the parent binary oxide (Cu_2O) despite the valence and conduction bands consisting mostly of Cu 3d character in both cases (the Fe 3d bands push the CB down). This helps to stabilise the material against the reduction of Cu^+ to Cu^0. Indeed, in stark contrast to Cu_2O, $CuFeO_2$ is also remarkably stable under PEC operation in basic electrolyte—operating without photocurrent decrease for days, making it a very promising photocathode material.[81]

A few different techniques have been used to produce $CuFeO_2$ photocathodes. Seminal reports using electrochemical deposition[80,82] or solid-state synthesis[83,84] yielded electrodes photoactive for water or CO_2 reduction, but offered only limited control over the morphology of the resulting electrode. More recent studies have employed a sol–gel approach[81,85,86] which offer improved tunability and scalability. Enhancements in performance have been made using TiO_2 as an electron-extracting overlayer[81] and $CuAlO_2$ as a hole-extracting underlayer,[85] promoting, respectively, better charge injection to the electrolyte, and better charge separation inside the absorbing layer. The electronic structure of the heterojunction $CuAlO_2/CuFeO_2$ based photocathode is shown in Figure 6.6(a). Photogenerated holes can transfer from the $CuFeO_2$ to the $CuAlO_2$ leaving the photogenerated electrons in the $CuFeO_2$ to drive the reduction reaction. For the case of sacrificial O_2 reduction in alkaline electrolyte, the solar photocurrent increased from about 1 mA cm^{-2} (in the $CuFeO_2$ only electrode) to 2.5 mA cm^{-2} in the heterojunction $CuAlO_2/CuFeO_2$ based electrode at 0 V *vs.* RHE (Figure 6.6(b)). Notably the photocurrent was not attributed to the $CuAlO_2$ given its poor light absorption. The quantum efficiency of the photoelectrodes show that the $CuAlO_2/CuFeO_2$ heterostructure increases the extraction of photogenerated carriers in $CuFeO_2$ to an IPCE of up to 25% (Figure 6.6(c)). The onset of productive light absorption in $CuAlO_2$ started at wavelengths less than 400 nm.

In addition to the heterostructure approach, a microwave treatment has also been reported to notably improve the performances of the material,[87] however, this technique also resulted in the presence of Cu(II) and the presence of secondary (corrosion-based) electrochemical processes. While the results to date with $CuFeO_2$ are promising, similar to the other ternary ferrites, the performance is far below the performance anticipated from the absorption spectrum of the material (which suggests a maximum possible solar photocurrent of *ca.* 15 mA cm^{-2}). Indeed, like many of the other oxide discussed above, more efforts are needed to understand the limitations of charge transport and especially the possible presence of surface states that limit the transfer of photogenerated electrons to the electrolyte for the fuel producing reactions.

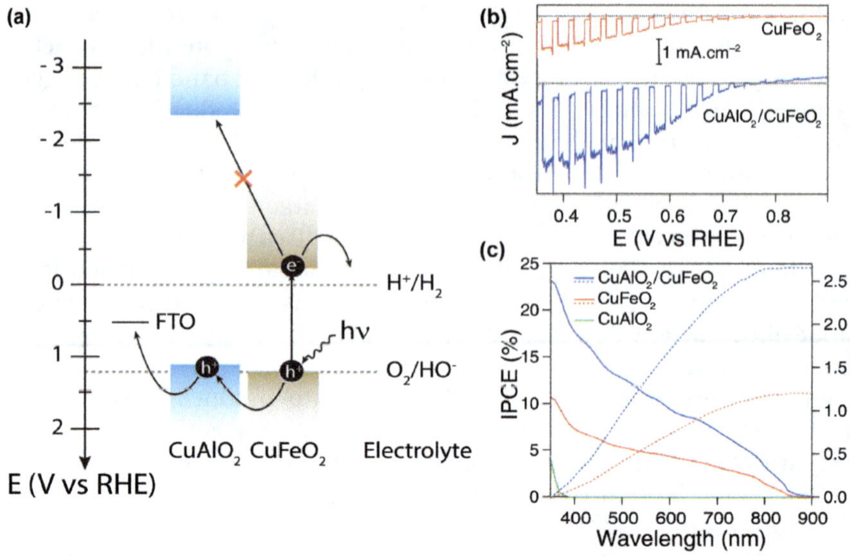

Figure 6.6 Performance of $CuFeO_2$ based photocathodes using a $CuAlO_2$ scaffold. (a) Simplified energy diagram of the electrode. Colourised areas represent energy bands, arrows indicate spontaneous charge transfer under illumination, and dashed lines mark the position of the water redox couples. (b) Comparison of *J–V* curves obtained for optimised $CuAlO_2$/$CuFeO_2$ and $CuFeO_2$ electrodes, under O_2 bubbling. Electrolyte: 1 M NaOH, scan rate: 10 mV s^{-1}, illumination: one sun. (b) IPCE spectra of $CuAlO_2$, $CuFeO_2$, and $CuFeO_2$/$CuAlO_2$ electrodes. Corresponding integrated photocurrents are represented by the dotted lines.
Reproduced from ref. 85 with permission from The Royal Society of Chemistry.

6.6 Strategies for Improving Multinary Oxides

One important theme regarding the development of multinary oxide semiconductor photoelectrodes emerges when taking into account the state of the art: the engineering of interfaces by nanostructuring[88] or interface engineering.[89] Both of these approaches are powerful tools to improve photoelectrode performance and also to gain important insights to direct the further development. Diverse interface-specific effects arise from interface treatments and have been recently categorised by Guijarro *et al.*[89] as (i) protecting against the photocorrosion, (ii) passivating surface trap states, (iii) tuning the band bending or band edge positions, and (iv) the selectively extracting carriers and accelerating catalytic activity. Importantly, extensive work has been performed on adding catalysts to the semiconductor–liquid junction to minimise surface recombination, especially if the system is limited by surface reaction kinetics. A large number of earth-abundant catalysts are available with water oxidation (reduction) overpotentials as low as 0.3–0.4 V (0.2 V) at 10 mA cm^{-2} in alkaline (acidic) solutions.[90] Combined interface strategies are possible to afford one or more of the beneficial effect of interface engineering. Importantly, the ability to tune the semiconductor at the liquid junction represents an important distinction between PEC devices and PV + electrolysis.

Approaches using nanostructuring and heterojunction formation are also now widely employed to improve the performance of oxide photoelectrodes. As previously mentioned nanostructured $WO_3/BiVO_4$ or $CuAlO_2/CuFeO_2$ heterojunctions both show significant enhancements over single semiconductor electrodes. However, it should be noted that while heterojunction formation and nanostructuring semiconductors in general can offer some improvement, intrinsic material properties such as carrier mobility, lifetime, stability, and absorption coefficient will ultimately define the performance limits of any particular semiconductor. These material properties need to be well studied in simple systems before too much stock is placed in any new material.

6.7 Outlook for Multinary Oxides

The number of oxide semiconductors for photoelectrochemical application has expanded dramatically from just a few key materials a decade ago—primarily the binary oxides TiO_2, WO_3, and Fe_2O_3—to a large number of potential multinary oxides that are relatively straightforward to prepare and simple to investigate for applications as photoanodes or photocathodes. New candidates for investigation have been chosen often by rationally engineering the band structure electronics to overcome the limitations of the classic binary oxides—*i.e.* a high band gap energy, poor transport properties or poor stability. While this approach has seen reasonable success, the field still lacks an ideal oxide that combines perfect characteristics for application in solar fuel production. Indeed, considering the effects of crystal structure, Jahn–Teller distortions, and intrinsic material defects, it remains a major challenge to accurately predict the optoelectronic properties of potential oxide semiconductors from first principle calculations.[91] Future advances in computational quantum mechanical modelling to predict the electronic structure of candidate oxide materials will eventually bring a revolution to the field of materials design. Indeed, the large number of untested atomic combinations gives hope that an ideal oxide can be still discovered.

In the meantime, experimentalists are still making contributions *via* combinatorial approaches and by studying and improving materials like $BiVO_4$ and $CuFeO_2$. Considering the development of new materials experimentally, an important task is to define selection criteria that can be applied to eliminate candidate oxide semiconductors from further time-consuming study. These criteria have been discussed in a recent article.[92] A first criterion is the band gap energy. To reach a possible solar energy conversion efficiency of 10%, the intrinsic band gap of the top absorber in a tandem cell should be smaller than 2.3 eV, assuming that all incident photons with energy $\geq E_g$ can be harvested. The stability of the semiconductor in direct contact with the electrolyte is a second major criterion for operation of a photoelectrochemical cell. Indeed, as the materials need to last on the order of 10 years under operation in direct contact with electrolyte, (assuming perfectly conformal and stable protection layers are not available in a scalable and cost-effective manner) there is little hope for developing materials that cannot operate for more than a few hours in the lab. A third criterion is the carrier diffusion length L_D. If L_D is only a few nm, chances are that the material is an intrinsically poor semiconductor or that it is susceptible to the formation of recombination centres. However, the origin of the poor diffusion length needs to be identified as if it stems from extrinsic factors (*e.g.* defects), amelioration may be possible. The fourth criterion is related to the band edge positions compared to the redox potentials of the desired electrochemical reaction, as these determine how much photovoltage the photoelectrode can develop. However, it should be noted that the band edges can be optimised with surface functionalisation as previously mentioned.

The future of the field should focus on the identification of promising multinary candidates by theoretical approaches followed by experimental efforts to realise these materials and study their properties using the above defined criteria to quickly eliminate hopeless candidates and focus on the most promising. Indeed, this approach will become more successful with the advancement of theoretical tools and also the large base of experimental material optimisation techniques developed over the decades of research developing the binary semiconducting oxides and more recently the promising ternary and multinary oxides for photoelectrochemical application.

References

1. M. G. Walter, E. L. Warren, J. R. McKone, S. W. Boettcher, Q. Mi, E. A. Santori and N. S. Lewis, *Chem. Rev.*, 2010, **110**, 6446–6473.
2. C. Janáky, K. Rajeshwar, N. R. de Tacconi, W. Chanmanee and M. N. Huda, *Catal. Today*, 2013, **199**, 53–64.
3. M. Kapilashrami, Y. Zhang, Y.-S. Liu, A. Hagfeldt and J. Guo, *Chem. Rev.*, 2014, **114**, 9662–9707.
4. V. Henrich and P. Cox, *The Surface Science of Metal Oxides*, Cambridge University Press, Cambridge, UK, 1996.
5. P. S. Bassi, Gurudayal, L. H. Wong and J. Barber, *Phys. Chem. Chem. Phys.*, 2014, **16**, 11834–11842.
6. K. Sivula, F. Le Formal and M. Grätzel, *ChemSusChem*, 2011, **4**, 432–449.
7. I. Sullivan, B. Zoellner and P. A. Maggard, *Chem. Mater.*, 2016, **28**, 5999–6016.
8. A. Paracchino, V. Laporte, K. Sivula, M. Grätzel and E. Thimsen, *Nat. Mater.*, 2011, **10**, 456–461.
9. M. Woodhouse and B. A. Parkinson, *Chem. Soc. Rev.*, 2009, **38**, 197–210.
10. J. A. McLeod, R. J. Green, E. Z. Kurmaev, N. Kumada, A. A. Belik and A. Moewes, *Phys. Rev. B: Condens. Matter*, 2012, **85**, 195201.
11. E. Thimsen, S. Biswas, C. S. Lo and P. Biswas, *J. Phys. Chem. C*, 2009, **113**, 2014–2021.

12. A. Walsh, Y. Yan, M. N. Huda, M. M. Al-Jassim and S.-H. Wei, *Chem. Mater.*, 2009, **21**, 547–551.
13. J. K. Cooper, S. Gul, F. M. Toma, L. Chen, P.-A. Glans, J. Guo, J. W. Ager, J. Yano and I. D. Sharp, *Chem. Mater.*, 2014, **26**, 5365–5373.
14. K. Ding, B. Chen, Z. Fang, Y. Zhang and Z. Chen, *Phys. Chem. Chem. Phys.*, 2014, **16**, 13465–13476.
15. Y. Liang, T. Tsubota, L. P. A. Mooij and R. van de Krol, *J. Phys. Chem. C*, 2011, **115**, 17594–17598.
16. F. F. Abdi, L. Han, A. H. Smets, M. Zeman, B. Dam and R. van de Krol, *Nat. Commun.*, 2013, **4**, 2195.
17. J. A. Seabold and K.-S. Choi, *J. Am. Chem. Soc.*, 2012, **134**, 2186–2192.
18. T. W. Kim and K.-S. Choi, *Science*, 2014, **343**, 990–994.
19. J. Su, L. Guo, S. Yoriya and C. A. Grimes, *Cryst. Growth Des.*, 2010, **10**, 856–861.
20. K. Sayama, A. Nomura, T. Arai, T. Sugita, R. Abe, M. Yanagida, T. Oi, Y. Iwasaki, Y. Abe and H. Sugihara, *J. Phys. Chem. B*, 2006, **110**, 11352–11360.
21. C. Laberty-Robert, S. Hilliard, D. Friedrich, H. Strubb, S. Kressman and V. Artero, *ChemPhotoChem*, 2017, **1**, 273–280.
22. A. Kudo, K. Ueda, H. Kato and I. Mikami, *Catal. Lett.*, 1998, **53**, 229–230.
23. M. Ziwritsch, S. Müller, H. Hempel, T. Unold, F. F. Abdi, R. van de Krol, D. Friedrich and R. Eichberger, *ACS Energy Lett.*, 2016, **1**, 888–894.
24. F. F. Abdi, N. Firet and R. van de Krol, *ChemCatChem*, 2013, **5**, 490–496.
25. S. K. Pilli, T. E. Furtak, L. D. Brown, T. G. Deutsch, J. A. Turner and A. M. Herring, *Energy Environ. Sci.*, 2011, **4**, 5028–5034.
26. L. Han, F. F. Abdi, R. van de Krol, R. Liu, Z. Huang, H.-J. Lewerenz, B. Dam, M. Zeman and A. H. M. Smets, *ChemSusChem*, 2014, **7**, 2832–2838.
27. S. J. Hong, S. Lee, J. S. Jang and J. S. Lee, *Energy Environ. Sci.*, 2011, **4**, 1781–1787.
28. P. M. Rao, L. Cai, C. Liu, I. S. Cho, C. H. Lee, J. M. Weisse, P. Yang and X. Zheng, *Nano Lett.*, 2014, **14**, 1099–1105.
29. X. Shi, I. Y. Choi, K. Zhang, J. Kwon, D. Y. Kim, J. K. Lee, S. H. Oh, J. K. Kim and J. H. Park, *Nat. Commun.*, 2014, **5**, 4775.
30. Y. Pihosh, I. Turkevych, K. Mawatari, J. Uemura, Y. Kazoe, S. Kosar, K. Makita, T. Sugaya, T. Matsui, D. Fujita, M. Tosa, M. Kondo and T. Kitamori, *Sci. Rep.*, 2015, **5**, 11141.
31. Y. Park, K. J. McDonald and K.-S. Choi, *Chem. Soc. Rev.*, 2013, **42**, 2321–2337.
32. F. F. Abdi and S. P. Berglund, *J. Phys. D: Appl. Phys.*, 2017, **50**, 193002.
33. P. Bornoz, F. F. Abdi, S. D. Tilley, B. Dam, R. van de Krol, M. Graetzel and K. Sivula, *J. Phys. Chem. C*, 2014, **118**, 16959–16966.
34. M. T. McDowell, M. F. Lichterman, J. M. Spurgeon, S. Hu, I. D. Sharp, B. S. Brunschwig and N. S. Lewis, *J. Phys. Chem. C*, 2014, **118**, 19618–19624.
35. R. Saito, Y. Miseki and K. Sayama, *Chem. Commun.*, 2012, **48**, 3833–3835.

36. M. F. Lichterman, M. R. Shaner, S. G. Handler, B. S. Brunschwig, H. B. Gray, N. S. Lewis and J. M. Spurgeon, *J. Phys. Chem. Lett.*, 2013, **4**, 4188–4191.
37. T. W. Kim and K.-S. Choi, *J. Phys. Chem. Lett.*, 2016, **7**, 447–451.
38. F. M. Toma, J. K. Cooper, V. Kunzelmann, M. T. McDowell, J. Yu, D. M. Larson, N. J. Borys, C. Abelyan, J. W. Beeman, K. M. Yu, J. Yang, L. Chen, M. R. Shaner, J. Spurgeon, F. A. Houle, K. A. Persson and I. D. Sharp, *Nat. Commun.*, 2016, **7**, 12012.
39. R. Dillert, D. H. Taffa, M. Wark, T. Bredow and D. W. Bahnemann, *APL Mater.*, 2015, **3**, 104001.
40. R. Dom, A. S. Chary, R. Subasri, N. Y. Hebalkar and P. H. Borse, *Int. J. Energy Res.*, 2015, **39**, 1378–1390.
41. W. Fan, M. Li, H. Bai, D. Xu, C. Chen, C. Li, Y. Ge and W. Shi, *Langmuir*, 2016, **32**, 1629–1636.
42. N. Helaili, G. Mitran, I. Popescu, K. Bachari, I.-C. Marcu and A. Boudjemaa, *J. Electroanal. Chem.*, 2015, **742**, 47–53.
43. A. Kezzim, N. Nasrallah, A. Abdi and M. Trari, *Energy Convers. Manage.*, 2011, **52**, 2800–2806.
44. J. H. Kim, J. H. Kim, J.-W. Jang, J. Y. Kim, S. H. Choi, G. Magesh, J. Lee and J. S. Lee, *Adv. Energy Mater.*, 2015, **5**, 1401933.
45. S. Ida, K. Yamada, T. Matsunaga, H. Hagiwara, Y. Matsumoto and T. Ishihara, *J. Am. Chem. Soc.*, 2010, **132**, 17343–17345.
46. M. I. Díez-García, T. Lana-Villarreal and R. Gómez, *ChemSusChem*, 2016, **9**, 1504–1512.
47. A. A. Tahir and K. G. U. Wijayantha, *J. Photochem. Photobiol., A*, 2010, **216**, 119–125.
48. M. G. Ahmed, T. A. Kandiel, A. Y. Ahmed, I. Kretschmer, F. Rashwan and D. Bahnemann, *J. Phys. Chem. C*, 2015, **119**, 5864–5871.
49. Y. Guo, Y. Fu, Y. Liu and S. Shen, *RSC Adv.*, 2014, **4**, 36967.
50. K. J. McDonald and K.-S. Choi, *Chem. Mater.*, 2011, **23**, 4863–4869.
51. Y. Gao, O. Zandi and T. W. Hamann, *J. Mater. Chem. A*, 2016, **4**, 2826–2830.
52. D. Hu, P. Diao, D. Xu, M. Xia, Y. Gu, Q. Wu, C. Li and S. Yang, *Nanoscale*, 2016, **8**, 5892–5901.
53. C. R. Lhermitte and B. M. Bartlett, *Acc. Chem. Res.*, 2016, **49**, 1121–1129.
54. P. K. Pandey, N. S. Bhave and R. B. Kharat, *Mater. Lett.*, 2005, **59**, 3149–3155.
55. W. Ye, F. Chen, F. Zhao, N. Han and Y. Li, *ACS Appl.* Mater. Interfaces, 2016, **8**, 9211–9217.
56. J. E. Yourey and B. M. Bartlett, *J. Mater. Chem.*, 2011, **21**, 7651–7660.
57. F. Zhan, J. Li, W. Li, Y. Liu, R. Xie, Y. Yang, Y. Li and Q. Chen, *Int. J. Hydrogen Energy*, 2015, **40**, 6512–6520.
58. H. Zhang, P. Yilmaz, J. O. Ansari, F. F. Khan, R. Binions, S. Krause and S. Dunn, *J. Mater. Chem. A*, 2015, **3**, 9638–9644.
59. K. J. Pyper, J. E. Yourey and B. M. Bartlett, *J. Phys. Chem. C*, 2013, **117**, 24726–24732.

60. T. Arai, Y. Konishi, Y. Iwasaki, H. Sugihara and K. Sayama, *J. Comb. Chem.*, 2007, **9**, 574–581.
61. E. W. Ong, G. H. Kwei, R. A. Robinson, B. L. Ramakrishna and R. B. Von Dreele, *Phys. Rev. B: Condens. Matter*, 1990, **42**, 4255–4262.
62. D. Kang, J. C. Hill, Y. Park and K.-S. Choi, *Chem. Mater.*, 2016, **28**, 4331–4340.
63. S. P. Berglund, F. F. Abdi, P. Bogdanoff, A. Chemseddine, D. Friedrich and R. van de Krol, *Chem. Mater.*, 2016, **28**, 4231–4242.
64. R. Patil, S. Kelkar, R. Naphade and S. Ogale, *J. Mater. Chem. A*, 2014, **2**, 3661–3668.
65. Y. Nakabayashi, M. Nishikawa and Y. Nosaka, *Electrochim. Acta*, 2014, **125**, 191–198.
66. U. A. Joshi, A. M. Palasyuk and P. A. Maggard, *J. Phys. Chem. C*, 2011, **115**, 13534–13539.
67. A. Kormányos, A. Thomas, M. N. Huda, P. Sarker, J. P. Liu, N. Poudyal, C. Janáky and K. Rajeshwar, *J. Phys. Chem. C*, 2016, **120**, 16024–16034.
68. S. Ouyang, Z. Li, Z. Ouyang, T. Yu, J. Ye and Z. Zou, *J. Phys. Chem. C*, 2008, **112**, 3134–3141.
69. S. Ouyang and J. Ye, *J. Am. Chem. Soc.*, 2011, **133**, 7757–7763.
70. J. Li, Q. Lu, J. Zhang and S. Zhao, *Mater. Sci. Semicond. Process.*, 2016, **49**, 54–60.
71. A. Renaud, L. Cario, P. Deniard, E. Gautron, X. Rocquefelte, Y. Pellegrin, E. Blart, F. Odobel and S. Jobic, *J. Phys. Chem. C*, 2014, **118**, 54–59.
72. A. Renaud, B. Chavillon, L. Le Pleux, Y. Pellegrin, E. Blart, M. Boujtita, T. Pauporté, L. Cario, S. Jobic and F. Odobel, *J. Mater. Chem.*, 2012, **22**, 14353.
73. M. Yu, G. Natu, Z. Ji and Y. Wu, *J. Phys. Chem. Lett.*, 2012, **3**, 1074–1078.
74. J. Ahmed, C. K. Blakely, J. Prakash, S. R. Bruno, M. Yu, Y. Wu and V. V. Poltavets, *J. Alloys Compd.*, 2014, **591**, 275–279.
75. A. Nattestad, *J. Photonics Energy*, 2011, **1**, 011103.
76. M. Yu, T. I. Draskovic and Y. Wu, *Phys. Chem. Chem. Phys.*, 2014, **16**, 5026.
77. H. Kawazoe, M. Yasukawa, H. Hyodo, M. Kurita, H. Yanagi and H. Hosono, *Nature*, 1997, **389**, 939–942.
78. I. Herraiz-Cardona, F. Fabregat-Santiago, A. Renaud, B. Julián-López, F. Odobel, L. Cario, S. Jobic and S. Giménez, *Electrochim. Acta*, 2013, **113**, 570–574.
79. J. Gu, Y. Yan, J. W. Krizan, Q. D. Gibson, Z. M. Detweiler, R. J. Cava and A. B. Bocarsly, *J. Am. Chem. Soc.*, 2013, **136**, 830–833.
80. U. Kang, S. K. Choi, D. J. Ham, S. M. Ji, W. Choi, D. S. Han, A. Abdel-Wahab and H. Park, *Energy Environ. Sci.*, 2015, **8**, 2638–2643.
81. M. S. Prévot, N. Guijarro and K. Sivula, *ChemSusChem*, 2015, **8**, 1359–1367.
82. C. G. Read, Y. Park and K.-S. Choi, *J. Phys. Chem. Lett.*, 2012, **3**, 1872–1876.
83. J. Gu, A. Wuttig, J. W. Krizan, Y. Hu, Z. M. Detweiler, R. J. Cava and A. B. Bocarsly, *J. Phys. Chem. C*, 2013, **117**, 12415–12422.

84. A. Wuttig, J. W. Krizan, J. Gu, J. J. Frick, R. J. Cava and A. B. Bocarsly, *J. Mater. Chem. A*, 2017, **5**, 165–171.
85. M. S. Prevot, Y. Li, N. Guijarro and K. Sivula, *J. Mater. Chem. A*, 2016, **4**, 3018–3026.
86. Y. J. Jang, Y. B. Park, H. E. Kim, Y. H. Choi, S. H. Choi and J. S. Lee, *Chem. Mater.*, 2016, **28**, 6054–6061.
87. Y. J. Jang, Y. B. Park, H. E. Kim, Y. H. Choi, S. H. Choi and J. S. Lee, *Chem. Mater.*, 2016, **28**, 6054–6061.
88. F. E. Osterloh, *Chem. Soc. Rev.*, 2013, **42**, 2294–2320.
89. N. Guijarro, M. S. Prevot and K. Sivula, *Phys. Chem. Chem. Phys.*, 2015, **17**, 15655–15674.
90. C. C. L. McCrory, S. Jung, I. M. Ferrer, S. M. Chatman, J. C. Peters and T. F. Jaramillo, *J. Am. Chem. Soc.*, 2015, **137**, 4347–4357.
91. A. Walsh and K. T. Butler, *Acc. Chem. Res.*, 2014, **47**, 364–372.
92. K. Sivula and R. van de Krol, *Nat. Rev. Mater.*, 2016, **1**, 15010.

CHAPTER 7

Particulate Photocatalyst Sheets for Efficient and Scalable Water Splitting

TAKASHI HISATOMI AND KAZUNARI DOMEN*

Department of Chemical System Engineering, School of Engineering, The University of Tokyo, 7-3-1 Hongo, Bunkyo-ku, Tokyo 113-8656, Japan
*Email: domen@chemsys.t.u-tokyo.ac.jp

7.1 Introduction

The efficient conversion of solar energy into storable, transportable chemical fuels is believed to have the potential to solve many energy and environmental challenges. This is because the amount of solar energy irradiating the Earth is almost four orders of magnitude larger than the current level of global primary energy consumption.[1] Water splitting into hydrogen and oxygen ($H_2O \rightarrow H_2 + 1/2\ O_2$) driven by solar energy is one possible means of achieving large-scale solar-to-chemical conversion. The standard Gibbs energy change for this reaction is $+237\ kJ\ mol^{-1}$ of water. Hydrogen, the reduction product of this reaction, can be used as a chemical fuel for hydrogen engines and fuels cells and generates only water as a by-product, which may be reused in the solar water splitting reaction. In addition, hydrogen is an important feedstock in chemical industries. At present, hydrogen is produced primarily *via* the reforming of fossil fuels; thus, the mass production of inexpensive solar hydrogen may have huge impacts on the economics and sustainability of chemical industries.

Water splitting over particulate photocatalysts has been studied extensively as a technology potentially capable of generating large amounts of renewable

Energy and Environment Series No. 20
Advances in Photoelectrochemical Water Splitting: Theory, Experiment and Systems Analysis
Edited by S. David Tilley, Stephan Lany and Roel van de Krol

Published by the Royal Society of Chemistry, www.rsc.org

solar hydrogen inexpensively, partly because these catalysts can be readily extended to large areas.[1–5] A review of photocatalytic systems for water splitting under visible light[6] and guiding principles for the correct measurement and reporting of photocatalytic efficiency[7] have recently been published. According to a technoeconomical analysis in 2013, a solar-to-hydrogen energy conversion efficiency (STH) of 5% or higher is required to make photocatalytic solar hydrogen production economically feasible.[8] However, the STH for water splitting using particulate photocatalysts is only approximately 1% at present. It is therefore essential to develop photocatalytic systems capable of splitting water efficiently under visible light irradiation to achieve such high STH values. Designing systems suitable for scaling-up is another important goal, because of the low areal density of solar energy. Recent analyses have suggested that a hydrogen generation facility would require an area of 25–180 km,[2,9] although efficient and inexpensive systems that are scalable over such large areas have yet to be demonstrated. In fact, the water splitting activities of particulate photocatalysts have been primarily evaluated in the form of suspensions in conventional laboratory studies,[7,10] and it is challenging to build a reactor capable of holding extremely large amounts of powder suspensions inexpensively. The use of suspensions also causes difficulties during the replacement of used photocatalysts for maintenance purposes. For all these reasons, it is likely more feasible to process particulate photocatalysts into panels and array these panels repeatedly, similar to photovoltaic applications.[11] Therefore, it is important to develop photocatalyst panels that combine high activity with good scalability and to understand the reactions that proceed in such devices.

Recently, the authors have studied particulate photocatalyst sheets based on a hydrogen evolution photocatalyst (HEP) and an oxygen evolution photocatalyst (OEP) embedded in a conductive layer using a particle transfer technique.[12–16] In particular, particulate photocatalyst sheets based on La- and Rh-co-doped $SrTiO_3$ ($SrTiO_3$:La,Rh) as the HEP and $BiVO_4$ or Mo-doped $BiVO_4$ (denoted as $BiVO_4$(:Mo) for simplicity) as the OEP embedded in gold (Au) or carbon (C) layers have been found to split water into hydrogen and oxygen efficiently *via* a Z-scheme mechanism based on two-step excitation, with STH values exceeding 1%.[13,15] These sheets can also be scaled up without any loss in their intrinsic activity. In this chapter, the structures, reaction properties, and working mechanisms of such photocatalyst sheets are described. In addition, the development of various particulate photocatalysts that are responsive to long wavelength visible light is presented, along with a discussion of the potential future applications of these materials in photocatalyst sheets.

7.2 Photocatalyst Sheets Based on $SrTiO_3$:La,Rh and $BiVO_4$

7.2.1 Preparation and Structure

The preparation of both $SrTiO_3$:La,Rh and $BiVO_4$(:Mo) powders has been described previously.[17–19] $SrTiO_3$:La,Rh, $BiVO_4$, and $BiVO_4$:Mo have particle sizes

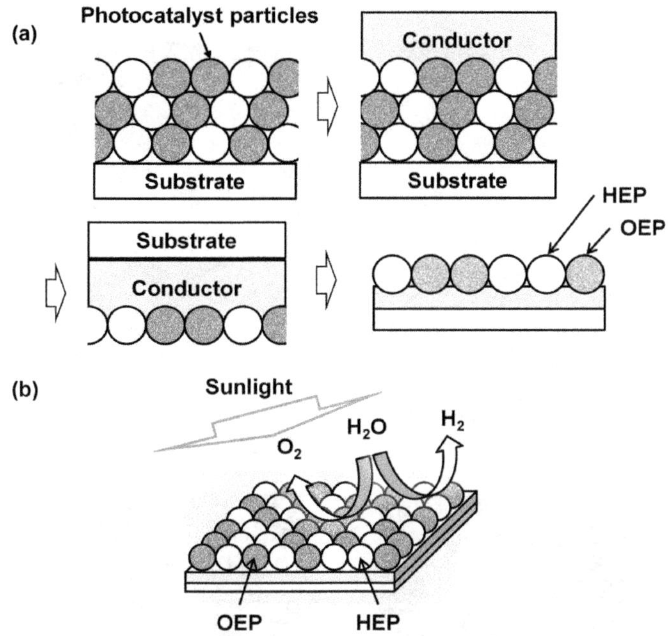

Figure 7.1 (a) Fabrication of a particulate photocatalyst sheet for Z-scheme water splitting by particle transfer, and (b) a schematic of water splitting on a particulate photocatalyst sheet. HEP and OEP stand for hydrogen evolution photocatalyst and oxygen evolution photocatalyst, respectively. Reproduced from ref. 1 with permission from The Royal Society of Chemistry.

ranging from several hundred nanometres to a few micrometres and tend to form aggregates to some extent. Particulate photocatalyst sheets are prepared by particle transfer, as illustrated by the schematic in Figure 7.1.[1,12–15] Briefly, a mixture of $SrTiO_3$ and $BiVO_4$(:Mo) powders suspended in a solvent is deposited onto a glass substrate. After the solvent evaporates, a thin layer of a conductive material, such as Au or C, having a nominal thickness of 350 nm is deposited by vacuum evaporation, electron beam evaporation, or direct current magnetron sputtering, depending on the material. The resulting composite consisting of the photocatalyst particles and the conductive layer is subsequently peeled off, fixed onto another glass substrate, and ultrasonicated to remove photocatalyst particles that are not directly fixed onto the conductive layer. The above process produces $SrTiO_3$:La,Rh/Au/$BiVO_4$(:Mo) and $SrTiO_3$:La,Rh/C/$BiVO_4$(:Mo) photocatalyst sheets.

Top-view scanning electron microscopy-energy dispersive X-ray spectroscopy (SEM-EDX) images of $SrTiO_3$:La,Rh/Au/$BiVO_4$ sheets are presented in Figure 7.2(a) as representative examples.[12] Both $SrTiO_3$:La,Rh and $BiVO_4$ particles are present on the photocatalyst sheet, and the morphologies of the individual photocatalysts are preserved. These photocatalyst particles are seen to cover approximately 80% of the underlying Au layer. Figure 7.2(b)

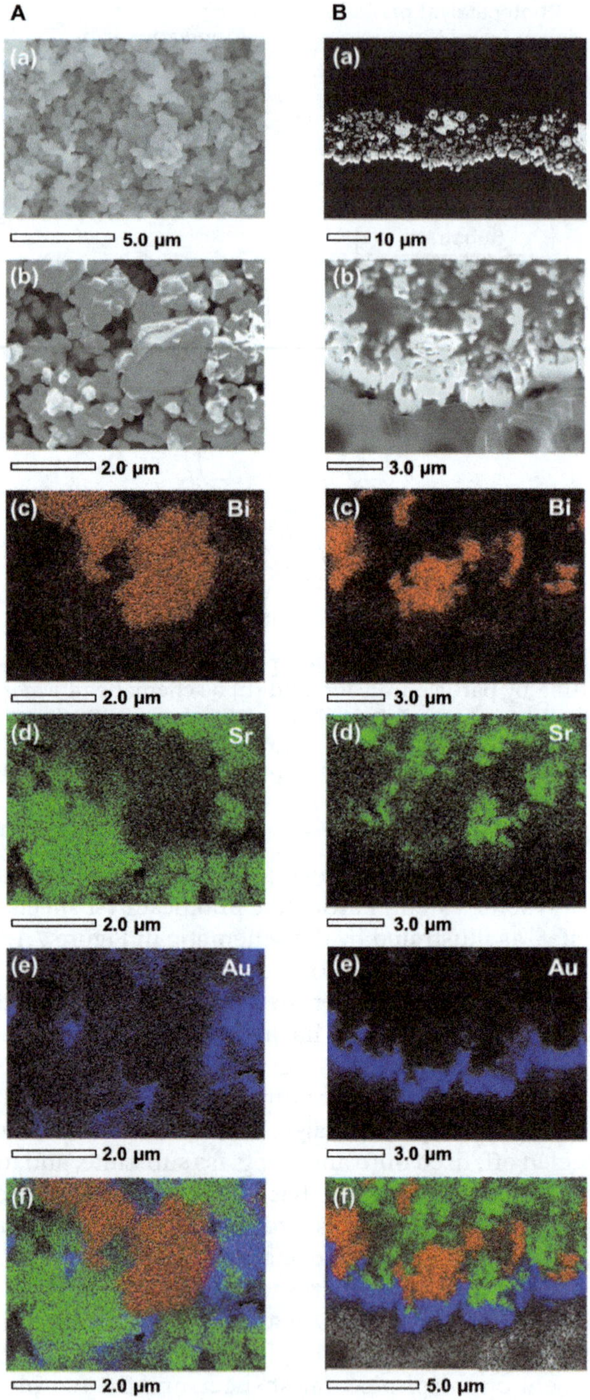
A
B
(a)
5.0 µm
(a)
10 µm
(b)
2.0 µm
(b)
3.0 µm
(c)
Bi
2.0 µm
(c)
Bi
3.0 µm
(d)
Sr
2.0 µm
(d)
Sr
3.0 µm
(e)
Au
2.0 µm
(e)
Au
3.0 µm
(f)
2.0 µm
(f)
5.0 µm

shows cross-sectional images of a $SrTiO_3$:La,Rh/Au/$BiVO_4$ sheet embedded within an epoxy resin in preparation for cross-sectional polishing.[12] It should be noted that the photocatalyst particle layer in this specimen is thicker than those in actual samples because of the penetration of the soft epoxy. However, it is evident that both $SrTiO_3$:La,Rh and $BiVO_4$ particles are directly and firmly embedded into the Au layer at the junction between this layer and the particles. As a result, both the HEP and OEP particles are in contact with the conductive layer and electrons can be transferred between these particles *via* the conductive material.

7.2.2 Z-Scheme Water Splitting Based on Electron Transfer *via* an Underlying Conductor

In experimental work, appropriate amounts of Ru species are loaded as co-catalysts on the photocatalyst sheet by photodeposition.[12] Thin layers of amorphous oxides such as Cr_2O_3 (ref. 20 and 21) and TiO_2 (ref. 22) can also be deposited to suppress backward reactions. A Xe lamp equipped with a long-pass filter ($\lambda > 420$ nm) and a solar simulator (AM 1.5 G) are used as light sources. The quantities of various gases evolved by photocatalytic water splitting are measured using a gas chromatograph directly connected to a closed circulation system made of Pyrex glass.

Table 7.1 shows the water splitting activity of $SrTiO_3$:La,Rh and $BiVO_4$, respectively, used as the HEP and OEP, in the forms of both photocatalyst sheets and powder suspensions.[12] The photocatalyst powder suspension exhibits water splitting activity even in the absence of electron mediators, because interparticle electron transfer[23] occurs *via* physical contact in this system. The selection of an optimum pH value is typically essential for successful interparticle electron transfer, because the HEP and OEP particles must carry opposing charges in the reaction solution so that they will aggregate in response to electrostatic attraction. A particulate photocatalyst sheet based on the same materials embedded in a Au layer showed an activity level during the water splitting reaction more than five times that of the photocatalyst suspension, while the activity of a sheet fabricated without any conductive layer was even lower than that of the suspension. This result demonstrates that electron transfer between the OEP and HEP particles occurs *via* the underlying Au layer in the sheet system, and that this electron transfer is more efficient than interparticle electron transfer.

Figure 7.2 (A) Top-view and (B) cross-sectional images of $SrTiO_3$:La,Rh/Au/$BiVO_4$ photocatalyst sheets, showing (a) and (b) SEM images, and (c)–(f) SEM-EDX mapping images for (c) Bi, (d) Sr, (e) Au, and (f) a superimposition of (c)–(e).
Reprinted from *J. Catal.*, **328**, Q. Wang, Y. Li, T. Hisatomi, M. Nakabayashi, N. Shibata, J. Kubota and K. Domen, Z-scheme water splitting using particulate semiconductors immobilised onto metal layers for efficient electron relay, 308–315,[12] Copyright (2015), with permission from Elsevier.

Table 7.1 Water splitting activity levels under visible light irradiation ($\lambda > 420$ nm) using $SrTiO_3$:La,Rh and $BiVO_4$ in various systems.[a]

Sample	$SrTiO_3$:La,Rh (HEP)	$BiVO_4$ (OEP)	Au	Activity/μmol h^{-1} cm^{-2} H_2	O_2
Suspension[b]	O[d]	O[d]		0.8	0.4
Sheet[c]	O[d]	O[d]	O[d]	4.5	2.2
	O[d]	O[d]		0.2	0.1
	O[d]		O[d]	<0.007	<0.02
		O[d]	O[d]	<0.007	<0.02

[a]Reactions were performed at 288 K and 5 kPa.
[b]Photocatalysts: 10 mg each; irradiation area: 13 cm^2.
[c]Sample area: 9 cm^2.
[d]'O' means that the material was used.

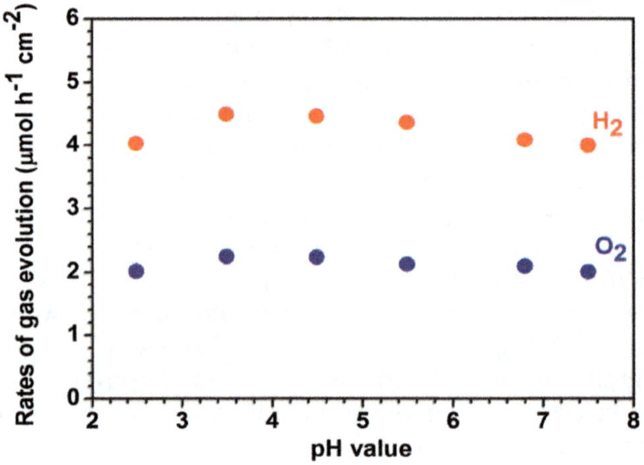

Figure 7.3 Dependence of the photocatalytic activity of a $SrTiO_3$:La,Rh/Au/$BiVO_4$ sheet on pH under visible light irradiation ($\lambda > 420$ nm). Reactions were performed at 288 K and 5 kPa.
Reprinted from *J. Catal.*, **328**, Q. Wang, Y. Li, T. Hisatomi, M. Nakabayashi, N. Shibata, J. Kubota and K. Domen, Z-scheme water splitting using particulate semiconductors immobilised onto metal layers for efficient electron relay, 308–315,[12] Copyright (2015), with permission from Elsevier.

Sheets made without either the HEP or the OEP do not show any water splitting activity. Therefore, it is clear that Z-scheme water splitting based on two-step excitation of the HEP and OEP occurs on the photocatalyst sheets.

Figure 7.3 plots variations in the water splitting activity of a $SrTiO_3$:La,Rh/Au/$BiVO_4$ sheet modified with Ru co-catalysts as a function of pH. The sheet is seen to have maintained activity at various pH values and even in pure water. This behaviour is characteristic of photocatalyst sheets and is distinct from the water splitting characteristics of photocatalyst powder suspensions and conventional electrochemical and photoelectrochemical (PEC) cells, in that the activity is not sensitive to pH values and highly-concentrated electrolytes and buffering reagents are not needed.

7.2.3 Comparison with Powder Suspensions and Photoelectrode Systems

Table 7.2 compares the water splitting activities of suspensions, photoelectrodes, and sheets based on using $SrTiO_3$:La,Rh as the HEP and $BiVO_4$:Mo as the OEP in various reaction solutions. These samples were modified with Ru co-catalysts by photodeposition. Here, the $SrTiO_3$:La,Rh and $BiVO_4$:Mo photoelectrodes were fabricated by embedding the respective photocatalysts individually in a conductive Au layer by particle transfer.[14] These photoelectrodes were connected with a conductive wire and placed side-by-side to construct a parallel p/n PEC cell. This sheet exhibits a level of water splitting activity more than ten times that of a powder suspension or of photoelectrodes in pure water. The addition of iron chloride, which works as an electron mediator,[24] has been found to enhance the activity of the powder suspension. Therefore, it is considered that the efficiency of the interparticle electron transfer in pure water is intrinsically low in such systems. Photoelectrodes also demonstrate higher water splitting activities when sodium sulphate and potassium phosphate (working as supporting electrolytes and buffering reagents) are added, indicating that the activity in pure water is limited because of both solution resistance and the presence of a pH gradient.[14] Specifically, the pH value is increased and decreased near the photocathode and photoanode, respectively, during the water splitting reaction. As a result, a pH gradient develops between the photocathode and the photoanode, generating a backward bias in accordance with the Nernst equation when the mass transfer is insufficient. Nevertheless, powder suspensions and photoelectrodes in these reaction solutions still show lower water splitting activities than sheets employed in pure water. As noted above, the

Table 7.2 Water splitting activity under visible light irradiation ($\lambda > 420$ nm) using $SrTiO_3$:La,Rh and $BiVO_4$:Mo.[a]

System	Reaction solution	Activity/μmol h^{-1} cm^{-2} H$_2$	O$_2$
Suspension	Distilled water	0.4	0.2
	Iron chloride aq. (2 mM, pH 2.4)	6.0	2.8
Photoelectrodes	Distilled water	0.4	0.2
	Sodium sulfate aq. (0.1 M, pH 6.8)	4.8	2.4
	Potassium phosphate buffer aq. (0.1 M, pH 6.8)	6.0	3.0
Sheet	Distilled water	11	5.6
	Diluted sulfuric acid (pH 3.5)	13	6.4

[a]Reactions were performed at 288 K and 5 kPa.

photocatalyst sheets are unaffected by the pH value or by pH gradients or the resistance of the reaction solution, because the photocatalyst particles are directly fixed into a conductive film and hydrogen and oxygen are generated in close proximity.[13] Therefore, the photocatalyst sheets can be directly scaled up without any special consideration given to the electron and mass transfer between the HEP and OEP particles. It is important to note that the applicable reaction solutions are not limited to pure water, and that the water splitting activity of these sheets is actually somewhat improved by slightly acidifying the reaction solution. Interestingly, the addition of a redox mediator or buffering reagent decreases the activity, because such ionic species can absorb incident light, promote reversible backward reactions, and/or strongly adsorb onto the photocatalyst particles.[14,15]

7.2.4 Influence of the Reaction Conditions on the Water Splitting Activity

Figure 7.4 plots the dependence of the photocatalytic activity of $SrTiO_3$:La,Rh/Au/$BiVO_4$:Mo sheets modified with Ru co-catalysts on the background pressure in a reactor.[14] The water splitting activity is seen to decrease drastically with increasing background pressure. However, the hydrogen evolution activity in aqueous methanol under oxygen-free conditions is almost independent of the background pressure. It is therefore believed that backward reactions, namely the oxygen reduction reaction and

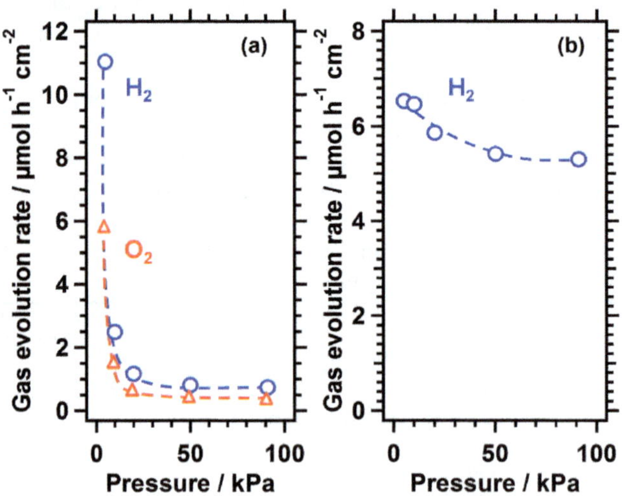

Figure 7.4 The gas evolution rates over Ru-loaded $SrTiO_3$:La,Rh/Au/$BiVO_4$:Mo sheets under visible light irradiation ($\lambda > 420$ nm) as functions of the background Ar pressure in (a) pure water and (b) aqueous methanol (10 vol%). Reactions were performed at 288 K.

water formation from hydrogen and oxygen, readily occur during the water splitting reaction on a photocatalyst sheet, because hydrogen and oxygen are generated in close proximity. As the background pressure increases, the growth rate of gas bubbles becomes slower, so the bubble residence time on the photocatalyst sheet is prolonged. As a consequence, the backward reactions become predominant, most likely on the Ru co-catalyst and the underlying Au layer.

It is known that thin layers of some amorphous oxides, such as Cr_2O_3 (ref. 20 and 21) and TiO_2 (ref. 22), function as a kind of molecular sieve to selectively inhibit the penetration of molecular oxygen and thereby decrease backward reactions during photocatalytic water splitting. In fact, a Ru-loaded $SrTiO_3$:La,Rh/Au/$BiVO_4$:Mo sheet exhibits higher water splitting activity following the photodeposition of Cr_2O_3, as shown in Figure 7.5.[13] Importantly, the negative effect of the background pressure on the water splitting activity is significantly suppressed in this scenario. The additional photodeposition of TiO_2 further reduces the pressure dependence of the activity. Therefore, suppression of the backward reactions can effectively enhance the water splitting activity of these photocatalyst sheets.

The water splitting reaction over a $SrTiO_3$:La,Rh/Au/$BiVO_4$:Mo sheet modified with a Cr_2O_3/Ru co-catalyst under simulated sunlight irradiation at 5 kPa was found to increase monotonically with increasing reaction temperature over the range from 279 to 318 K. The apparent activation energy of the overall water splitting reaction was determined to be 18 kJ mol^{-1}. Figure 7.6 shows the time course of gas evolution during the water splitting reaction using a $SrTiO_3$:La,Rh/Au/$BiVO_4$:Mo sheet modified with a Cr_2O_3/Ru co-catalyst at 331 K and 10 kPa under simulated sunlight irradiation.[13] Under simulated sunlight irradiation, bubbles were vigorously evolved on

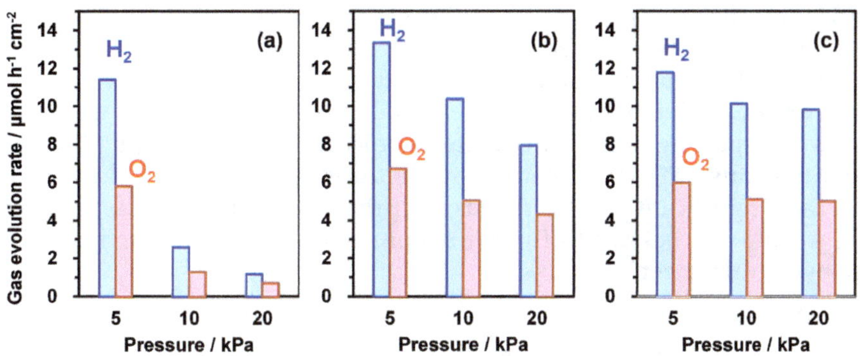

Figure 7.5 The water splitting rates over $SrTiO_3$:La,Rh/Au/$BiVO_4$:Mo sheets modified with (a) a Ru co-catalyst, (b) a Ru co-catalyst and Cr_2O_3, and (c) a Ru co-catalyst, Cr_2O_3, and amorphous TiO_2 under visible light illumination ($\lambda > 420$ nm) as functions of the background Ar pressure. Reactions were performed at 288 K.
Adapted by permission from Macmillan Publishers Ltd: *Nature* Materials (ref. 13), copyright (2016).

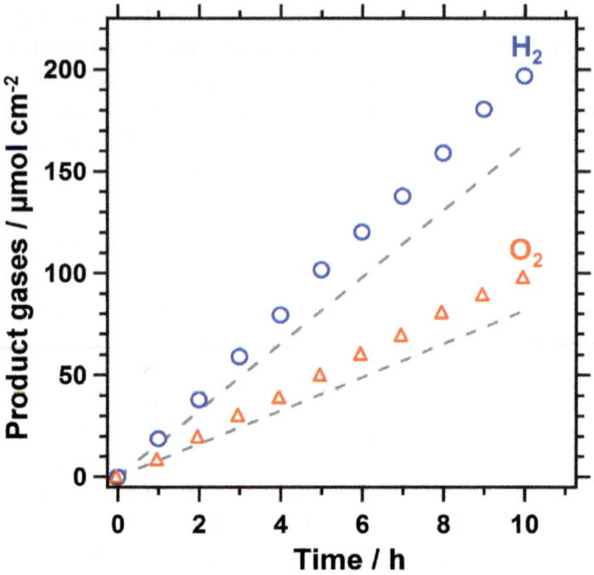

Figure 7.6 Time course of gas evolution during the water splitting reaction using a $SrTiO_3$:La,Rh/Au/$BiVO_4$:Mo sheet modified with a Ru co-catalyst and Cr_2O_3 under simulated sunlight (AM 1.5 G). Reactions were performed at 331 K and 10 kPa. The dashed lines show gas evolution rates corresponding to a solar-to-hydrogen energy conversion efficiency of 1%.

the photocatalyst sheet, and the evolution of bubbles was observed to cease immediately upon termination of the irradiation. The gaseous products consisted of a mixture of hydrogen and oxygen at a molar ratio of two to one, demonstrating the occurrence of the overall water splitting reaction and confirming that the photocatalyst sheet was able to split pure water in a stable manner at the elevated temperature. The apparent quantum efficiency and the STH of this reaction were 33% at 419 nm and 1.1%, respectively. These represent the highest values reported as of 2016 for steady, unassisted pure water splitting using particulate photocatalysts. However, it should be noted that this STH was recorded at a reduced pressure of 10 kPa, and it decreased significantly with increases in the background pressure. Therefore, it will be necessary to further suppress the backward reactions when employing this sheet.

7.2.5 Carbon Conductor-based Sheets Operable at Ambient Pressure

Recently, it has been demonstrated that the water splitting activity of photocatalyst sheets consisting of $SrTiO_3$:La,Rh and $BiVO_4$:Mo can be largely maintained even at ambient pressure by using C as a conductive layer.[15] Figure 7.7 summarises the pressure dependence of the water

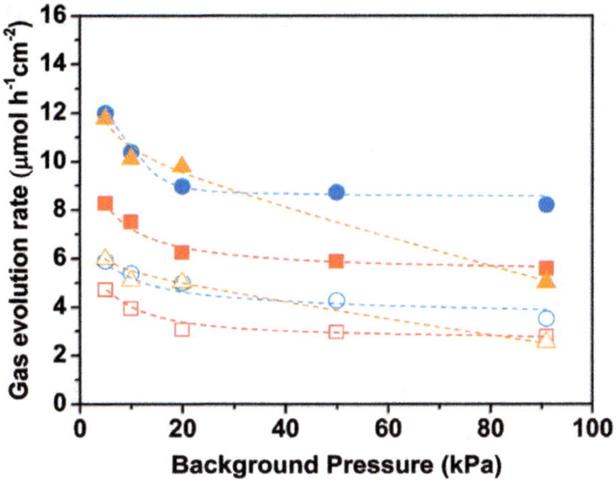

Figure 7.7 Gas evolution rates over photocatalyst sheets under visible light irradiation ($\lambda > 420$ nm) as functions of background pressure. Squares (□), circles (○), and triangles (△) indicate Ru-loaded $SrTiO_3$:La,Rh/C/Bi-VO_4:Mo, Cr_2O_3/Ru-loaded $SrTiO_3$:La,Rh/C/$BiVO_4$:Mo, and amorphous TiO_2/Cr_2O_3/Ru-loaded $SrTiO_3$:La,Rh/Au/$BiVO_4$:Mo, respectively. Closed and open symbols represent hydrogen and oxygen, respectively. The reaction was carried out at 288 K.
Reprinted with permission from Q. Wang, T. Hisatomi, Y. Suzuki, Z. Pan, J. Seo, M. Katayama, T. Minegishi, H. Nishiyama, T. Takata, K. Seki, A. Kudo, T. Yamada and K. Domen, *J. Am. Chem. Soc.*, 2017, **139**, 1675.[15] Copyright © 2017 American Chemical Society.

splitting rates over Au-based and C-based photocatalyst sheets under visible irradiation.[15] The Ru-loaded $SrTiO_3$:La,Rh/C/$BiVO_4$:Mo sheet evidently retained approximately 70% of its base performance at 5 kPa at an elevated background pressure of 91 kPa without modifications with Cr_2O_3 or TiO_2. This is in contrast to the behaviour of a $SrTiO_3$:La,Rh/Au/$BiVO_4$:Mo sheet (see Figure 7.4(a)). The water splitting activity of the Ru-loaded $SrTiO_3$:La,Rh/C/$BiVO_4$:Mo sheet can also be slightly enhanced by additional modification with Cr_2O_3, although this has minimal effect on the sensitivity of the sheet to the background pressure. As a result, the photocatalyst sheet incorporating a C conductor exhibits greater water splitting activity at ambient pressure than a sheet based on a Au conductor and optimally modified with Cr_2O_3 and TiO_2 layers. The distinct sensitivity of the water splitting activity to the background pressure originates from the activity of the conductive materials with regard to the oxygen reduction reaction. This occurs because C is less active than Au during oxygen reduction to hydrogen peroxide. The $SrTiO_3$:La,Rh/C/$BiVO_4$:Mo sheet incorporating a C conductor deposited by sputtering at 373 K and modified with a Ru co-catalyst and Cr_2O_3 exhibited unassisted pure water splitting with a STH of 1.2% at 331 K and 10 kPa, and retained 80% of this efficiency even at 91 kPa.[15]

7.2.6 Simulation of Band Diagrams and Carrier Density Distributions

To date, photocatalyst sheets consisting of $SrTiO_3$:La,Rh and $BiVO_4$(:Mo) have shown high water splitting activities when conductive materials with large work functions and high durability, such as Rh (5.0 eV), Au (5.1 eV) and C (5.2 eV), are employed.[12,15] In contrast, sheets using conductive materials with small work functions, such as Ag (4.3 eV), do not split water into hydrogen and oxygen efficiently.[12] Conductors that are unstable during the water splitting reaction, such as Al or Ni, are not suitable for use, either.[12] It therefore appears that the junction properties at the interfaces of the photocatalysts and conductor materials in photocatalyst sheets have a significant impact on the performance of the sheet.

Band diagrams and carrier density distributions of photocatalyst sheets based on C and Ag conductors have been determined using a semiconductor device simulator (AFORS-HET), as shown in Figures 7.8 and 7.9.[15] In the case of the C-based sheet, $SrTiO_3$:La,Rh forms Schottky-type barriers at the interfaces with both water and C (Figure 7.8(b)), whilst $BiVO_4$:Mo does not form appreciable Schottky-type barriers because the depletion layer width exceeds the dimensions of the particles owing to the low electron density ($\sim 10^{14}$ cm^{-3}). Under illumination (Figure 7.8(c)), electrons close to the H_2O/$SrTiO_3$:La,Rh interface can migrate to the water phase by following the electric field, whilst holes accumulate in the $SrTiO_3$:La,Rh and most likely migrate to the C layer across the Schottky-type barrier by concentration diffusion (Figure 7.8(d)). In contrast, in $BiVO_4$:Mo, electrons and holes are able to migrate to the C and water, respectively, by diffusion because the band profile is virtually flat and does not have significant energy barriers. However, the use of Ag as a conductive layer generates a steeper Schottky-type barrier at the interface between the $SrTiO_3$:La,Rh and the Ag as a result of the small work function of this metal. This leads to the accumulation of photo-excited electrons at the interface (Figure 7.9). Therefore, the migration of holes to the Ag layer by concentration diffusion is highly unfavourable. In fact, the hole current density from $SrTiO_3$:La,Rh to Ag is predicated to be almost zero (Figure 7.9(e)), suggesting that photogenerated holes in $SrTiO_3$:La,Rh contacting Ag are lost by recombination.

In general, therefore, conductive materials with large work functions are favourable in this sheet system because the Schottky barrier is moderated and hole migration from $SrTiO_3$:La,Rh to the conductive material becomes easier. The present model of the photocatalysis mechanism entails some uncertainty because defect states at all the interfaces are ignored and some semiconductor parameters are missing. Nevertheless, this model can differentiate the behaviour of photo-excited carriers in C- and Ag-based photocatalyst sheets and provide some understanding of the working mechanism.

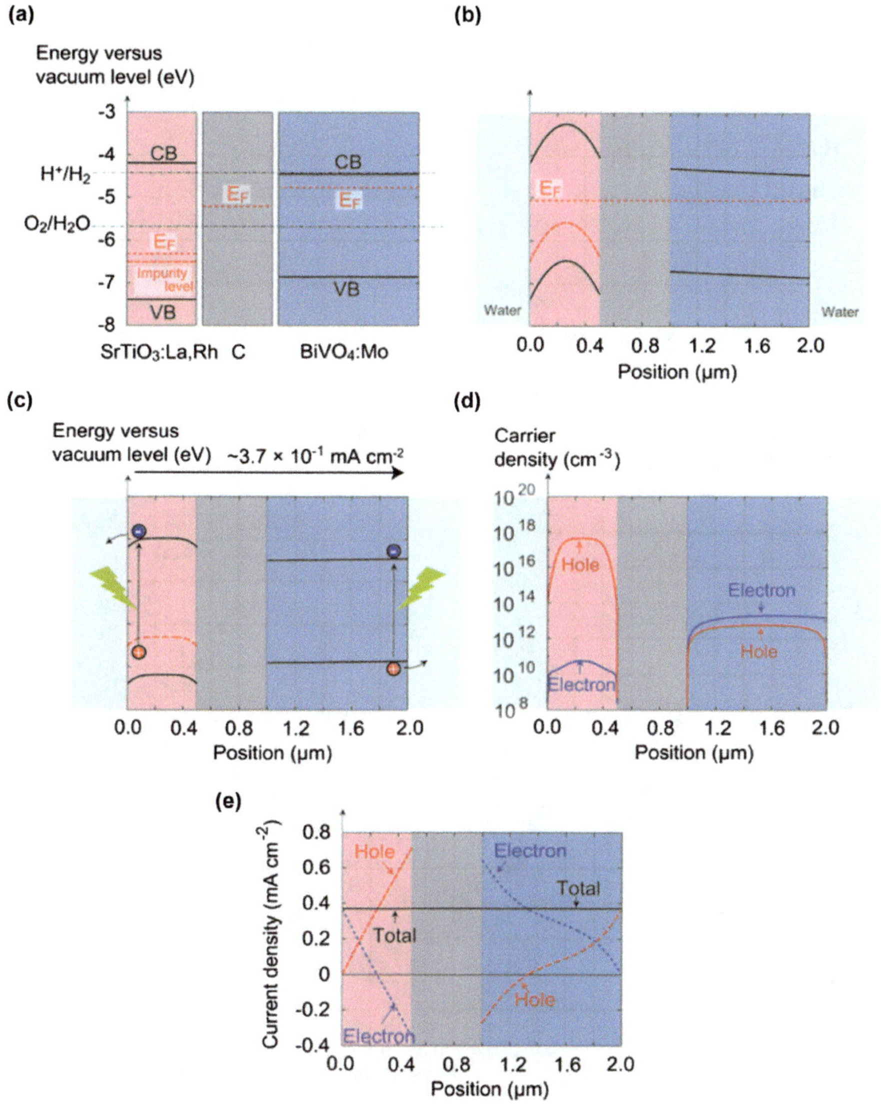

Figure 7.8 Band diagrams and carrier density distributions for $SrTiO_3$:La,Rh/C/$BiVO_4$:Mo sheets. (a)–(c), Band diagrams (a) before contact, (b) under darkness after contact, and (c) under illumination. (d) Carrier density distribution under illumination with a current of 3.7×10^{-1} mA cm^{-2}. (e) Current density distribution. The total, electron, and hole current densities are shown by solid black, dotted blue, and dashed red curves, respectively. Positive electron and hole current values indicate the flow of electrons and holes toward the left and right, respectively.
Adapted with permission from Q. Wang, T. Hisatomi, Y. Suzuki, Z. Pan, J. Seo, M. Katayama, T. Minegishi, H. Nishiyama, T. Takata, K. Seki, A. Kudo, T. Yamada and K. Domen, *J. Am. Chem. Soc.*, 2017, **139**, 1675.[15]

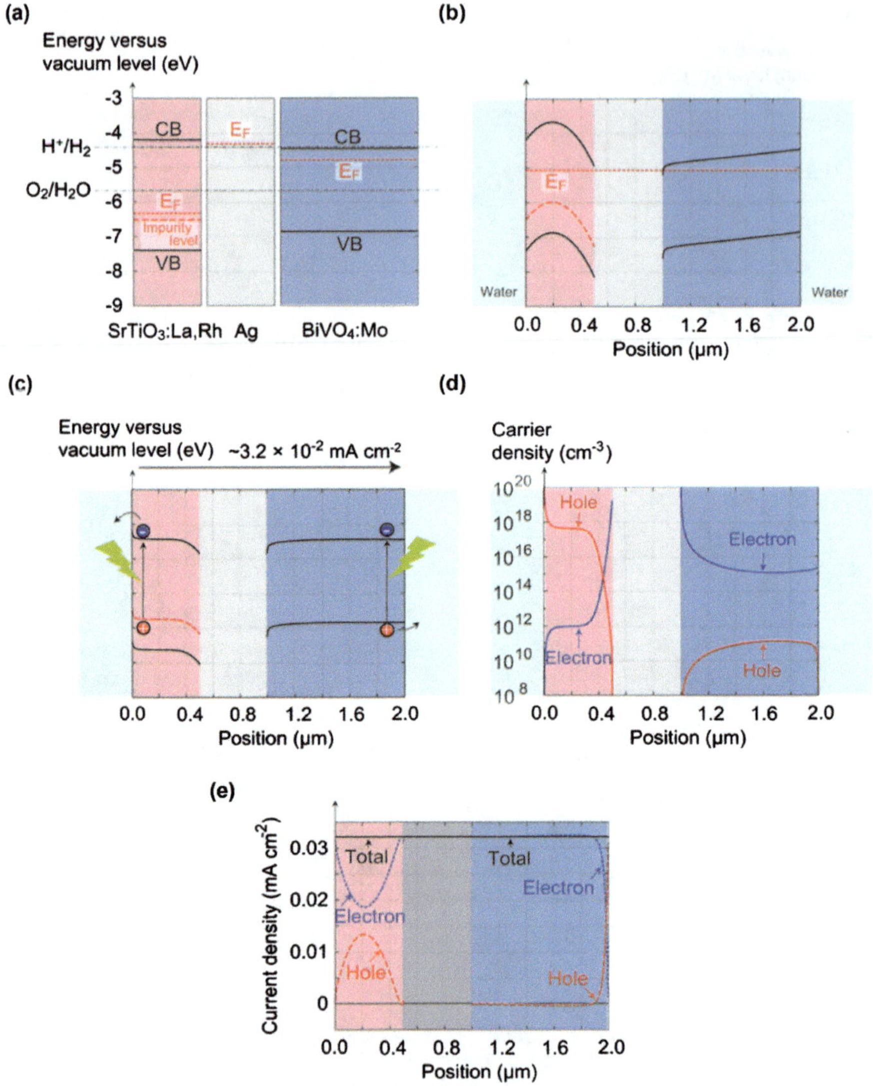

Figure 7.9 Band diagrams and carrier density distributions for $SrTiO_3$:La,Rh/Ag/$BiVO_4$:Mo sheets. (a)–(c) Band diagrams (a) before contact, (b) under darkness after contact, and (c) under illumination. (d) Carrier density distribution under illumination with a current of 3.2×10^{-2} mA cm^{-2}. (e) Current density distribution. The total, electron, and hole current densities are indicated by solid black, dotted blue, and dashed red curves, respectively. Positive electron and hole current values indicate the flow of electrons and holes toward the left and right, respectively.
Reprinted with permission from Q. Wang, T. Hisatomi, Y. Suzuki, Z. Pan, J. Seo, M. Katayama, T. Minegishi, H. Nishiyama, T. Takata, K. Seki, A. Kudo, T. Yamada and K. Domen, *J. Am. Chem. Soc.*, 2017, **139**, 1675.[15]

7.3 Approaches to the Development of Photocatalyst Sheets Based on Narrow Band Gap Photocatalysts

The $SrTiO_3$:La,Rh and $BiVO_4$:Mo photocatalysts used to prepare these sheets are capable of absorbing visible light up to 520–540 nm.[12] However, these wavelengths are too short to achieve the STH of 5% *via* two-step excitation water splitting.[4] Some particulate (oxy)nitride and oxysulfide photocatalysts have suitable band structures for water splitting under visible light.[5,25] In fact, $LaMg_{1/3}Ta_{2/3}O_2N$,[26–28] $BaTaO_2N$,[29] $La_5Ti_2CuS_5O_7$,[30] and certain other materials whose absorption edge wavelengths are situated in the range of approximately 600–700 nm have been applied to photocatalytic and PEC water splitting in the forms of powder suspensions or parallel p/n PEC cells. Such photocatalysts are potentially applicable to the sheet approach as well.

7.3.1 $LaMg_{1/3}Ta_{2/3}O_2N$ as a Hydrogen Evolution Photocatalyst

Recent studies have shown that a suspension of a $LaMg_{1/3}Ta_{2/3}O_2N$ powder with appropriate surface modifications is active in the overall water splitting reaction *via* one-step excitation under irradiation up to 600 nm.[26–28] $LaMg_{1/3}Ta_{2/3}O_2N$, originally reported by Kim in 2007,[31] is a perovskite-type n-type semiconductor consisting of a solid solution of $LaMg_{2/3}Ta_{1/3}O_3$ and $LaTaON_2$. $LaTaON_2$ loaded with rhodium chromium oxide ($RhCrO_x$) as a hydrogen evolution co-catalyst can evolve hydrogen (although only very little oxygen) from pure water under irradiation. In contrast, $LaMg_{1/3}Ta_{2/3}O_2N$ generates both hydrogen and oxygen from pure water, although this is accompanied by the production of nitrogen owing to self-oxidation.[26] The modification of $LaMg_{1/3}Ta_{2/3}O_2N$ with amorphous TiO_2 layer by a photo-deposition method results in water splitting into hydrogen and oxygen at the expected stoichiometric ratio. The amorphous TiO_2 layer functions as a surface protective layer to suppress self-oxidation and also as a molecular sieve to inhibit diffusion of molecular oxygen to the photocatalyst surface, and therefore suppresses the backward reactions (Figure 7.10).[26] Moreover, additional modification with amorphous SiO_2 enhances the water splitting activity of $LaMg_{1/3}Ta_{2/3}O_2N$, presumably because the density of the amorphous oxide layer is modified.

The conduction band edge and valence band edge of $LaMg_{1/3}Ta_{2/3}O_2N$ are located at −0.66 and +1.40 V *vs.* RHE. These values are both more negative than those of the (oxy)nitride photocatalysts, such as $LaTiO_2N$ and Ta_3N_5, which have similar band gap energies but do not exhibit overall water splitting *via* one-step excitation.[28] Therefore, the band gap potential of $LaMg_{1/3}Ta_{2/3}O_2N$ favours the hydrogen evolution reaction, although the apparent quantum efficiency of overall water splitting with this material is at most 0.18%.[27] This is in part because $LaMg_{1/3}Ta_{2/3}O_2N$

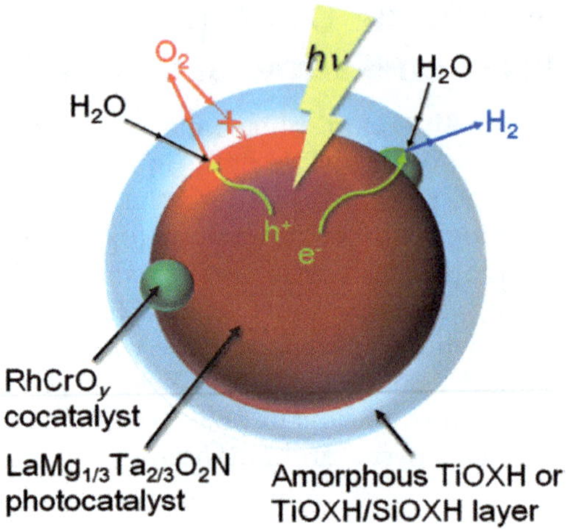

Figure 7.10 Reaction mechanism for water splitting on a surface-coated photocatalyst.
Reproduced with permission from C. Pan, T. Takata, M. Nakabayashi, T. Matsumoto, N. Shibata, Y. Ikuhara and K. Domen, *Angew. Chem. Int. Ed.*, 2015, **54**, 2955,[26] John Wiley and Sons, © 2015 Wiley-VCH Verlag GmbH & Co. KGaA, Weinheim.

as synthesised contains a high density of anion defects. It is therefore necessary to improve the synthesis of this material. In addition, the driving force for the oxygen evolution reaction is believed to be low as a result of the small offset between the valence band edge and the oxygen evolution potential. It is generally believed that the oxygen evolution reaction requires a larger overpotential than the hydrogen evolution reaction.

Based on its band gap potential, it is reasonable to apply $LaMg_{1/3}Ta_{2/3}O_2N$ as the HEP in a photocatalyst sheet system. As such, photocatalyst sheets employing $LaMg_{1/3}Ta_{2/3}O_2N$ as the HEP and $BiVO_4$(:Mo) as the OEP, both embedded in a Au layer, have been demonstrated to split water into hydrogen and oxygen under visible light irradiation.[32,33] This photocatalyst sheet also evolves nitrogen as a result of the self-oxidation of $LaMg_{1/3}Ta_{2/3}O_2N$, although this can be effectively suppressed by applying an amorphous TiO_2 layer of the appropriate thickness,[32,34] similar to the case for the powder suspension.[26] In addition, the incorporation of reduced graphene oxide (RGO) enhances the charge transfer within the particulate photocatalyst layer on the sheet (Figure 7.11).[33] However, $LaMg_{1/3}Ta_{2/3}O_2N$/(Au,RGO)/$BiVO_4$:Mo sheets show an STH of 3.5×10^{-3}% at most, which is more than three orders of magnitude lower than that obtained from sheets consisting of $SrTiO_3$:La,Rh and $BiVO_4$:Mo.[13,15] Such results suggest that the performance of $LaMg_{1/3}Ta_{2/3}O_2N$ as the HEP is inferior to that of $SrTiO_3$:La,Rh, presumably because of the high densities of anion defects and grain boundaries in $LaMg_{1/3}Ta_{2/3}O_2N$ particles.

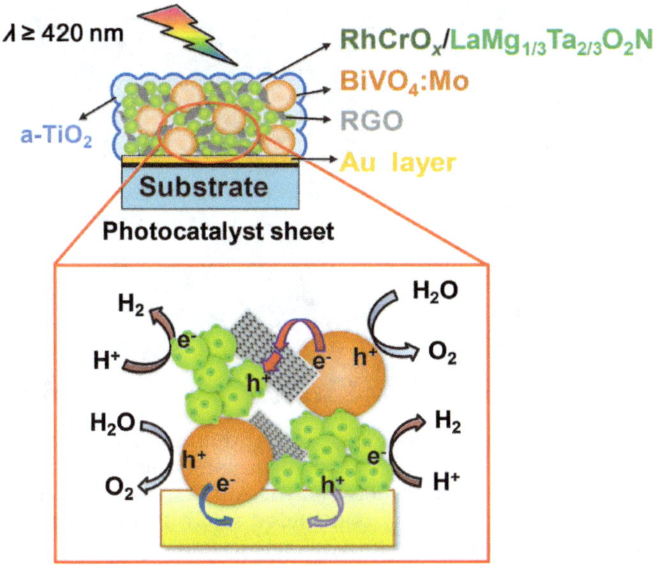

Figure 7.11 Schematic showing the mechanism by which an amorphous TiO_2-coated ($RhCrO_x/LaMg_{1/3}Ta_{2/3}O_2N$)/(Au,RGO)/$BiVO_4$:Mo photocatalyst sheet functions.
Reproduced with permission from Z. Pan, T. Hisatomi, Q. Wang, S. Chen, A. Iwase, M. Nakabayashi, N. Shibata, T. Takata, M. Katayama, T. Minegishi, A. Kudo and K. Domen, *Adv. Funct. Mater.*, 2016, **26**, 7011,[33] John Wiley and Sons, © 2016 WILEY-VCH Verlag GmbH & Co. KGaA, Weinheim.

7.3.2 Unassisted Photoelectrochemical Water Splitting Using a $La_5Ti_2(Cu,Ag)S_5O_7$ Photocathode and a $BaTaO_2N$ Photoanode

$La_5Ti_2CuS_5O_7$ (LTC) and $La_5Ti_2AgS_5O_7$ (LTA), semiconducting materials with absorption edge wavelengths of 650 and 570 nm, respectively, were originally reported by Meignen *et al.* in 2004 (ref. 35). These compounds have been shown to be active during photocatalytic hydrogen and oxygen evolution from aqueous solutions containing appropriate sacrificial electron donors and acceptors under visible light irradiation.[36] Both materials can be obtained as rod-shaped particles by a solid state reaction and exhibit a unique one-dimensional electron conductivity.[37] In particular, LTC functions as a photocathode for PEC water splitting when embedded in a Au film having a large work function (Figure 7.12(a)) as a back electrode.[30,38] Interestingly, this compound also functions as a photoanode when a Ti film (with a small work function) is used as the back electrode.[38] The polarity of the surface photovoltage also depends on the back electrode material. This is because the carrier diffusion length in LTC is on the order of micrometres, and charge separation at the LTC/back electrode interface can determine the PEC properties.[38,39] The most important parameters of LTC as a photocathode

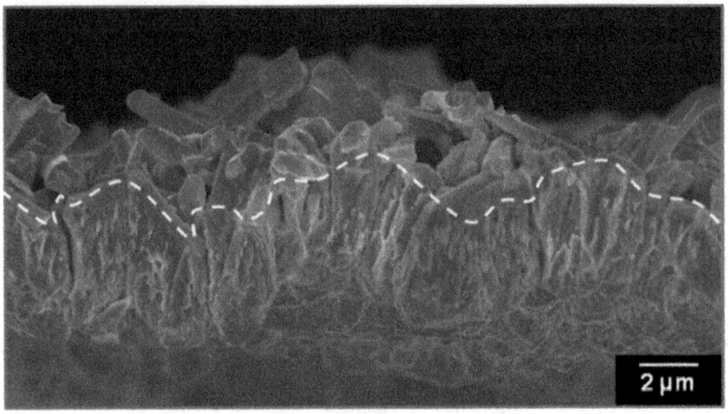

Figure 7.12 A cross-sectional SEM image of an Al-doped $LTC_{0.9}A_{0.1}$/Au photoelectrode fabricated by the particle transfer method. The dashed path indicates the interface between the Al-$LTC_{0.9}A_{0.1}$ particle layer and the underlying Au layer and is included solely as a guide for the eye. Reproduced from ref. 30 with permission from The Royal Society of Chemistry.

material are its positive onset potential of *ca.* +0.9 V *vs.* RHE for the photocathodic current and the ability to generate hydrogen at +0.7 V *vs.* RHE,[40] both of which are well suited for use in p/n PEC cells. The photocathodic current can be significantly enhanced by p-type doping into the Ti sites of LTC,[40–42] by the formation of LTC–LTA solid solutions ($LTC_{1-x}A_x$),[30] and by surface modifications with CdS and certain oxides (Figure 7.13).[43,44] $LTC_{1-x}A_x$ electrodes are not particularly stable in aqueous phosphate buffer solutions or in highly basic solutions because of the specific adsorption of ions and degradation.[43] This is undesirable with respect to the construction of p/n PEC cells because photoanodes and oxygen evolution catalysts generally work more efficiently and with greater stability in highly buffered alkaline solutions. However, it has been demonstrated that modifications with oxide layers improve the stability of $LTC_{1-x}A_x$ electrodes such that they can be employed in alkaline solutions up to pH 13.[43]

$BaTaO_2N$, an n-type semiconducting material with an absorption edge wavelength of 660 nm, has been found to work as a photocatalyst for hydrogen evolution in the presence of sacrificial electron donors under visible light (ref. 45) and has been applied as the HEP for Z-scheme water splitting.[46,47] This material has also been considered as a photoanode material since 2012.[48,49] In 2015, photoanodes composed of $BaTaO_2N$ embedded in a Ta/Ti bilayer by particle transfer *via* radio-frequency magnetron sputtering were reported.[29] The PEC properties of $BaTaO_2N$/Ta/Ti photoanodes can be improved by depositing a Ta layer at elevated temperatures. It has been confirmed by Mott–Schottky analysis that an increase in the deposition temperature decreases the donor concentration in $BaTaO_2N$. Therefore, the depletion layer thickness is increased and charge

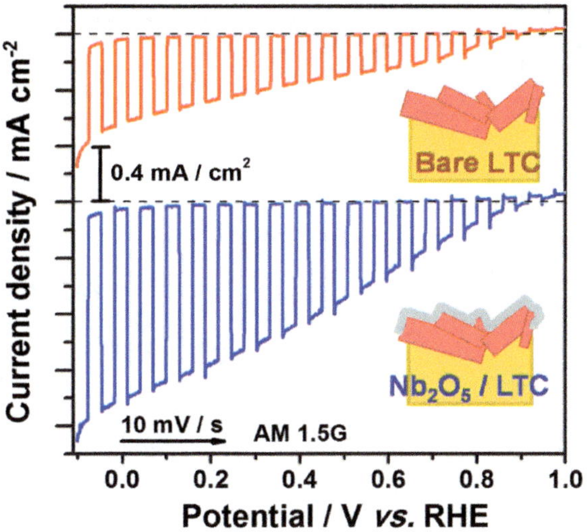

Figure 7.13 Current–potential plots obtained from (a) a bare Mg–LTC photocathode, and (b) a photocathode surface-modified with Nb_2O_5. The measurements were carried out in a 0.1 M aqueous Na_2SO_4 solution (pH 10) under chopped simulated sunlight irradiation. The photocathodes had been modified by the photodeposition of Pt.
Reprinted with permission from J. Liu, T. Hisatomi, D. H. K. Murthy, M. Zhong, M. Nakabayashi, T. Higashi, Y. Suzuki, H. Matsuzaki, K. Seki, A. Furube, N. Shibata, M. Katayama, T. Minegishi and K. Domen, *J. Phys. Chem. Lett.*, 2017, **8**, 375.[44] Copyright © 2016, American Chemical Society.

separation is facilitated, while the series resistance is raised. As a consequence, a $BaTaO_2N$ photoanode deposited with a Ta layer at 873 K has demonstrated the highest PEC properties yet reported. The $BaTaO_2N$ photoanode generates a steady photoanodic current at +0.7 V *vs.* RHE and more positive potentials. This current is attributable solely to oxygen evolution from water at the relatively negative potential of 0.8 V *vs.* RHE (Figure 7.14).[29] Furthermore, the onset potential of the photocurrent of a $BaTaO_2N$ photoanode can be improved by using Zr as a contact layer and by depositing Ir species *via* magnetron sputtering as the second catalyst following electrodeposition of the Co species.[46]

Because the $LTC_{1-x}A_x$ photocathode and the $BaTaO_2N$ photoanode generate photocurrents at the same electrode potential, series-connected photoelectrodes are capable of splitting water under irradiation up to 660 nm without any applied voltage.[30,43] Figure 7.15 presents the time course of gas evolution during unassisted PEC water splitting using an Al-doped $LTC_{0.9}A_{0.1}$/Au photocathode modified with CdS, TiO_2, and Pt and a $BaTaO_2N$/Zr/Ti photoanode modified with Co and Ir species.[43] Hydrogen and oxygen are produced at the expected stoichiometric ratio at an initial STH of 0.14%. During the water splitting, the potential of the electrodes is maintained at approximately +0.7 V *vs.* RHE, and the Faradaic efficiencies of the hydrogen

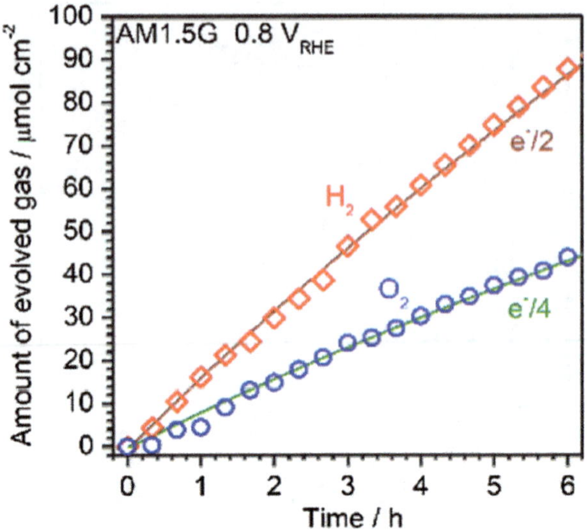

Figure 7.14 Amounts of hydrogen and oxygen evolved over a Co/$BaTaO_2N$/Ta/Ti electrode at +0.8 V *vs.* RHE under simulated AM1.5G light. A 0.2 M aqueous potassium phosphate solution (pH 13) and a CrO_x-coated Pt mesh were used as the electrolyte and counter electrode, respectively. The symbols represent the gas quantities determined by gas chromatography, while the solid lines indicate the values calculated from the total charge recorded as the photocurrent.
Adapted with permission from K. Ueda, T. Minegishi, J. Clune, M. Nakabayashi, T. Hisatomi, H. Nishiyama, M. Katayama, N. Shibata, J. Kubota, T. Yamada and K. Domen, *J. Am. Chem. Soc.*, 2015, **137**, 2227.[29]

and oxygen evolution reactions are virtually unity, as expected based on the performance of the individual photoelectrodes. However, the STH of the p/n PEC cell is far less than the values obtained during photocatalytic water splitting using sheets[13,15] or photovoltaic-powered water electrolysis.[50] The primary reason for this low STH is the late photocurrent onset potential, which gives a low photocurrent at the operating potential. Therefore, it is important to improve the onset potential for the photocurrent. In addition, the photoelectrodes from which these data were obtained were prepared individually and subsequently combined. This was done so as to employ the optimal back contact materials for the respective photoelectrodes and thus to generate a suitable level of PEC performance. Generally, it is preferable to use low and high work function materials for photocathodes and photoanodes, respectively, to establish ohmic contacts between the semiconductors and the back electrode. In the future, the development of efficient monolithic photocatalyst sheets will require a better understanding of the associated band diagrams and current distributions as well as modification of the semiconducting properties and sheet fabrication processes.

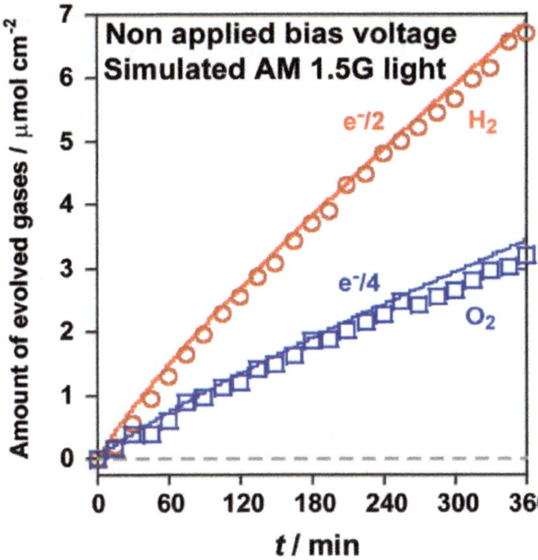

Figure 7.15 Gas generation over time for a PEC cell composed of a Pt/TiO_2/CdS/Al-doped $LTC_{0.9}A_{0.1}$ photocathode with a geometrical electrode area of 0.44 cm^2 and an Ir/Co/$BaTaO_2N$ photoanode with a geometrical electrode area of 0.46 cm^2. The experiment was carried out in an aqueous 0.1 M Na_2HPO_4 solution with a pH value adjusted to 13 by NaOH addition, under simulated AM1.5G light irradiation without the application of an external bias voltage. The symbols represent the gas quantities determined by gas chromatography, while the solid lines indicate the values calculated from the total charge recorded as the photocurrent.
Adapted with permission from T. Higashi, Y. Shinohara, A. Ohnishi, J. Liu, K. Ueda, S, Okamura, T. Hisatomi, M. Katayama, H. Nishiyama, T. Yamada, T. Minegishi, K. Domen, *ChemPhotoChem*, 2017, **1**, 167,[43] John Wiley and Sons, © 2016 Wiley-VCH Verlag GmbH & Co. KGaA, Weinheim.

7.4 Summary and Future Prospects

This chapter summarised the development and reaction mechanisms of state-of-the-art particulate photocatalyst sheet systems. These units are capable of exhibiting higher photocatalytic activity during pure water splitting than conventional powder suspension or (photo)electrochemical approaches because HEP and OEP particles are rigidly embedded in conductive layers and the hydrogen and oxygen evolution reactions take place in close proximity. This feature enables efficient electron transfer between HEP and OEP particles, reduces the solution resistance, and inhibits the formation of a pH gradient arising from mass transfer issues between the hydrogen and oxygen evolution sites. The water splitting activities of these photocatalyst sheets are sensitive to the oxygen reduction reaction and decrease with increasing background pressure. However, these issues can be avoided by modifying the surfaces of the sheets with thin oxide layers that inhibit the access of molecular oxygen to active sites, and by applying a carbon conductor layer that is relatively unaffected by the oxygen reduction reaction. A STH value for pure water splitting at ambient pressure as high as 1.0% can be obtained by modifying $SrTiO_3$:La,Rh/C/$BiVO_4$:Mo sheets with a Ru-based co-catalyst and a Cr_2O_3 layer.

The primary challenge associated with achieving more efficient sunlight-driven water splitting by photocatalyst sheets lies in the effective utilisation of narrow-band-gap photocatalysts. Photocatalyst sheets incorporating $LaMg_{1/3}Ta_{2/3}O_2N$ as the HEP and p/n PEC cells based on $LTC_{1-x}A_x$ photocathodes and $BaTaO_2N$ photoanodes are presented herein as attempts to do so. These systems are able to harvest longer wavelengths of visible light than sheets incorporating $SrTiO_3$:La,Rh and $BiVO_4$:Mo, although the associated STH values are still low. Based on the similarity between the charge transfer processes in sheet and photoelectrode systems, it will be vital to upgrade the photocatalytic materials to improve the photocurrent and the onset potentials after fixation onto specific conductive materials. An in-depth understanding of the band diagrams and current distributions of these compounds, and engineering of the semiconducting properties and sheet fabrication processes, should promote the development of future efficient monolithic photocatalyst sheets. Spatially-resolved analyses of photovoltages (associated with charge separation) and photocurrents (associated with activity) may also play important roles in the development of particulate photocatalyst sheets. This is because these are highly inhomogeneous systems that are not conducive to observations of the photocurrent or photovoltage of the entire device using conventional bulk or wide-range techniques.[51]

Another important point that needs to be addressed to allow the practical implementation of solar hydrogen production by photocatalyst sheets is the development of scalable, inexpensive reaction systems that allow for the safe and efficient separation of hydrogen from oxygen in the products. This requirement stems from the mixture of hydrogen and oxygen evolved over the photocatalyst sheets. Gas separation is largely a chemical engineering issue that will require balancing a broad range of factors in conjunction with the operation of photocatalyst sheets capable of splitting pure water into hydrogen and oxygen at elevated pressures. It is expected that such processes will eventually be achieved by integrating membrane separators into the panel reactors.

Acknowledgements

This work was financially supported by Grants-in-Aids for Scientific Research (A) (No. 16H02417) and for Young Scientists (A) (No. 15H05494) from the Japan Society for the Promotion of Science (JSPS) and by the Artificial Photosynthesis Project of the New Energy and Industrial Technology Development Organisation (NEDO).

References

1. T. Hisatomi and K. Domen, *Faraday Discuss.*, 2017, **198**, 11.
2. K. Maeda and K. Domen, *J. Phys. Chem. Lett.*, 2010, **1**, 2655.
3. T. Hisatomi, J. Kubota and K. Domen, *Chem. Soc. Rev.*, 2014, **43**, 7520.
4. T. Hisatomi, K. Takanabe and K. Domen, *Catal. Lett.*, 2015, **145**, 95.
5. K. Maeda and K. Domen, *Bull. Chem. Soc. Jpn.*, 2016, **89**, 627.
6. M. Fabian, S. Hu, N. Singh, F. A. Houle, T. Hisatomi, K. Domen, F. E. Osterloh and S. Ardo, *Energy Environ. Sci.*, 2015, **8**, 2825.
7. M. Qureshi and K. Takanabe, *Chem. Mater.*, 2017, **29**, 158.
8. B. A. Pinaud, J. D. Benck, L. C. Seitz, A. J. Forman, Z. Chen, T. G. Deutsch, B. D. James, K. N. Baum, G. N. Baum, S. Ardo, H. Wang, E. Miller and T. F. Jaramillo, *Energy Environ. Sci.*, 2013, **6**, 1983.
9. R. Sathre, J. B. Greenblatt, K. Walczak, I. D. Sharp, J. C. Stevens, J. W. Ager III and F. A. Houle, *Energy Environ. Sci.*, 2016, **9**, 803.
10. A. Kudo and Y. Miseki, *Chem. Soc. Rev.*, 2009, **38**, 253.
11. A. Xiong, G. Ma, K. Maeda, T. Takata, T. Hisatomi, T. Setoyama, J. Kubota and K. Domen, *Catal. Sci. Technol.*, 2014, **4**, 325.
12. Q. Wang, Y. Li, T. Hisatomi, M. Nakabayashi, N. Shibata, J. Kubota and K. Domen, *J. Catal.*, 2015, **328**, 308.
13. Q. Wang, T. Hisatomi, Q. Jia, H. Tokudome, M. Zhong, C. Wang, Z. Pan, T. Takata, M. Nakabayashi, N. Shibata, Y. Li, I. D. Sharp, A. Kudo, T. Yamada and K. Domen, *Nat. Mater.*, 2016, **15**, 611.
14. Q. Wang, T. Hisatomi, M. Katayama, T. Takata, T. Minegishi, A. Kudo, Taro Yamada and K. Domen, *Faraday Discuss.*, 2017, **197**, 491.
15. Q. Wang, T. Hisatomi, Y. Suzuki, Z. Pan, J. Seo, M. Katayama, T. Minegishi, H. Nishiyama, T. Takata, K. Seki, A. Kudo, T. Yamada and K. Domen, *J. Am. Chem. Soc.*, 2017, **139**, 1675.
16. T. Minegishi, N. Nishimura, J. Kubota and K. Domen, *Chem. Sci.*, 2013, **4**, 1120.
17. Q. Wang, T. Hisatomi, S. S. K. Ma, Y. Li and K. Domen, *Chem. Mater.*, 2014, **26**, 4144.
18. A. Iwase, Y. H. Ng, Y. Ishiguro, A. Kudo and R. Amal, *J. Am. Chem. Soc.*, 2011, **133**, 11054.
19. A. Kudo, K. Omori and H. Kato, *J. Am. Chem. Soc.*, 1999, **121**, 11459.
20. K. Maeda, K. Teramura, D. Lu, N. Saito, Y. Inoue and K. Domen, *Angew. Chem., Int. Ed.*, 2006, **45**, 7806.
21. M. Yoshida, K. Takanabe, K. Maeda, A. Ishikawa, J. Kubota, Y. Sakata, Y. Ikezawa and K. Domen, *J. Phys. Chem. C*, 2009, **113**, 10151.
22. T. Takata, C. Pan, M. Nakabayashi, N. Shibata and K. Domen, *J. Am. Chem. Soc.*, 2015, **137**, 9627.
23. Y. Sasaki, H. Nemoto, K. Saito and A. Kudo, *J. Phys. Chem. C*, 2009, **113**, 17536.
24. H. Kato, Y. Sasaki, N. Shirakura and A. Kudo, *J. Mater. Chem. A*, 2013, **1**, 12327.
25. K. Maeda and K. Domen, *J. Phys. Chem. C*, 2007, **111**, 7851.

26. C. Pan, T. Takata, M. Nakabayashi, T. Matsumoto, N. Shibata, Y. Ikuhara and K. Domen, *Angew. Chem., Int. Ed.*, 2015, **54**, 2955.
27. C. Pan, T. Takata and K. Domen, *Chem. – Eur. J.*, 2016, **22**, 1854.
28. C. Pan, T. Takata, K. Kumamoto, S. S. K. Ma, K. Ueda, T. Minegishi, M. Nakabayashi, T. Matsumoto, N. Shibata, Y. Ikuhara and K. Domen, *J. Mater. Chem. A*, 2016, **4**, 4544.
29. K. Ueda, T. Minegishi, J. Clune, M. Nakabayashi, T. Hisatomi, H. Nishiyama, M. Katayama, N. Shibata, J. Kubota, T. Yamada and K. Domen, *J. Am. Chem. Soc.*, 2015, **137**, 2227.
30. T. Hisatomi, S. Okamura, J. Liu, Y. Shinohara, K. Ueda, T. Higashi, M. Katayama, T. Minegishi and K. Domen, *Energy Environ. Sci.*, 2015, **8**, 3354.
31. Y.-I. Kim and P. M. Woodward, *J. Solid State Chem.*, 2007, **180**, 3224.
32. Z. Pan, T. Hisatomi, Q. Wang, S. Chen, M. Nakabayashi, N. Shibata, C. Pan, T. Takata, M. Katayama, T. Minegishi, A. Kudo and K. Domen, *ACS Catal.*, 2016, **6**, 7188.
33. Z. Pan, T. Hisatomi, Q. Wang, S. Chen, A. Iwase, M. Nakabayashi, N. Shibata, T. Takata, M. Katayama, T. Minegishi, A. Kudo and K. Domen, *Adv. Funct. Mater.*, 2016, **26**, 7011.
34. Z. Pan, T. Hisatomi, Q. Wang, M. Nakabayashi, N. Shibata, C. Pan, T. Takata and K. Domen, *Appl. Catal., A*, 2016, **521**, 26.
35. V. Meignen, L. Cario, A. Lafond, Y. Moëlo, C. Guillot-Deudon and A. Meerschaut, *J. Solid State Chem.*, 2004, **177**, 2810.
36. T. Suzuki, T. Hisatomi, K. Teramura, Y. Shimodaira, H. Kobayashi and K. Domen, *Phys. Chem. Chem. Phys.*, 2012, **14**, 15475.
37. G. Ma, J. Liu, T. Hisatomi, T. Minegishi, Y. Moriya, M. Iwase, H. Nishiyama, M. Katayama, T. Yamada and K. Domen, *Chem. Commun.*, 2015, **51**, 4302.
38. G. Ma, Y. Suzuki, R. B. Singh, A. Iwanaga, Y. Moriya, T. Minegishi, J. Liu, T. Hisatomi, H. Nishiyama, M. Katayama, K. Seki, A. Furube, T. Yamada and K. Domen, *Chem. Sci.*, 2015, **6**, 4513.
39. Y. Suzuki, R. B. Singh, H. Matsuzaki, A. Furube, G. Ma, T. Hisatomi, K. Domen and K. Seki, *Chem. Phys.*, 2016, **476**, 9.
40. J. Liu, T. Hisatomi, G. Ma, A. Iwanaga, T. Minegishi, Y. Moriya, M. Katayama, J. Kubota and K. Domen, *Energy Environ. Sci.*, 2014, **7**, 2239.
41. J. Liu, T. Hisatomi, M. Katayama, T. Minegishi, J. Kubota and K. Domen, *J. Mater. Chem. A*, 2016, **4**, 4848.
42. E. Sakai, N. Nagamura, J. Liu, T. Hisatomi, T. Yamada, K. Domen and M. Oshima, *Nanoscale*, 2016, **8**, 18893.
43. T. Higashi, Y. Shinohara, A. Ohnishi, J. Liu, K. Ueda, S. Okamura, T. Hisatomi, M. Katayama, H. Nishiyama, T. Yamada, T. Minegishi and K. Domen, *ChemPhotoChem*, 2017, **1**, 167.
44. J. Liu, T. Hisatomi, D. H. K. Murthy, M. Zhong, M. Nakabayashi, T. Higashi, Y. Suzuki, H. Matsuzaki, K. Seki, A. Furube, N. Shibata, M. Katayama, T. Minegishi and K. Domen, *J. Phys. Chem. Lett.*, 2017, **8**, 375.

45. G. Hitoki, T. Takata, J. N. Kondo, M. Hara, H. Kobayashi and K. Domen, *Electrochemistry*, 2002, **70**, 463.
46. M. Higashi, R. Abe, K. Teramura, T. Takata, B. Ohtani and K. Domen, *Chem. Phys. Lett.*, 2008, **452**, 120.
47. M. Higashi, R. Abe, T. Takata and K. Domen, *Chem. Mater.*, 2009, **21**, 1543.
48. K. Maeda and K. Domen, *Angew. Chem., Int. Ed.*, 2012, **51**, 9865.
49. M. Higashi, K. Domen and R. Abe, *J. Am. Chem. Soc.*, 2013, **135**, 10238.
50. J. Luo, J.-H. Im, M. T. Mayer, M. Schreier, M. K. Nazeeruddin, N.-G. Park, S. D. Tilley, H. J. Fan and M. Grätzel, *Science*, 2014, **345**, 1593.
51. T. Hisatomi and K. Domen, *Curr. Opin. Electrochem.*, 2017, **2**, 148.

CHAPTER 8

Applications of Bipolar Membranes for Electrochemical and Photoelectrochemical Water Splitting

DAVID A. VERMAAS*[a,b] AND WILSON A. SMITH*[c]

[a] Transport Phenomena, Department of Chemical Engineering, Delft University of Technology, 2629 HZ Delft, The Netherlands;
[b] AquaBattery B.V., Lijnbaan 3C, 2352CK Leiderdorp, The Netherlands;
[c] Materials for Energy Conversion and Storage, Department of Chemical Engineering, Delft University of Technology, 2629 HZ Delft, The Netherlands
*Email: d.a.vermaas@tudelft.nl; w.smith@tudelft.nl

8.1 Introduction

The conversion and storage of solar energy has the potential to provide a sustainable supply of clean energy to power the world of the future. The key to unlocking this energy infrastructure is to utilise abundant resources to power the transition, in terms of both feedstocks and active materials. Water and CO_2 are readily available across most of the globe, and can act as the storage mediums through various technologies, most notably electrochemical and photoeletrochemical conversion. If these applications can be accomplished with cheap, earth abundant, selective and stable materials, then there is an opportunity to develop cost effective alternatives to todays energy supply.

Energy and Environment Series No. 20
Advances in Photoelectrochemical Water Splitting: Theory, Experiment and Systems Analysis
Edited by S. David Tilley, Stephan Lany and Roel van de Krol

Published by the Royal Society of Chemistry, www.rsc.org

Among the many ways to convert and store solar energy, solar driven water splitting (either PV-powered electrolysers or direct photoelectrochemical cells) offers a direct technology that can feasibly contribute to this energy transition. Many designs have been proposed to develop solar water splitting systems, with a strong emphasis on catalyst optimisation (*i.e.* increasing the activity and concentration of active sites), and with an increasing focus on the effects of the catalyst support, the electrolyte, and overall cell design. However, a practical system has yet to be realised that can contribute in a substantial way to storing excess renewable electricity.

8.1.1 Challenges in (Solar Driven) Water Splitting

Two key challenges in the rational design of a (photo)electrochemical water splitting reactor are (i) the separation of products (hydrogen and oxygen) and (ii) the development of an overall system with a low operating potential. The overall potential required to drive an electrochemical cell is the sum of many components, shown in eqn (8.1):

$$V_{total} = E_{red} + E_{ox} + \eta_{red} + \eta_{ox} + iR \quad (8.1)$$

Where V_{total} is the total applied potential, E_{red} and E_{ox} are the equilibration potentials for the reduction and oxidation of water, η_{red} and η_{ox} are the overpotentials for the respective reduction and oxidation reactions, and iR is the resistance of the electrolyte, membrane and wires between electrodes. While E_{red} and E_{ox} are governed solely by thermodynamics, catalytic overpotentials (η_{red} and η_{ox}) are determined by the activity of the catalyst/support system. In general, iR drops across the solution are compensated for half-cell reaction measurements (*i.e.* in a three-electrode measurement), and not given as much consideration as the electrode materials themselves.

The first device/reactor challenge, gas separation, can be addressed directly by introducing a membrane between the electrodes. Membranes can offer pathways to allow ions to transfer between the two electrodes, but should be able to block the passage of gaseous products such as oxygen and hydrogen. For this, monopolar membranes such as Nafion have been readily used. However, the use of monopolar membranes limits the choice of electrolyte based on which ionic species is dominant in the corresponding electrochemical reactions (*i.e.* H^+ for acidic electrolytes require a cation exchange membrane, and OH^- for basic electrolytes require an anion exchange membrane).

8.1.2 Effect of pH

For (photo)electrochemical water splitting systems, this interplay of electrolyte and membranes for a given reaction are further complicated when assessing the overall system. Specifically, it has been demonstrated that the oxygen evolution reaction (OER) is able to be driven at a lower η_{ox} than in neutral or acidic environment.[1–4] In fact, for the OER, not only can the system evolve oxygen at a lower potential in a basic environment, but this also is achieved using earth abundant $NiFeO_x$ catalysts, where only IrO_x can be used to evolve oxygen in an acidic environment (with a larger overpotential), see Figure 8.1(a). This shows that the OER can be done more efficiently and with cheaper materials when performed in a basic electrolyte. Likewise, the hydrogen evolution reaction (HER) prefers an acidic environment (Figure 8.1(b)), and the sensitivity of precious metal catalysts also becomes more apparent.[5–7]

To find the lowest overpotential combination of OER and HER catalysts and electrolytes, the optimum pairing would be to perform OER in base and HER in acid. These conditions cannot be maintained by a monopolar membrane, and thus new solutions are needed to be able to maintain this high pH gradient, which has the potential to drive the overall system at a lower voltage than previous attempts. Recently, the introduction of bipolar membranes to the electrolysis field has shown that such a pH gradient can be maintained over long periods of time, without adding any significant overpotentials to the system, thus offering a solution to the aforementioned problems of both gas separation and electrode/electrolyte compatibility.[8,9]

8.1.3 Membranes in (Solar) Water Splitting

For both electrochemical and photoelectrochemical water splitting, the overall reaction will split water into hydrogen and oxygen. Like any chemical process, it is important to separate the product streams to prevent the need for further processing/separation downstream, as well as to avoid unwanted cross-over for safety reasons. Several attempts have been made for membrane-less systems,[10,11] but these often result in higher ohmic losses for large-scale operation because of the large electrode spacing required.[12] Therefore, introducing a membrane into a (photo)electrochemical cell can intrinsically make such a device safe by physically separating the product streams of hydrogen and oxygen while not compromising the yield at larger scale.

Traditionally, ion-exchange membranes with a single charge (either cation exchange membranes, such as a proton exchange membrane, or anion exchange membranes) are used in (photo)electrochemical systems.[13] Membranes with both cation and anion exchange layer, *i.e.*, bipolar membranes, have been recently introduced and have drawn significant attention for further development. In the next section, we will discuss both monopolar and bipolar membranes for the use in (photo)electrochemical water splitting.

8.2 Monopolar and Bipolar Ion-exchange Membranes

Among the different types of membranes, ion-exchange membranes are the most suitable to apply in (photo)electrochemical water splitting, as they allow charge transport and confine the produced (dissolved) gasses.

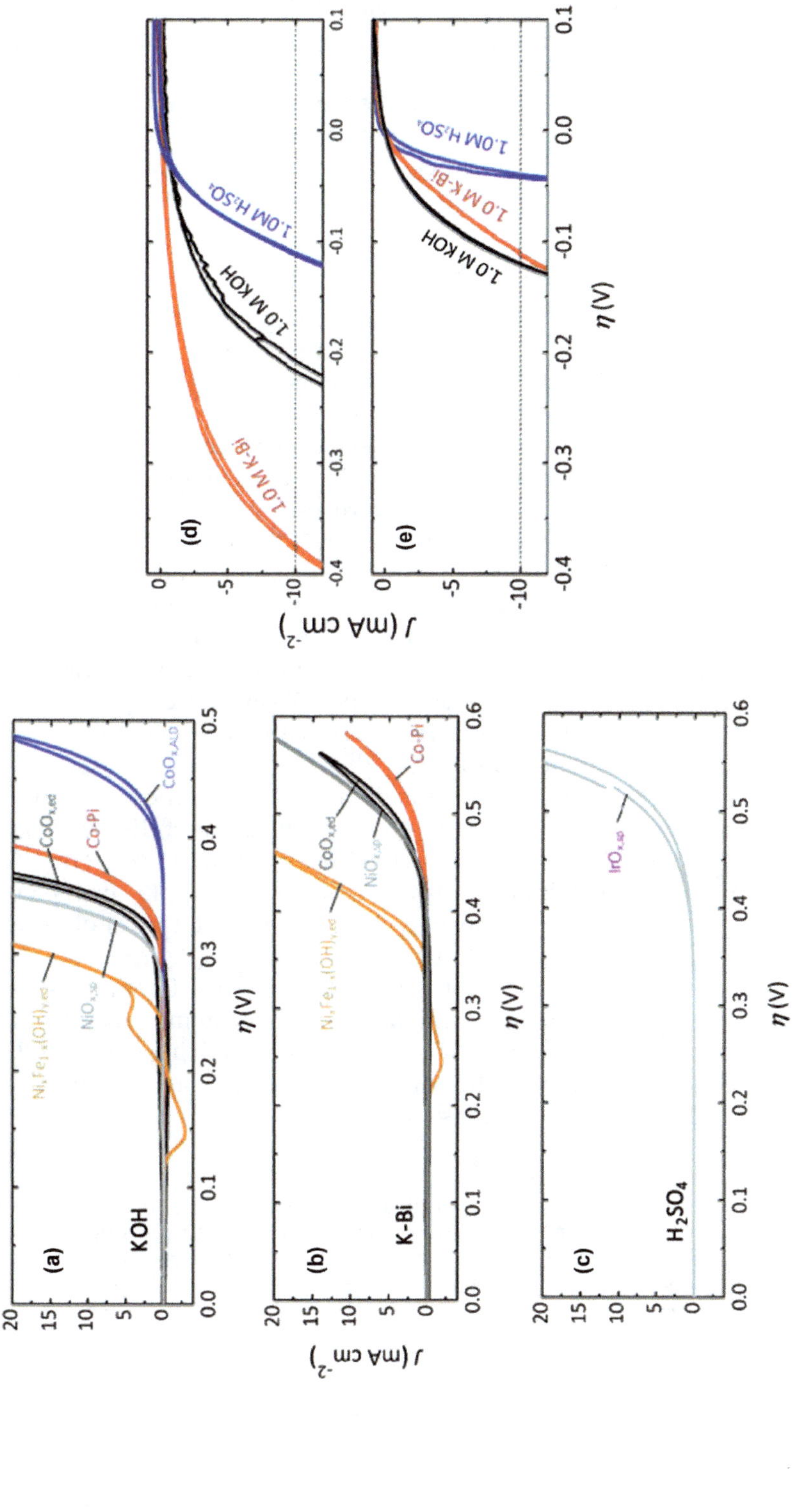

Figure 8.1 Current–voltage curves for various catalysts for the oxygen evolution reaction performing in KOH (a), K-Borate (b) and H_2SO_4 (c). Hydrogen evolution is tested in the same environments for Ni_xMo_y catalyst (d) and Pt catalyst (e). Reproduced from ref. 4 with permission from The Royal Society of Chemistry.

Ion-exchange membranes are dense membranes, which means that they do not contain pores for the transport of species.[14–16] Instead, charged groups in the membrane polymer provide selectivity for either cationic or anionic species to pass the membrane.

Monopolar membranes, which allow either cations or anions to pass, will be discussed first in this section. Subsequently, we will discuss the use of bipolar membranes for (photo)electrochemical water splitting.

8.2.1 Monopolar Membranes

Monopolar membranes are the most common group of ion exchange membranes, allowing only cationic species (cation exchange membrane, CEM) or anionic species (anion exchange membrane, AEM) to pass the membrane. The dense membrane structure ensures that not only gas bubbles, but also dissolved gasses are rejected. The confinement of dissolved gasses is increasingly important at low current densities, as the concentration of dissolved oxygen and hydrogen is in the same order of magnitude as that of gas bubbles for 10 mA cm^{-2} (Table 8.1).

Because the selectivity for either cations or anions depends on the membrane charge polarity, standard CEMs and AEMs allow all cations or all anions to pass. At the same time, H^+ or OH^- are involved in the electrode reactions. Hence, transport of H^+ or OH^- through the ion-exchange membrane is required to close the mass balance. To avoid transport of other (co-)ions, CEMs should be used for acidic cells, while AEMs should be used for cells operating in alkaline conditions.[17,18]

The imbalance in H^+/OH^- transport is demonstrated in the case of water splitting at near-neutral pH. At this pH, the concentration of dissolved salts (required to make the electrolyte conductive) far exceeds the concentration of H^+ and OH^-, which results in transport of salt ions (*e.g.*, K^+ or $H_2PO_4^-$) through the membrane. As a consequence, the pH of the anolyte will increase and the pH of the catholyte will decrease (Figure 8.2). This phenomenon has been observed for several membranes for operation at near-neutral pH,[9,19] and even at membrane-less systems operating at neutral pH.[11] Modestino *et al.* proposed to use a partial mixing of the anolyte and catholyte, which neutralises the pH polarisation, at least in small scale devices, with only very small gas cross-over.[20]

Table 8.1 Concentrations of hydrogen and oxygen as gas and dissolved species in water. Calculated for 20 °C and at atmospheric pressure, using a current density of 10 mA cm^{-2}, 30 s residence time of the gas bubbles, 1 cm electrode-membrane spacing, and at gas saturation condition.

Gas	Present as gas/mmol L^{-1}	Dissolved in water/mmol L^{-1}
Hydrogen	1.6	0.8
Oxygen	0.8	0.3

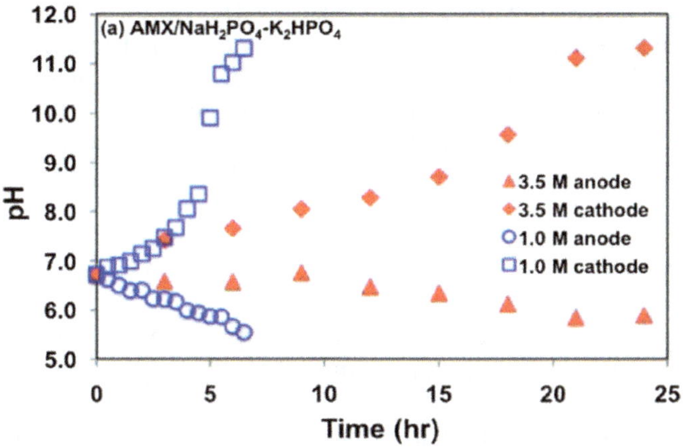

Figure 8.2 pH in anolyte and catholyte over time, when using an AEM as separator. Reproduced from ref. 19 with permission from The Royal Society of Chemistry.

8.2.1.1 Cation-exchange Membranes (CEMs)

8.2.1.1.1 Principle. Ion-exchange membranes were already reported in the late 19th century by W. Ostwald (1890) and later extensively investigated by F. G. Donnan.[21,22] However, the first industrial-scale application of ion-exchange membranes were developed in the 1950s, initially for water purification *via* electrodialysis (ED).[23] The main hurdle for the chemical industry was to be able to develop ion exchange materials in a thin film structure.

In CEMs, the membrane polymer contains fixed negatively charged groups, for example $-SO_3^-$ or $-COO^-$ groups.[24] The negatively charged groups are neutralised by cations and excludes anions (Donnan exclusion).[15,16] The cations are mobile, *i.e.* they can be exchanged with the surrounding electrolyte at either side of the membrane, while the negative charged groups are covalently bonded to the polymer. This creates a selective barrier for cations to pass the membrane.

The selectivity of the CEM is mainly determined by the charge density of the membrane, which is typically in the order of 1–10 mole per kg water absorbed in the membrane.[25–27] The selectivity is also dependent on the ion concentration in the electrolyte, as the charge density of the membrane should be large compared to the ion concentration in the electrolyte, and the (difference in) ion mobility.[16] The Teorell–Meyer–Sievers theory estimates the corresponding electrical potential difference over the membrane that separates two compartments, here denoted as compartment A and B. When neglecting mobility differences between anionic and cationic species, and neglecting activity coefficients, this potential difference φ is expressed as eqn (8.2):[25,28]

$$\varphi = \frac{\mathrm{RT}}{zF}\left(\ln\left(\frac{c_\mathrm{A}}{c_\mathrm{B}}\right)\frac{X+\sqrt{X^2+4{c_\mathrm{B}}^2}}{X+\sqrt{X^2+4{c_\mathrm{A}}^2}}\right) \tag{8.2}$$

in which R is the universal gas constant (8.314 J K^{-1} mol^{-1}), T is the absolute temperature (K), z is the ion valance (–), F is the Faraday constant (96 485 C mol^{-1}), c is the ion concentration in the electrolyte (in compartment A and B) (mol m^{-3}) and X is the fixed charge density of the membrane (mol m^{-3}). When assuming that $X \gg c_A$ and c_B, and using monovalent ions ($z=1$), this equation simplifies to the Nernst law (eqn (8.3)):

$$\varphi = \frac{RT}{F}\ln\left(\frac{c_A}{c_B}\right) \tag{8.3}$$

Hence, the equilibrium potential over a monopolar ion-exchange membrane equals zero when $c_A = c_B$. When the applied electrical potential is larger than the membrane potential, cations are transported through the CEM. In case of equal cation concentrations at either side (*i.e.*, equal chemical potentials, $c_A = c_B$), the driving force is simply the electrical potential, but in the case where different electrolytes are used at either side, the driving force is equal to the electrochemical potential, which is the sum of the electrical potential difference and the chemical potential difference over the membrane.

In addition to the driving force, the electrical resistance of the membrane determines the current density through the membrane. The electrical resistance of the membrane is usually measured as an area resistance in Ω cm^{-2}, because a larger membrane gives a lower resistance. This property depends on the membrane conductivity and membrane thickness.[29] A typical CEM has a resistance of 1–5 Ω cm^{-2},[24,30,31] which implies a voltage loss of 10–50 mV at 10 mA cm^{-2}.

8.2.1.1.2 Membrane Structure. CEMs can either be made by embedding ion exchange resin particles in a polymer matrix (heterogeneous membranes), or by functionalising the polymer itself with a charged group (homogeneous membranes). At present, homogeneous CEMs outperform the heterogeneous membranes in terms of selectivity and resistance, due to the thinner membrane structure and higher charge density.[30]

Even homogeneous membranes show voids at the microscopic level (Figure 8.3), which is believed to promote the membrane conductivity as ions can 'hop' from one charge group to another,[32] similar to the Grotthuss mechanism for proton conduction. For the same reason, the membrane resistance appears to be dependent on the ion concentration in the electrolyte.[33,34]

8.2.1.1.3 PEM. A special type of CEM is the proton-exchange membrane (PEM), sometimes referred to as polymer electrolyte membrane. PEMs are applied at an industrial scale in electrolysers and fuel-cells and are often made of fluoropolymers,[35] from which DuPont's Nafion® is the most well-known membrane. The selectivity of PEMs relies—similar as for CEMs—on fixed negative charges in the membrane polymer, and additionally features high transport numbers for H^+ and high chemical stability for

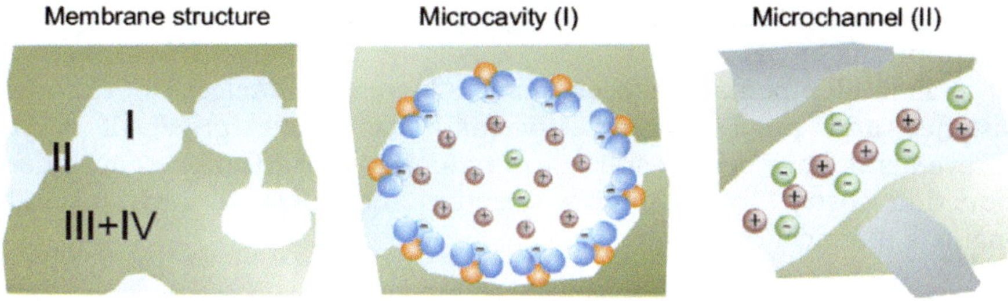

Figure 8.3 Membrane structure, composed of charged microcavities (I), poorly charged microchannels (II) and polymer material (III + IV), which comprises charged polymer groups and reinforcement.
Reprinted from *Journal of Membrane Science*, **467**, A. H. Galama, D. A. Vermaas, J. Veerman, M. Saakes, H. H. M. Rijnaarts, J. W. Post and K. Nijmeijer, Membrane resistance: The effect of salinity gradients over a cation exchange membrane, 279–291,[33] Copyright (2014), with permission from Elsevier.

strongly acidic conditions and often elevated temperatures.[35] The high transport number for H^+ provides a very low electrical resistance ($\ll 1\ \Omega\,cm^{-2}$) in strongly acidic conditions,[36] and similar electrical resistance as regular CEMs in near-neutral conditions.[37]

For practical water splitting, it needs to be considered that proton exchange membranes (such as Nafion) are good at transporting protons and do not or hardly allow the transport of OH^-. Therefore, the application of proton exchange membranes limits the overall device to operate in an overall acidic environment. This constraint limits the use of OEC's to be IrO_x, which still has a relatively high overpotential. Examples of using PEM for (solar driven) water splitting are given in Section 8.2.1.2.2.

8.2.1.2 Anion-exchange Membranes (AEMs)

The rapid development of earth abundant hydrogen and oxygen evolving catalysts, for example Ni-based oxides, and their requirement for alkaline environments, has increased the need to use AEMs for (photo)electrochemical water splitting. The principles and examples of these membranes are discussed below.

8.2.1.2.1 Principles. In analogy to a CEM, an AEM comprises fixed positive charges, which are compensated by mobile anions. The most common fixed positive charge groups are quaternary ammonium groups ($-NH_3^+$).[38] Compared to CEMs, the selectivity of AEMs is in general lower due to, among other reasons, a typically larger swelling degree of the membrane material, and hence a lower charge density.[30] Also the low chemical resistance to alkaline environments is a challenge for AEMs.[17] This incompatibility with alkaline environments applies to many CEMs as well, but are more inconvenient for AEMs because they are intended for use with large concentrations of OH^-.

Several developments have been made to increase the selectivity, chemical resistance and process complexity (and hence the price) of AEMs. Improvements have been made to tune the AEM specifications and reduce production costs,[39] but a highly selective and chemical resistant AEM remains a practical challenge.[38]

8.2.1.2.2 Examples of CEM, PEM and AEM. The large availability of selective and chemically stable CEMs/PEMs has given rise to many examples of (photo)electrochemical water splitting. First of all, obviously, commercial PEM electrolysers comprise a PEM as a key component.[40,41] Also many solar driven PEC and PV-EC cells include a CEM, if any membrane,[42] in which Nafion is most commonly used. In fact, the use of a CEM or PEM (Nafion) is the reference case for (photo)electrochemical water splitting.

However, very little research has been performed using AEMs in photoelectrochemical systems. Hernandez-Pagan *et al.* (2012)[19] investigated both anion- and cation-exchange membranes in PEC systems for near neutral pH (4.7–9.3). For both membrane types polarisation of the pH occurs during operation. Dependent on the electrolyte and membrane type, neutralisation can partly occur during down-time, reflecting night-time for a PEC-system. For example, when using acetate–acetic acid and an AEM, the undesired polarisation due to acetate cross-over can be partly neutralised by back-diffusion of the neutral acetic acid when no current is applied. Similar mechanisms can occur for CEMs, for example using an imidazole-based electrolyte.[19] Nevertheless, in all cases a pH gradient of approximately four units, corresponding to an additional 240 mV, remains after 24 h. In addition, Vargas-Barbosa *et al.* (2014) observed the polarisation for AEMs[9] for near-neutral pH.

For strongly alkaline pH, as used in industrial alkaline electrolysers, usually porous materials are used instead of ion exchange membranes, to provide the high chemical stability.[17,38,43]

8.2.1.3 *Special Membrane Types*

Within the monopolar membranes, some less well known membrane types can be used in (photo)electrochemical water splitting. Xu[24] used amphoteric membranes for ion exchange, which contain weak acid and weak basic groups in the membrane. Because the charge of these groups is dependent on the protonation state, amphoteric membranes can either exchange cations or anions dependent on the pH. Although the amphoteric membranes are not industrially mature yet, and envisioned mainly for biomedical applications,[24] these membranes may provide a negative feedback to pH changes in electrolysis systems. When, for whatever reason, the pH in the electrolysis system becomes too low (or too high), the membrane charge becomes more positive (or more negative), which makes the membrane more selective for anions (or cations).

In addition, monovalent ion exchange membranes may find an application in electrolysis systems. As the name suggests, monovalent ion-exchange membranes are selectively allowing transport of monovalent ions.[16,44] This is established by either an additional layer of with the same charge as the ions (repelling multivalent ions stronger than monovalent)[45] or increased cross-linking to create steric hindrance for the usually larger multivalent ions.[46] In electrolysis, this will better confine salt co-ions when using multivalent salts, compared to monovalent protons and hydroxide ions.

Third, nanoporous membranes have been suggested to be used in (photo-)electrochemical water splitting as well, for example made from a glass frit or micro- or nanoporous membranes,[47,48] and are commercially applied for alkaline electrolysers. Nanoporous membranes are not specifically selective for ions, and therefore can show larger gas cross-over (as discussed at Table 8.1). Nevertheless, the wide availability and chemical resistance makes them easy to apply, also in electrolysis.

8.2.2 Bipolar Membranes (BPM)

8.2.2.1 Principle

Bipolar membranes comprise both a cation exchange layer (CEL) and an anion exchange layer (AEL). As the CEL blocks the transport of cations from the solution, and the AEL blocks anions, a bipolar membrane should not be conductive to ions in the electrolyte. Hence, when an electrical current is applied in reverse bias (see Figure 8.4), a large electrical membrane potential can be established, which facilitates the dissociation of water into H^+ and OH^- at the interfacial layer between the AEL and CEL. This allows the ability to produce acid at one side of the bipolar membrane while also producing base at the other side.

In addition to the operation in reverse bias, BPMs may be operated in forward bias mode. In that case, protons and hydroxides will be associated into water. This mode has been demonstrated for electrolysis[9] and fuel cells,[49] but is less practical than the reverse bias for electrolysis purpose. A forward bias implies either an alkaline catholyte and acidic anolyte, which is not compatible with Earth abundant catalyst materials, or (in combination

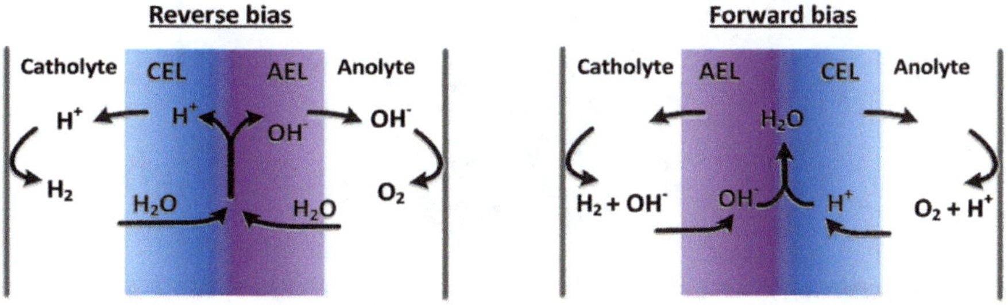

Figure 8.4 Bipolar membrane in reverse bias (left) and forward bias (right). A reverse bias is the default operation for electrolysis purpose.

with an acidic catholyte and alkaline anolyte) a situation where the membrane layers will transport salt ions, leading to non-steady electrolyte pH. Another issue with the forward bias is that the produced water molecules at the interface layer may delaminate the CEL and AEL, leading to water blisters in the membrane. Hence, we limit ourselves to discussing the reverse bias of BPMs for electrolysis purpose.

The working mechanism of BPM-assisted water dissociation requires not only the blockage of co-ions, but also a small (nanoscopic) distance between the anion and cation exchange layers. Water dissociation could technically also be established by oppositely charged monopolar membranes that are kept a macroscopically large distance away from each other. When the salt concentration near the membrane interface approaches zero, for example due to concentration polarisation (*i.e.* depletion of the diffusive boundary layer, see Figure 8.5(a)), no more ions can be transported, which is observed as the limiting current density.[50,51] When even more potential difference is applied over the monopolar membranes, water dissociation is observed (Figure 8.5(b)), which causes a further increase in current density, which is the so-called overlimiting current.[52–56] However, when water dissociation performed with monopolar membranes, a large overpotential is required

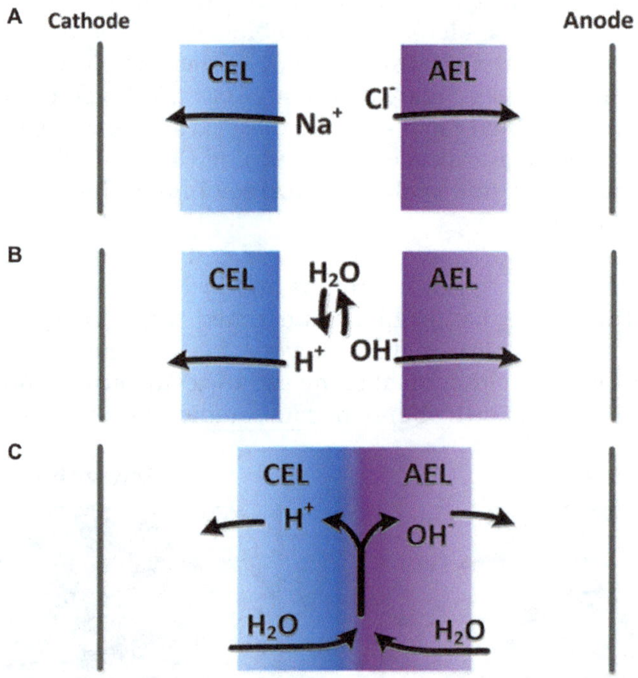

Figure 8.5 Principle of water dissociation when desalinating a thin compartment between a cation-exchange membrane (CEM) and anion-exchange membrane (AEM) (step A and B). When the distance between the membrane layers is zero (step C), this is a bipolar membrane (BPM).

because the zero salt concentration in the diffusive boundary layer causes a large ohmic loss. A bipolar membrane directly solves this problem by bringing the distance between the CEL and AEL to zero, which effectively brings the ohmic loss of this water layer to zero (Figure 8.5(c)). Hence, water dissociation can be established at a high energy efficiency.

An interesting note is that not all materials allow water dissociation at low overpotentials; a catalyst for water dissociation within the interfacial layer is often required. Known catalysts for water dissociation in the interfacial layer are polyacrylic acid/polyvinylpyridine, metal alkoxides, aluminosilicates and Fe(III) or Cr(III) hydroxides.[8,58,59] The behaviour of catalytic water dissociation is also observed for overlimiting currents at monopolar membranes: AEMs facilitate water dissociation at much lower potentials than CEMs, possibly due to the catalytic function of ammonium groups in AEMs.[52,53,56,60]

8.2.2.2 *Chemical, Electrochemical, and Electric Potential*

In contrast to a monopolar membrane, a bipolar membrane can maintain a constant pH difference over the membrane when used in water electrolysis, as the H^+ and OH^- produced by the BPM are consumed in equal rates at the catalytically active electrodes. The pH difference over the BPM is of interest for water electrolysis, as the hydrogen evolution and oxygen evolution potentials are dependent on the pH of the electrolyte. Hence, the energy analysis for BPM water splitting should include the *electrical* potential difference over the membrane as well as the *chemical* potential difference over the membrane. The sum of the electrical and chemical potential is the electrochemical potential, $\bar{\mu}$ (in J mol^{-1}) (eqn (8.4)):

$$\bar{\mu} = \mu + zF\varphi \tag{8.4}$$

in which μ is the chemical potential (J mol^{-1}), z is the ion valence (–), F is the Faraday constant (96 485 C mol^{-1}) and φ is the electric potential (V).

The chemical potential μ (in J mol^{-1}) as a function of the concentration c for a certain species (*e.g.* H^+) is defined as eqn (8.5):

$$\mu = \mu_0 + \mathrm{RT}\ \ln(\gamma c) \tag{8.5}$$

in which is the standard chemical potential for a species at standard conditions (1 M), R is the universal gas constant (8.314 J mol^{-1} K^{-1}), T is the temperature (K), γ is the activity coefficient and c is the concentration.

When solutions at both sides of a membrane are in equilibrium, the difference in electrochemical potential is zero. Hence, for a perfectly selective membrane, the electrical potential difference over the membrane, in equilibrium state, is dependent on the ion concentration at either side of the membrane. When substituting eqn (8.5) in eqn (8.4), and taking the difference between side a and side b, the electrical potential difference is given by:

$$\varphi = \frac{\mathrm{RT}}{zF} \ln\left(\frac{\gamma_a c_a}{\gamma_b c_b}\right) \tag{8.6}$$

which is the Nernst law. Eqn (8.6) is similar to eqn (8.3), but now includes the activity coefficients and ion valence. In this equation, the subscripts a or b indicate the side a and b of the membrane, respectively. The concentrations c_a and c_b should be considering the same ion species at either side. For a bipolar membrane, which should allow no other ions than protons and hydroxides, we can use the proton concentration for c_a and c_b. In that case, this equation can be rewritten as eqn (8.7):

$$\varphi = \frac{\mathrm{RT}\ln(10)}{zF}(\mathrm{pH}_a - \mathrm{pH}_b) = 0.0591 \cdot \Delta\mathrm{pH} \tag{8.7}$$

Hence, the ideal thermodynamic potential over a (perfectly selective) BPM is dependent on the pH gradient over the membrane.

When using an ideal (perfectly selective, infinitely conductive) bipolar membrane, the electrochemical potential over the BPM is zero, even though the electrical potential is non-zero. This may seem counter-intuitive, as a voltage is measured over the BPM when a pH gradient is apparent, for example when using an acidic catholyte and alkaline anolyte. However, the chemical potential difference, arising from the difference in pH and therefore affecting the redox potentials, has exactly the same magnitude and opposite sign. In fact, the BPM creates a chemical potential difference at the expense of an electrical potential difference. While the electric potential over the BPM requires energy, the different pH at either side lowers the energy for the chemical reaction at exactly the same amount of energy. This becomes also clear from a Pourbaix diagram for the hydrogen evolution and oxygen evolution reactions (Figure 8.6).

Hence, when considering the electrolysis cell as a whole, the theoretical total cell voltage required to split water into H_2 and O_2 is still the same 1.23 V, independent of the electrolyte pH, as illustrated in Figure 8.6.

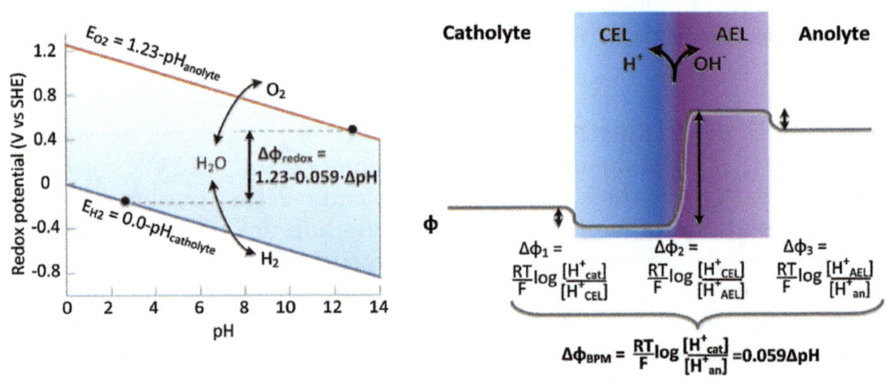

Figure 8.6 Pourbaix diagram (left) and potential differences at BPM (right). According to the thermodynamic equilibria, both the redox potential and the BPM potential are a function of the pH difference over the membrane. The sum $\varphi_{\mathrm{redox}} + \varphi_{\mathrm{BPM}} = 1.23$ V, independent of ΔpH.

In reality, the electric potential over the BPM may differ from the Nernst potential. At low current densities, measured membrane voltages are close to, sometimes even lower than, the Nernst potential that is expected from the thermodynamic analysis. Voltages even lower than the Nernst potential are possible because the membrane is not perfectly selective, and demonstrates some ion cross-over. Although the ion cross-over decreases the energy consumption, it also causes a slow change in pH difference, and thereby is not sustainable. However, at higher current densities, the membrane potentials approaches or exceeds 0.83 V independent of the pH gradient,[8,61] which is the potential difference for an extreme pH-gradient, *i.e.* pH 0 to pH 14. We hypothesised that this behaviour is related to a local pH adjacent to the membrane, in a diffusive boundary layer.[61] Sun *et al.* (2016)[62] have observed similar behaviour in a BPM study using pH 0 *vs.* pH 9.3. Modelling indicated that the local pH at the AEL-electrolyte interface was approaching pH 13 for 20 $mA\,cm^{-2}$, while the pH in the bulk was only 9.3. This high pH near the membrane surface increases the membrane voltage, as described by eqn (8.7).

In addition, at low current density, relatively large amounts of salt ions cross-over through the BPM. This non-ideal behaviour lowers the membrane voltage at low current density (as the salt permeation does not require a driving voltage for water dissociation). For higher current density, where the salt ion cross-over is relatively small,[62] the membrane potential has less benefit from this effect, which effectively increases the membrane potential for higher current densities. At the extreme, for poorly selective BPMs, there is no water dissociation at all for low current densities, and a threshold in current density (*i.e.* the limiting current density) needs to be exceeded before water dissociation starts.[50,57,63,64] However, modern commercial BPMs show that even at very low current densities (<1 $mA\,cm^{-2}$) 70–90% of the current is carried by water dissociation.[61,62] Hence, the salt ion cross-over does play a role in the low membrane voltage at low current density, but the other way around, a low membrane voltage does not infer exclusive salt ion cross-over.

In addition, for very high current densities (>100 $mA\,cm^{-2}$), ohmic resistance of the BPM and water diffusion into the BPM interface layer plays a significant role and limits the current density.[57] Bejerano *et al.* (2014)[64] and recently Shen *et al.* (2017)[65] demonstrated that membrane modification can extend the maximum current density (with selective water dissociation) above 1000 $mA\,cm^{-2}$. The latter research uses a 3D structure at the interface layer to increase the CEL–AEL contact area.

8.2.2.3 Commercial Development

Despite the unique ability to dissociate water close to the thermodynamic potential difference, research into bipolar membrane materials is limited. Bipolar membranes (BPMs) were first introduced in the 1950s,[66] although most (commercial) development on BPMs, for the production of acid and base, was performed in the 1990s and 2000s.[57,63,67–69] Only recently have bipolar membranes been used for other applications, such as (photo)electrochemical water splitting,[8,9,61,70] CO_2 reduction[71–73] and (microbial) fuel cells.[74,75] Because all electrochemical conversion applications of BPMs are in their infancy, only one commercial BPM is available on the market (Fumasep FBM), which has built on the membrane technology developed in the former company Aqualytics, which was taken over by FumaTech in 1999. Other commercial bipolar membranes have been developed, and new materials are being investigated,[8,58,59,69,76,77] but they are currently not available for sale. Hence, the FumaTech bipolar membranes are applied in the majority of the solar fuel research papers. An exception is the recent demonstration of a bipolar membrane with transparent properties, developed by combining two existing monopolar membranes,[78] which allow photocatalytic device designs with front-illumination.

Bipolar membranes are manufactured from combining two membrane layers with opposite charge, which can be done by *e.g.* casting, gluing or hot-pressing two layers to each other[58] (*e.g.* Figure 8.7). Although many ionomer solutions are available for that process, two practical issues rise when using them in a bipolar membrane. First, the membrane layers should be resistant against extremely acidic/alkaline conditions, where most standard grade membranes degrade especially in strong alkaline conditions.[79] Second, both membrane layers should have a similar swelling degree in order to prevent curling or cracking of the membrane when wetted. In addition to these criteria, an efficient operation requires also a catalyst for water dissociation in the interface layer to minimise overpotentials,[8,58] and a high water diffusivity to allow high current densities without the diffusion of water towards the interface layer being limiting.

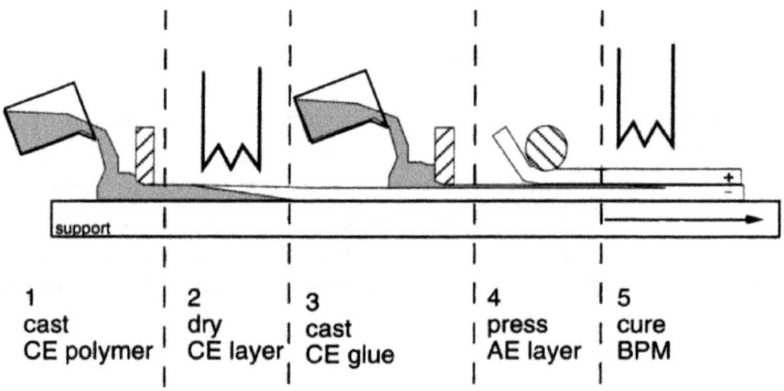

Figure 8.7 Example of bipolar membrane manufacturing, in this case with casting a first layer (step 1) and gluing a second layer on top (steps 3 and 4). Reprinted from *Journal of Membrane Science*, **182**, F. G Wilhelm, I Pünt, N. F. A van der Vegt, M. Wessling and H. Strathmann, Optimisation strategies for the preparation of bipolar membranes with reduced salt ion leakage in acid–base electrodialysis, 13–28,[58] Copyright (2001), with permission from Elsevier.

8.3 Membrane Performance

8.3.1 *I–V* Curves for Monopolar and Bipolar Membranes

Figure 8.8 shows current–voltage curves for BPMs in (1 M) acid–base media and near-neutral media. In case of (1 M) acid–base solutions, the 14 units pH gradient should give a BPM voltage of 0.83V, which is the case for small current densities (Figure 8.8(b)). For larger current densities, the membrane voltage increases, due to ohmic losses. The area resistance of BPMs are specified as approximately $3\ \Omega\,\mathrm{cm}^{-2}$,[80] but can be larger when including the diffusive boundary layers (approximately $8\ \Omega\,\mathrm{cm}^{-2}$ in Figure 8.8(a)[81]). As a reference, the resistance of $3\ \Omega\,\mathrm{cm}^{-2}$ for BPMs in acid–base solutions is equivalent to the resistance of an electrolyte compartment (*e.g.*, 1 M KOH, which has a conductivity of ∼200 $\mathrm{mS\,cm}^{-1}$) with approximately 6 mm thickness.

For smaller pH gradients (pH 0 *vs.* 7 in Figure 8.8(b)), the membrane voltage at low current density is significantly lower than 0.83 V, and is close to the thermodynamic voltage (which is 0.41 V in this case).[61] Also, at low concentrations (10 mM in Figure 8.8(a)), the membrane resistance is measured to be an order of magnitude larger, which may be associated with the lower conductivity of the electrolyte. Also, it has been recognised that the resistance of the membrane material increases for low salt concentrations in the water compartments.[33,34]

Overall, it is recognised that the practical use of commercial BPMs for non-extreme pH gradients (such as pH 0 *vs.* 7) is challenging, as the thermodynamic membrane voltage is only achieved at low current densities.

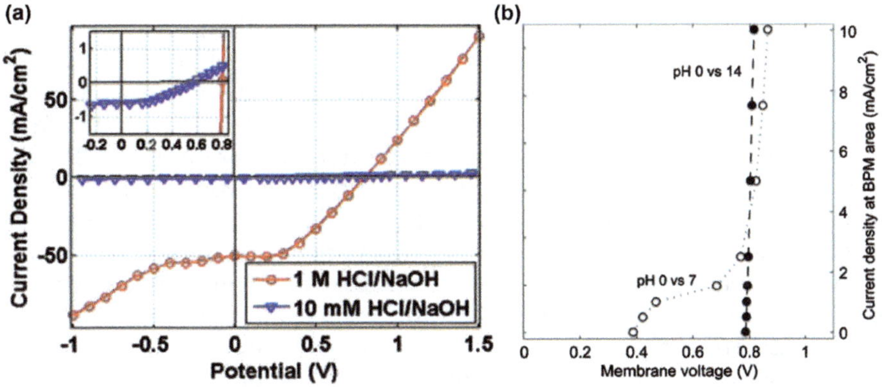

Figure 8.8 Examples of *I–V* curves for BPM. Left: *I–V* curve for 1 M HCl/NaOH and 0.01 M HCl/NaOH. Positive current densities represent a reverse bias. Reprinted with permission from *J. Electrochem. Soc.*, 2016, **163**, H3132–H3134,[81] http://dx.doi.org/10.1149/2.0201604jes. © The Author(s) 2015. Published under the terms of the CC BY 4.0 licence, https://creativecommons.org/licenses/by/4.0/. Right: *I–V* curve for 1 M H_2SO_4/KOH (pH 0 *vs.* 14) and 1 M H_2SO_4 *vs.* 1 M KPi (pH 0 *vs.* 7). Reproduced from ref. 61 with permission from The Royal Society of Chemistry.

For that reason, the highest energy efficiencies has been obtained by using BPMs for more extreme pH gradients.

8.3.2 Membrane Pricing

An underexposed facet is the ion exchange membranes costs. Monopolar (homogeneous) ion-exchange membranes cost typically 50 $€ m^{-2}$, while special grade membranes such as Nafion and BPMs are >100 $€ m^{-2}$. For solar fuel devices where the current density is low (typically 10 mA cm^{-2}), the membrane costs are significant (>10% of total costs).[18] One of the reasons for the high prices of ion-exchange membranes (much higher than for osmosis membranes or gas membranes) is that electrochemical processes are smaller than alternative routes (*e.g.*, hydrogen production *via* steam reforming) at the moment.

The total price per kg of hydrogen is very sensitive to membrane costs—more sensitive than to the cathode and anode costs.[18] Hence, reducing the membrane price is an important aspect for commercial application. Based on cost reduction when scaling up membrane production—similar to reverse osmosis membrane upscaling, which made prices drop by an order of magnitude over the past two decades[82]—an order of magnitude lower membrane price is possible.

8.4 Demonstration of BPM's in Water Splitting Systems

8.4.1 Single Electrolyte

McDonald *et al.*[8] and Vargas-Barbosa *et al.*[9] were simultaneously the first to show the use of a BPM in an electrochemical cell designed for overall water splitting. McDonald *et al.*[8] used a cell with 100 mL total volume (50 mL in each the anode and cathode compartments) and Pt electrodes were used for both hydrogen and oxygen evolution. The authors first examined a BPM in a system with 1 M $NaClO_4$ in both anodic and cathodic compartments. After Na^+ and ClO_4^- migrate from the BPM, the potential across the membrane is increased (>0.8 V), and electroneutrality across the membrane is maintained by the dissociation of water, where the OH^- ions migrate towards the positive electrode and H^+ migrate towards the negative electrode.

Vargas-Barbosa *et al.*[9] demonstrated the performance of an electrochemical system operating with a BPM and 3.6 M phosphate buffer (1 : 1 $H_2PO_4^-$ to HPO_4^{2-}) under both forward and reverse bias. This process of applying the potential across the system both ways shows how respective anions/cations can penetrate the individual AEL/CEL layers and *vice versa*.

Under forward bias conditions, the authors found a significant change in electrolyte pH over time, due to the non-closed mass balance (no water dissociation is present). This also drives salt ions into the BPM, causing a reduction in the buffer capacity of the catholyte, which in turn causes a gradual increase in the catholyte pH and decrease of the anolyte pH. Also running with a monopolar membrane gives a change in pH over time, causing a membrane potential.

These first studies[8,9] showed that BPMs could be used for stable operation (*i.e.*, creating minimal pH changes over time), but that the use of two neutral electrolytes brings in a significant voltage loss over the BPM.

8.4.2 Extreme pH Gradient Across BPM's

The use of an extreme pH gradient (*i.e.*, pH 0 *vs.* 14) brings in the advantage of two different environments that favor the HER and OER, and at the same time the BPM voltage is closer to the thermodynamic membrane voltage. To show the contribution of the anodic and cathodic half-cells in the overall water splitting system, McDonald *et al.*[8] measured the potential drop from the HER and OER simultaneously, and measured the membrane potential. Because the chemical potential difference of 14 pH-units is equal to 0.83 V and the membrane voltage is (at least for low current density) close to 0.8 V, the system operates close to the thermodynamic limit. The authors showed that when the pH gradient is largest, the total cell voltage with a BPM is similar to that when using Nafion®, while the BPM-case has the additional advantage that better (earth abundant) catalysts are available when using a pH gradient.

Here, the authors note that the primary challenges for future BPM optimisation for electrochemistry lie mainly in the ionic permselectivity of the membrane and ionic current resistance. This may be achieved through finding an AEL/CEL pair that maximise water dissociation in the interfacial layer and are preferential in anion/cation permselectivity. Furthermore, for applications in potential solar fuel devices, the optical properties of BPM's must also be taken into consideration.

Shortly after, Luo and Vermaas *et al.* leveraged this work to produce a highly efficient solar to hydrogen conversion device that employed a tandem perovskite solar cell with NiFe electrocatalysts in pH 0 and 14, to generate a STH of 12.7%. Using NaOH as the anolyte and H_2SO_4 as the catholyte, the authors showed electrochemical performance of this system in operation with a BPM, and also using photons to drive the reaction *via* a PV-coupled electrolysis, aided by the use of two perovskite solar cells (Figure 8.9). This achievement was remarkable not only for its high efficiency, but that it was also accomplished using only earth abundant materials (for both the solar cell and catalysts). The potential over the BPM was measured while using a tri-potentiostat where potentials are applied over the cathode compartment and anode compartment separately, and the BPM potential was measured by sensing the potential drop between the two compartments *via* two reference electrodes in each electrolyte compartment. The potential over the system was maintained up to 90 h of operation at a cell potential around 1.6 V (Figure 8.9(c)). Linear sweep voltammetry was applied before and after long term electrolysis, and show nearly identical *I*–*V* curves.

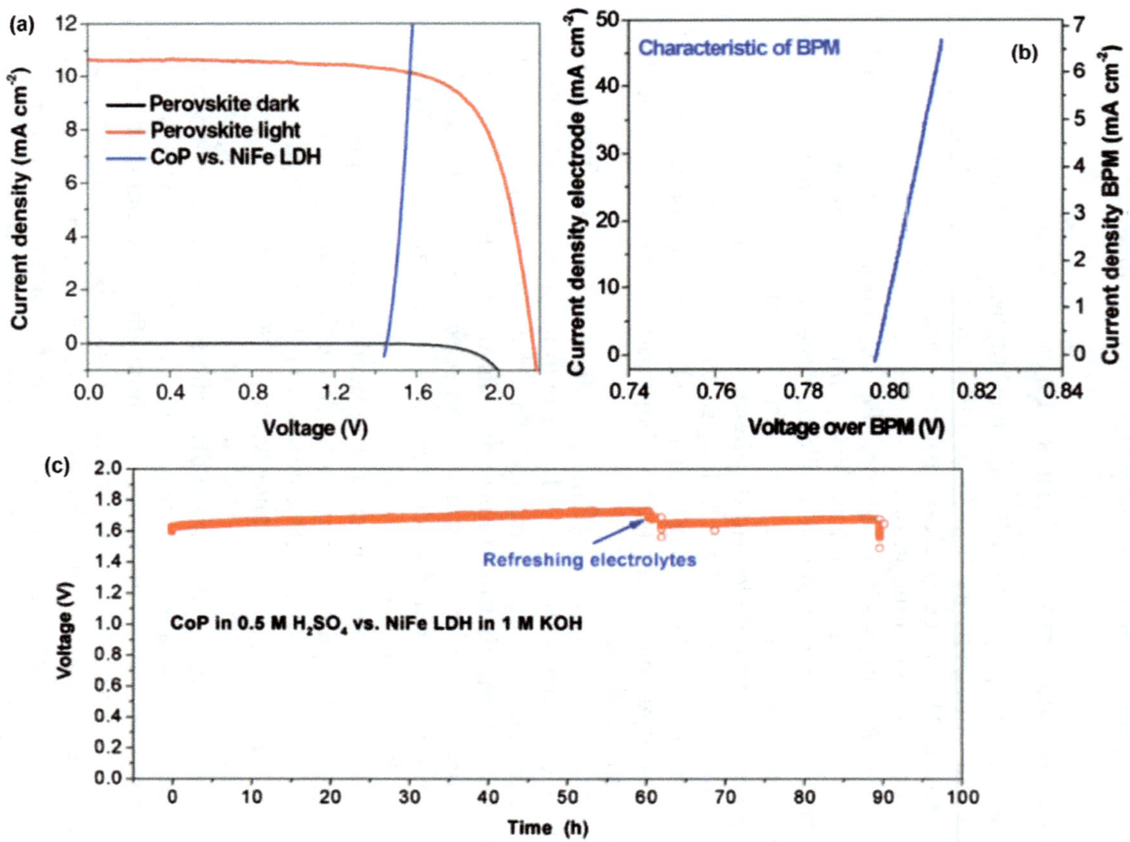

Figure 8.9 *I–V* curves for Perovskite solar cell and CoP–BPM–NiFeO$_x$ water splitting cell (a) and BPM (b). The voltage over the CoP–BPM–NiFeO$_x$ (at 10 mA cm^{-2}) is shown in panel (c).
Reproduced from Luo *et al.*[70] (2016) with permission from John Wiley & Sons, © 2016 WILEY-VCH Verlag GmbH & Co. KGaA, Weinheim.

8.4.3 Intermediate pH Gradient

A similar highly-efficient achievement of BPM water splitting was made using a photoelectrode, demonstrated by Sun *et al.* They showed a 10% solar-to-hydrogen efficiency using a GaAs/InGaP/TiO_2/Ni photoanode immersed in a 0.5 M KBi (pH 9.3) and a Pt cathode in 1 M H_2SO_4 (pH 0). The authors were able to show an unassisted solar to hydrogen conversion efficiency over 10% for 100 h operation. In this system, the role of the BPM was to maintain a strongly acidic catholyte and mildly alkaline anolyte. They also investigated the salt cross-over.

The concept of using a system with a BPM and an photoelectrode embedded in the same cell was shown for the first time a year earlier. Vermaas *et al.* used a $BiVO_4$ photoanode in pH 7 (KPi) against a pH 0 (H_2SO_4) compartment with a Pt cathode (Figure 8.10). The authors measured the potential across the anode compartment, the cathode compartment and the BPM simultaneously. Although the system offered a benefit compared to other $BiVO_4$ systems (freedom of choice for cathode), the photoanode still requires a bias to drive the HER and OER.

8.4.4 Co-ion Transport (at Non-extreme pH)

Although a perfectly selective BPM should dissociate water and allow no cations or anions to pass both layers of the BPM, practical membranes do leak some salt ions. This is referred to as co-ion transport, where the name 'co-' refers to the same charge as the membrane charge. Because a BPM has both a layer with positive and a layer with negative charges, all salt ions in the electrolyte that cross-over are referred to as co-ion transport. This ion

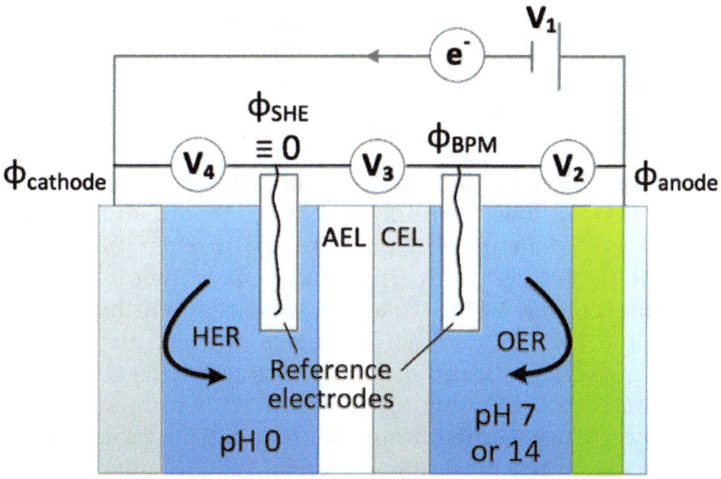

Figure 8.10 Photo-assisted water splitting using a BPM-facilitated pH gradient. Reproduced from ref. 61 with permission from The Royal Society of Chemistry.

transport satisfies the charge transport, but does not satisfy a steady state mass balance, as only protons and hydroxides react at the electrodes. Over time, the pH will change due to this ion cross-over.

8.4.4.1 Early Work on BPM for Acid/Base Production in BPMED

During the early development of bipolar membranes in the 90s, extensive research has been conducted on the use of bipolar membranes in neutral pH, in particular NaCl solutions (*i.e.*, pH 7 *vs.* 7), for the production of acid and base in bipolar membrane electrodialysis (BPMED).[50,57,67,79,83,84] The main interest was to find the limiting current density, *i.e.*, the current density at which water dissociation starts. It was postulated that at very low current density (<1 mA cm^{-2}), the main charge transport mechanism (for the membranes being investigated at that time) was *via* co-ions that permeate through the non-ideal BPM. At higher current densities, where the membrane voltages exceeds $\sim$0.6–0.9 V, water dissociation was thought to be responsible for the main charge transport mechanism.[50,84]

8.4.4.2 Direct Measurements of Co-ion Transport

Although several models have been developed for the contribution of co-ion transport and several experimental *I–V* curves are obtained, experimental results of co-ion cross-over are sparse. Moussaoui *et al.* (1994)[83] did demonstrate experimental results, indicating that co-ion cross-over is enhanced at low current density and high salt concentrations in the electrolyte. This can be understood from the Donnan exclusion principle; the amount of ion cross-over is limited due to the charge repulsion of ions in either the CEL or AEL.[28] This repulsion is determined by the ratio between the fixed charge density in the membrane material and the ion concentration. The fixed charged density is typically in the order of 1–10 M,[26,27] which means the BPM is more selective for water dissociation at low salt concentrations.

The work of Moussaoui *et al.*[83] shows clearly the dependency of co-ion transport as a function of the charge density, ion concentration, membrane thickness, current density and diffusion coefficients of ions in the membrane. However, the low membrane charge for the AEL (0.35 meq g^{-1}, compared to 1–10 M in present membranes) in their home-made BPMs make the absolute values from their home-made membranes not representative for present-day BPMs. In a way, this argument holds for all BPMs from the 90s and early 2000s: the *I–V* curves of most historic membranes (Figure 8.11) show significant different characteristics than the current commercial BPMs, possibly due to an improved membrane selectivity and interface catalyst. The AP6 membrane (from Aqualytics) comes closest to the current *I–V* curves for the present commercial BPM, obtained by Reiter *et al.* (Figure 8.11).[81]

Here, Reiter *et al.* have shown that, for non-extreme pH gradients, the most abundant ionic species in the electrolyte determines the *I–V* curve,

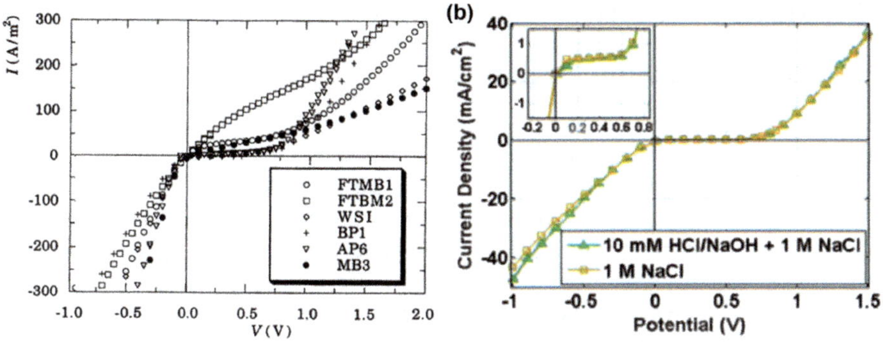

Figure 8.11 Left: *I–V* curves for commercial (not in production anymore) bipolar membranes in 0.5 M KCl solution. Reprinted from *Polymer*, **41**, A. Alcaraz, P. Ramırez, S. Mafé, H. Holdik and B. Bauer, Ion selectivity and water dissociation in polymer bipolar membranes studied by membrane potential and current–voltage measurements, 6627–6634,[67] Copyright (2000), with permission from Elsevier. Right: *I–V* curve for FumaTech BPM in NaCl. Reprinted from *J. Electrochem. Soc.*, 2016, **163**, H3132–H3134,[81] http://dx.doi.org/10.1149/2.0201604jes. © The Author(s) 2015. Published under the terms of the CC BY 4.0 licence, https://creativecommons.org/licenses/by/4.0/.

rather than the pH gradient itself.[81] They show that the *I–V* curve for a 1 M NaCl solution (*i.e.*, pH 7 *vs.* 7) is very similar to that of (1 M NaCl + 10 mM HCl)/(1 M NaCl + 10 mM NaOH) (*i.e.*, pH 2 *vs.* 12). They propose that at very low current densities (<0.5 mA cm^{-2}), co-ion transport carries charge, and hence explains the very low membrane potential in this case.

Direct measurements of co-ion transport have been presented in two papers. First, Vermaas *et al.* measured the ion cross-over from the molar fraction of potassium, sulphates, and phosphates in both the catholyte and anolyte compartments, as well as the pH from measurements carried out over four days at low current density (1 mA cm^{-2}). In these experiments the pH of the anolyte decreased from 7.01 to 6.15, while the pH of the catholyte increased from 0.08 to 0.14. The most significant ionic exchange was found for K^+, which increased from <0.02 to 0.29 M in the catholyte, while it decreased from 1.48 to 1.30 M in the anolyte after four days operation. This corresponds to approximately 10–15% of the transport charge, which means 85–90% of the charge is carried by water dissociation. Sun *et al.* found similar results, showing that 73–90% of the charge is carried by water dissociation at even lower current density (0.5 mA cm^{-2}), and much higher water dissociation contributions at higher current density.

These direct co-ion transport measurements show that the modern commercial BPMs do demonstrate water dissociation (in addition to some co-ion cross-over) at very low current density, in contrast to the prediction that all current is carried below a limiting current density. This finding is of major importance to allow separation of any combination of catholyte and anolyte with a BPM, without being limited to extreme pH gradients.

8.5 BPM's in Other Electrochemical Systems

8.5.1 Fuel Cells and Batteries

Before being used in electrolysis cells, the use of BPMs have been demonstrated in fuel cells,[75] microbial fuel cells[74] and batteries,[85] and is the subject of a European project named BAoBaB (Blue Acid/Base Battery),[86] where a BPM is reversibly used for water dissociation (during charging) and water recombination (thereby generating electricity). When using BPMs in fuel cells and batteries, the formation of water from H^+ and OH^- occurs, also referred to as forward bias. In contrast to water dissociation, this water production now consumes acid and base, and the membrane voltage is now contributing to the obtained cell voltage. Recombination of water in a BPM provides a new challenge for the membrane structure, as the CEL and AEL can delaminate when water is formed at the interlayer. As a result, blisters are sometimes observed in the BPM.[75]

8.5.2 CO_2 Reduction

The increasing interest in using bipolar membranes to benefit photodriven and electrochemical water splitting systems has also lead to the use of BPM's also in electrochemical CO_2 reduction. Many similar approaches have been taken (photoelectrochemical, photo-driven electrochemical, and electrochemical), and combine neutral electrolyte CO_2 electroreduction separated by a BPM to an alkaline water oxidation half-cell.

The first demonstration of CO_2 reduction with a BPM cell was by Zhou *et al.*[73] The authors use a GaAs/InGaP/TiO_2/Ni photoanode in 1.0 M KOH (pH = 13.7) for OER, and a Pd/C nanoparticle coated Ti mesh cathode in 2.8 M $KHCO_3$ (pH = 8.0) for the CO_2 reduction reaction (CO_2RR). This system was able to produce oxygen efficiently in the anode reaction, while having a high yield of formic acid in the catholyte and demonstrated a lower overall two-electrode potential than any previous CO_2RR system. A 10% solar-to-fuel efficiency was obtained in this research, demonstrated stable for three hours (see Figure 8.12).

Similarly, Vermaas and Smith[71] demonstrated an alkaline water oxidation system made with a NiFe anode in NaOH separated by a BPM from a $KCHO_3$ solution and a Ag cathode. While this system similarly produced oxygen in the anolyte, the choice of a Ag electrocatalyst gave a high faradaic efficiency (FE) for CO evolution, with small amounts of H_2 also being produced in the catholyte. Compared to a Nafion system, the BPM system operated at a significant lower bias voltage (up to 1 V lower), and kept the pH much more stable, thereby causing a more stable production of CO.

In another approach using a BPM for CO_2 reduction systems, Schreier *et al.*[87] used an ALD grown layer of SnO_2 on CuO nanowires to evolve CO (with small amounts of H_2) and oxygen. The authors were able to photodrive

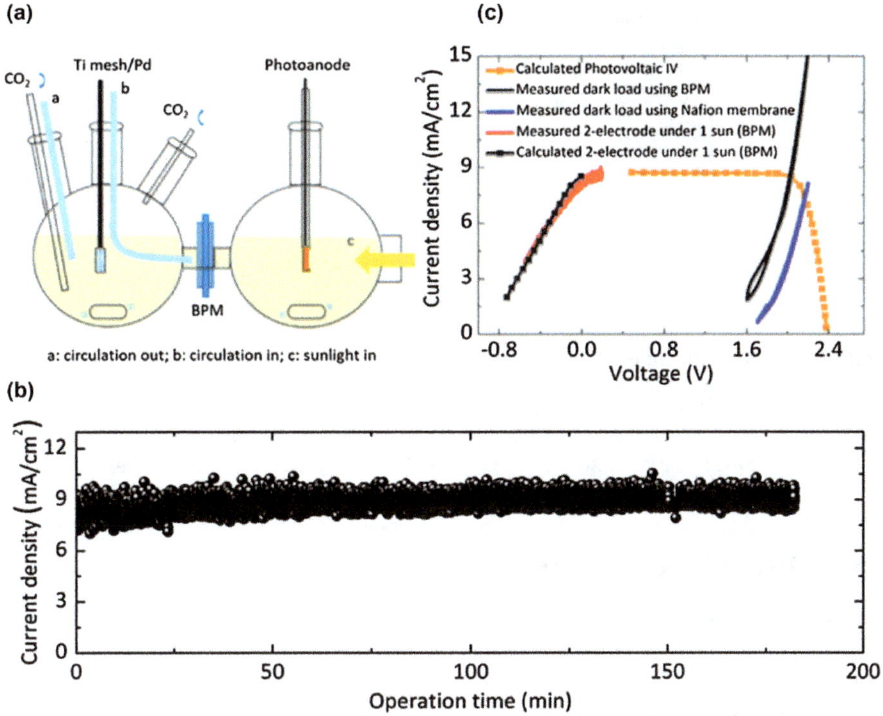

Figure 8.12 (A) Schematic illustration of two-electrode electrochemical setup for CO_2 reduction. (B) Current density as a function of time, using a GaAs/InGaP/TiO_2/Ni photoanode and a Pd/C-coated Ti mesh cathode under 100 mW cm^{-2} of simulated AM 1.5 illumination. (C) The measured (red) and calculated (black) two-electrode current – voltage behaviour of the GaAs/InGaP/TiO_2/Ni photoanode wired to a Pd/C coated Ti mesh cathode was measured under 100 mW cm^{-2} of simulated AM 1.5 illumination, together with calculated photovoltaic *I–V* curve (orange) and under loads with BPM and Nafion system (black and blue line). Reprinted with permission from X. Zhou, R. Liu, K. Sun, Y. Chen, E. Verlage, S. A. Francis, N. S. Lewis and C. Xiang, *ACS Energy Letters*, 2016, **1**, 764–770.[73] Copyright 2016 American Chemical Society.

this reaction using a GaInP/GaInAs/Ge solar cell, which resulted in a solar to fuel conversion efficiency of 13.4%.

A significant next step was taken by Li *et al.*,[72] using a BPM to separate an alkaline water oxidation half-cell from a gas diffusion electrode (GDE) and bismuth-based catalyst that was converting gas phase CO_2 to different products. It was demonstrated that a BPM allowed an order of magnitude higher current densities than when using a monopolar AEM, and similar to using a GDE with Nafion® (Figure 8.13).

In all cases, the use of a BPM shows high potential for CO_2 reduction, offering high energy efficiencies and creating a stable environment for both electrode reactions.

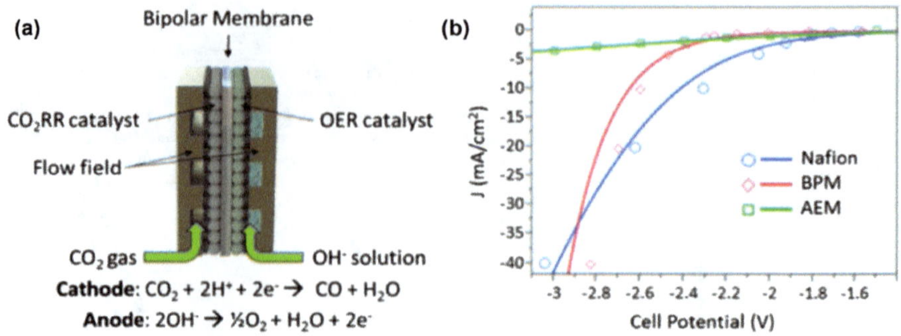

Figure 8.13 Schematic drawing of a gas-fed CO_2 electrolyser (a) and $I - V$ curves for the gas-fed cell comparing BPM, AEM, and Nafion® membranes (b). Reprinted with permission from Y. C. Li, D. Zhou, Z. Yan, R. H. Gonçalves, D. A. Salvatore, C. P. Berlinguette and T. E. Mallouk, *ACS Energy Letters*, 2016, **1**, 1149–1153.[72] Copyright 2016 American Chemical Society.

8.6 Outlook/Conclusions

The use of bipolar membranes offers a strong potential for separating environments for anode and cathode. A pH gradient, maintained by water dissociation in a BPM, creates a chemical potential difference at the expense of an electrical voltage. For extreme pH gradients (*i.e.*, pH 0 *vs.* 14), results have been presented with small overpotential and therefore high electrolysis efficiencies. A challenge remains to operate energy efficient and without salt cross-over for smaller pH gradients. The growing interest in BPM's should warrant further exploration into membrane technology, the importance of mass/ion fluxes in electrochemical systems, and overall systems considerations.

References

1. C. C. L. McCrory, S. Jung, J. C. Peters and T. F. Jaramillo, Benchmarking Heterogeneous Electrocatalysts for the Oxygen Evolution Reaction, *J. Am. Chem. Soc.*, 2013, **135**(45), 16977–16987.
2. M. D. Merrill and R. C. Dougherty, Metal Oxide Catalysts for the Evolution of O2 from H2O, *J. Phys. Chem. C*, 2008, **112**(10), 3655–3666.
3. L. Trotochaud, S. L. Young, J. K. Ranney and S. W. Boettcher, Nickel–Iron Oxyhydroxide Oxygen-Evolution Electrocatalysts: The Role of Intentional and Incidental Iron Incorporation, *J. Am. Chem. Soc.*, 2014, **136**(18), 6744–6753.
4. K. Sun, et al., A comparison of the chemical, optical and electrocatalytic properties of water-oxidation catalysts for use in integrated solar-fuel generators, *Energy Environ. Sci.*, 2017, **10**(4), 987–1002.
5. C. C. L. McCrory, S. Jung, I. M. Ferrer, S. M. Chatman, J. C. Peters and T. F. Jaramillo, Benchmarking Hydrogen Evolving Reaction and Oxygen

Evolving Reaction Electrocatalysts for Solar Water Splitting Devices, *J. Am. Chem. Soc.*, 2015, **137**(13), 4347–4357.

6. F. H. Saadi, A. I. Carim, E. Verlage, J. C. Hemminger, N. S. Lewis and M. P. Soriaga, CoP as an Acid-Stable Active Electrocatalyst for the Hydrogen-Evolution Reaction: Electrochemical Synthesis, Interfacial Characterization and Performance Evaluation, *J. Phys. Chem. C*, 2014, **118**(50), 29294–29300.
7. P. C. K. Vesborg, B. Seger and I. Chorkendorff, Recent Development in Hydrogen Evolution Reaction Catalysts and Their Practical Implementation, *J. Phys. Chem. Lett.*, 2015, **6**(6), 951–957.
8. M. B. McDonald, S. Ardo, N. S. Lewis and M. S. Freund, Use of Bipolar Membranes for Maintaining Steady-State pH Gradients in Membrane-Supported, Solar-Driven Water Splitting, *ChemSusChem*, 2014, **7**(11), 3021–3027.
9. N. M. Vargas-Barbosa, G. M. Geise, M. A. Hickner and T. E. Mallouk, Assessing the Utility of Bipolar Membranes for use in Photoelectrochemical Water-Splitting Cells, *ChemSusChem*, 2014, **7**(11), 3017–3020.
10. S. M. H. Hashemi, M. A. Modestino and D. Psaltis, Membrane-less microelectrolyzer for hydrogen production across the pH scale, *Energy Environ. Sci.*, 2015, **8**, 2003–2009.
11. J. Jin, K. Walczak, M. R. Singh, C. Karp, N. S. Lewis and C. Xiang, An experimental and modeling/simulation-based evaluation of the efficiency and operational performance characteristics of an integrated, membrane-free, neutral pH solar-driven water-splitting system, *Energy Environ. Sci.*, 2014, **7**, 3371–3380.
12. C. Xiang, *et al.*, Modeling, Simulation, and Implementation of Solar-Driven Water-Splitting Devices, *Angew. Chem., Int. Ed.*, 2016, **55**(42), 12974–12988.
13. S. L. McFarlane, B. A. Day, K. McEleney, M. S. Freund and N. S. Lewis, Designing electronic/ionic conducting membranes for artificial photosynthesis, *Energy Environ. Sci.*, 2011, **4**(5), 1700–1703.
14. G. M. Geise, H.-S. Lee, D. J. Miller, B. D. Freeman, J. E. McGrath and D. R. Paul, Water purification by membranes: The role of polymer science, *J. Polym. Sci., Part B: Polym. Phys.*, 2010, **48**(15), 1685–1718.
15. T. Sata, *Ion Exchange Membranes: Preparation, Characterization, Modification and Application*, RSC, Cambridge, 2004, p. 321.
16. H. Strathmann, *Ion Exchange Membrane Separation Processes*, Elsevier, 2004, vol. 9.
17. J. R. McKone, N. S. Lewis and H. B. Gray, Will Solar-Driven Water-Splitting Devices See the Light of Day? *Chem. Mater.*, 2014, **26**(1), 407–414.
18. C. Rodriguez, M. A. Modestino, C. Moser and D. Psaltis, Design and cost considerations for practical solar-hydrogen generators, *Energy Environ. Sci.*, 2014, **7**, 3828–3835.
19. E. A. Hernandez-Pagan, N. M. Vargas-Barbosa, T. Wang, Y. Zhao, E. S. Smotkin and T. E. Mallouk, Resistance and polarization losses in

aqueous buffer-membrane electrolytes for water-splitting photoelectrochemical cells, *Energy Environ. Sci.*, 2012, **5**(6), 7582–7589.
20. M. A. Modestino, K. A. Walczak, A. Berger, C. M. Evans, S. Haussener, C. Koval, J. S. Newman, J. W. Ager and R. A. Segalman, Robust production of purified H2 in a stable, self-regulating, and continuously operating solar fuel generator, *Energy Environ. Sci.*, 2013, **7**(1), 297–301.
21. F. G. Donnan, The theory of membrane equilibrium in oresence of a non-dialyzable electrolyte, *Z. Electrochem.*, 1911, **17**, 572.
22. W. Ostwald, Elektrische eigenschaften halbdurchlassiger scheidewande, *Z. Phys. Chem.*, 1890, **6**, 71.
23. J. R. Wilson, *Demineralization by Electrodialysis*, Butterworths Scientific Publications, 1960.
24. T. Xu, Ion exchange membranes: State of their development and perspective, *J. Membr. Sci.*, 2005, **263**(1–2), 1–29.
25. A. H. Galama, J. W. Post, H. V. M. Hamelers, V. V. Nikonenko and P. M. Biesheuvel, On the Origin of the Membrane Potential Arising Across Densely Charged Ion Exchange Membranes: How Well Does the Teorell-Meyer-Sievers Theory Work? *J. Membr. Sci. Res.*, 2016, **2**(3), 128–140.
26. A. H. Galama, J. W. Post, M. A. Cohen Stuart and P. M. Biesheuvel, Validity of the Boltzmann equation to describe Donnan equilibrium at the membrane–solution interface, *J. Membr. Sci.*, 2013, **442**, 131–139.
27. G. M. Geise, L. P. Falcon, B. D. Freeman and D. R. Paul, Sodium chloride sorption in sulfonated polymers for membrane applications, *J. Membr. Sci.*, 2012, **423–424**, 195–208.
28. V. M. Barragán, C. Rueda and C. Ruiz-Bauzá, On the Fixed Charge Concentration and the Water Electroosmotic Transport in a Cellulose Acetate Membrane, *J. Colloid Interface Sci.*, 1995, **172**(2), 361–367.
29. E. Güler, R. Elizen, D. A. Vermaas, M. Saakes and K. Nijmeijer, Performance-determining membrane properties in reverse electrodialysis, *J. Membr. Sci.*, 2013, **446**, 266–276.
30. P. Długołęcki, K. Nymeijer, S. Metz and M. Wessling, Current status of ion exchange membranes for power generation from salinity gradients, *J. Membr. Sci.*, 2008, **319**(1–2), 214–222.
31. J. Veerman, R. M. de Jong, M. Saakes, S. J. Metz and G. J. Harmsen, Reverse electrodialysis: Comparison of six commercial membrane pairs on the thermodynamic efficiency and power density, *J. Membr. Sci.*, 2009, **343**(1–2), 7–15.
32. V. I. Zabolotsky and V. V. Nikonenko, Effect of structural membrane inhomogeneity on transport properties, *J. Membr. Sci.*, 1993, **79**(2), 181–198.
33. A. H. Galama, D. A. Vermaas, J. Veermas, M. Saakes, H. H. M. Rijnaarts, J. W. Post and K. Nijmeijer, Membrane resistance: The effect of salinity gradients over a cation exchange membrane, *J. Membr. Sci.*, 2014, **467**, 279–291.
34. G. M. Geise, A. J. Curtis, M. C. Hatzell, M. A. Hickner and B. E. Logan, Salt Concentration Differences Alter Membrane Resistance in Reverse Electrodialysis Stacks, *Environ. Sci. Technol. Lett.*, 2014, **1**(1), 36–39.

35. M. A. Hickner, H. Ghassemi, Y. S. Kim, B. R. Einsla and J. E. McGrath, Alternative Polymer Systems for Proton Exchange Membranes (PEMs), *Chem. Rev.*, 2004, **104**(10), 4587–4612.
36. L. Wang, A. Husar, T. Zhou and H. Liu, A parametric study of PEM fuel cell performances, *Int. J. Hydrogen Energy*, 2003, **28**(11), 1263–1272.
37. M. J. Kelly, B. Egger, G. Fafilek, J. O. Besenhard, H. Kronberger and G. E. Nauer, Conductivity of polymer electrolyte membranes by impedance spectroscopy with microelectrodes, *Solid State Ionics*, 2005, **176**(25–28), 2111–2114.
38. J. R. Varcoe, et al., Anion-exchange membranes in electrochemical energy systems, *Energy Environ. Sci.*, 2014, **7**(10), 3135–3191.
39. E. Guler, Y. Zhang, M. Saakes and K. Nijmeijer, Tailor-Made Anion-Exchange Membranes for Salinity Gradient Power Generation Using Reverse Electrodialysis, *ChemSusChem*, 2012, **5**(11), 2262–2270.
40. M. Ni, M. K. H. Leung and D. Y. C. Leung, Energy and exergy analysis of hydrogen production by a proton exchange membrane (PEM) electrolyzer plant, *Energy Convers. Manage.*, 2008, **49**(10), 2748–2756.
41. J. M. Spurgeon and N. S. Lewis, Proton exchange membrane electrolysis sustained by water vapor, *Energy Environ. Sci.*, 2011, **4**(8), 2993–2998.
42. J. W. Ager, M. Shaner, K. Walczak, I. D. Sharp and S. Ardo, Experimental Demonstrations of Spontaneous, Solar-Driven Photoelectrochemical Water Splitting, *Energy Environ. Sci.*, 2015, DOI: 10.1039/C5EE00457H.
43. K. Zeng and D. Zhang, Recent progress in alkaline water electrolysis for hydrogen production and applications, *Prog. Energy Combust. Sci.*, 2010, **36**(3), 307–326.
44. E. Güler, W. van Baak, M. Saakes and K. Nijmeijer, Monovalent-ion-selective membranes for reverse electrodialysis, *J. Membr. Sci.*, 2014, **455**, 254–270.
45. Y. Mizutani, Ion exchange membranes with preferential permselectivity for monovalent ions, *J. Membr. Sci.*, 1990, **54**(3), 233–257.
46. J. W. Post, H. V. M. Hamelers and C. J. N. Buisman, Influence of multivalent ions on power production from mixing salt and fresh water with a reverse electrodialysis system, *J. Membr. Sci.*, 2009, **330**(1–2), 65–72.
47. S. Haussener, C. Xiang, J. M. Spurgeon, S. Ardo, N. S. Lewis and A. Z. Weber, Modeling, simulation, and design criteria for photoelectrochemical water-splitting systems, *Energy Environ. Sci.*, 2012, **5**(12), 9922–9935.
48. B. A. Pinaud, et al., Technical and economic feasibility of centralized facilities for solar hydrogen production via photocatalysis and photoelectrochemistry, *Energy Environ. Sci.*, 2013, **6**(7), 1983–2002.
49. S. Peng, S. Lu, J. Zhang, P.-C. Sui and Y. Xiang, Evaluating the interfacial reaction kinetics of the bipolar membrane interface in the bipolar membrane fuel cell, *Phys. Chem. Chem. Phys.*, 2013, **15**(27), 11217–11220.
50. H. Strathmann, J. J. Krol, H. J. Rapp and G. Eigenberger, Limiting current density and water dissociation in bipolar membranes, *J. Membr. Sci.*, 1997, **125**(1), 123–142.

51. N. W. Rosenberg and C. E. Tirrell, Limiting Currents in Membrane Cells, *Ind. Eng. Chem.*, 1957, **49**(4), 780–784.
52. M. B. Andersen, M. van Soestbergen, A. Mani, H. Bruus, P. M. Biesheuvel and M. Z. Bazant, Current-Induced Membrane Discharge, *Phys. Rev. Lett.*, 2012, **109**(10), 108301.
53. J. Balster, M. H. Yildirim, D. F. Stamatialis, R. Ibanez, R. G. H. Lammertink, V. Jordan and M. Wessling, Morphology and Microtopology of Cation-Exchange Polymers and the Origin of the Overlimiting Current, *J. Phys. Chem. B*, 2007, **111**(9), 2152–2165.
54. J. J. Krol, M. Wessling and H. Strathmann, Concentration polarization with monopolar ion exchange membranes: current–voltage curves and water dissociation, *J. Membr. Sci.*, 1999, **162**(1), 145–154.
55. Y. Tanaka, Concentration polarization in ion-exchange membrane electrodialysis—the events arising in a flowing solution in a desalting cell, *J. Membr. Sci.*, 2003, **216**(1), 149–164.
56. V. V. Nikonenko, A. V. Kovalenko, M. K. Urtenov, N. D. Pismenskaya, J. Han, P. Sistat and G. Pourcelly, Desalination at overlimiting currents: State-of-the-art and perspectives, *Desalination*, 2014, **342**(Supplement C), 85–106.
57. J. J. Krol, Monopolar and bipolar ion exchange membranes. Bipolar membrane, PhD thesis, Universiteit Twente, 1997.
58. F. Wilhelm, N. Vegt, M. Wessling, and H. Strathmann, *Bipolar Membrane Preparation*, 2000.
59. A. M. Rajesh, T. Chakrabarty, S. Prakash and V. K. Shahi, Effects of metal alkoxides on electro-assisted water dissociation across bipolar membranes, *Electrochim. Acta*, 2012, **66**(Supplement C), 325–331.
60. J. C. de Valença, R. M. Wagterveld, R. G. H. Lammertink and P. A. Tsai, Dynamics of microvortices induced by ion concentration polarization, *Phys. Rev. E*, 2015, **92**(3), 031003.
61. D. A. Vermaas, M. Sassenburg and W. A. Smith, Photo-assisted water splitting with bipolar membrane induced pH gradients for practical solar fuel devices, *J. Mater. Chem. A*, 2015, **3**, 19556–19562.
62. K. Sun, R. Liu, Y. Chen, E. Verlage, N. S. Lewis and C. Xiang, A Stabilized, Intrinsically Safe, 10% Efficient, Solar-Driven Water-Splitting Cell Incorporating Earth-Abundant Electrocatalysts with Steady-State pH Gradients and Product Separation Enabled by a Bipolar Membrane, *Adv. Energy Mater.*, 2016, **6**(13), 1600379.
63. F. G. Wilhelm, I. Pünt, N. F. A. van der Vegt, M. Wessling and H. Strathmann, Optimisation strategies for the preparation of bipolar membranes with reduced salt ion leakage in acid–base electrodialysis, *J. Membr. Sci.*, 2001, **182**(1–2), 13–28.
64. T. T. Bejerano, E. Korngold and R. Messalem, New bipolar membrane setup for high current densities, *Desalin. Water Treat.*, 2014, **56**, 1421–1426.
65. C. Shen, R. Wycisk and P. N. Pintauro, High performance electrospun bipolar membrane with a 3D junction, *Energy Environ. Sci.*, 2017, **10**(6), 1435–1442.

66. V. J. Frilette, Preparation and characterization of bipolar ion exchange membranes, *J. Phys. Chem.*, 1956, **60**(4), 435–439.
67. A. Alcaraz, P. Ramírez, S. Mafé, H. Holdik and B. Bauer, Ion selectivity and water dissociation in polymer bipolar membranes studied by membrane potential and current–voltage measurements, *Polymer*, 2000, **41**(17), 6627–6634.
68. X. Tongwen, Electrodialysis processes with bipolar membranes (EDBM) in environmental protection—a review, *Resour., Conserv. Recycl.*, 2002, **37**(1), 1–22.
69. F. G. Wilhelm, N. F. A. van der Vegt, H. Strathmann and M. Wessling, Comparison of bipolar membranes by means of chronopotentiometry, *J. Membr. Sci.*, 2002, **199**(1–2), 177–190.
70. J. Luo, D. A. Vermaas, D. Bi, A. Hagfeldt, W. A. Smith and M. Grätzel, Bipolar Membrane Assisted Solar Water Splitting in Optimal pH, *Adv. Energy Mater.*, 2016, **6**, 1600100.
71. D. A. Vermaas and W. A. Smith, Synergistic Electrochemical CO2 Reduction and Water Oxidation with a Bipolar Membrane, *ACS Energy Lett.*, 2016, **1**(6), 1143–1148.
72. Y. C. Li, D. Zhou, Z. Yan, R. H. Gonçalves, D. A. Salvatore, C. P. Berlinguette and T. E. Mallouk, Electrolysis of CO2 to Syngas in Bipolar Membrane-Based Electrochemical Cells, *ACS Energy Lett.*, 2016, **1**(6), 1149–1153.
73. X. Zhou, R. Liu, K. Sun, Y. Chen, E. Verlage, S. A. Francis, N. S. Lewis and C. Xiang, Solar-Driven Reduction of 1 atm of CO2 to Formate at 10% Energy-Conversion Efficiency by Use of a TiO2-Protected III–V Tandem Photoanode in Conjunction with a Bipolar Membrane and a Pd/C Cathode, *ACS Energy Lett.*, 2016, **1**(4), 764–770.
74. A. ter Heijne, H. V. M. Hamelers, V. de Wilde, R. A. Rozendal and C. J. N. Buisman, A Bipolar Membrane Combined with Ferric Iron Reduction as an Efficient Cathode System in Microbial Fuel Cells, *Environ. Sci. Technol.*, 2006, **40**(17), 5200–5205.
75. M. Ünlü, J. Zhou and P. A. Kohl, Hybrid Anion and Proton Exchange Membrane Fuel Cells, *J. Phys. Chem. C*, 2009, **113**(26), 11416–11423.
76. S. Graillon, F. Persin, G. Pourcelly and C. Gavach, Development of electrodialysis with bipolar membrane for the treatment of concentrated nitrate effluents, *Desalination*, 1996, **107**(2), 159–169.
77. R. Simons, Preparation of a high performance bipolar membrane, *J. Membr. Sci.*, 1993, **78**(1), 13–23.
78. S. Chabi, A. G. Wright, S. Holdcroft and M. S. Freund, Transparent Bipolar Membrane for Water Splitting Applications, *ACS Appl. Mater. Interfaces*, 2017, **9**(32), 26749–26755.
79. B. Bauer, F. J. Gerner and H. Strathmann, Development of bipolar membranes, *Desalination*, 1988, **68**(2), 279–292.
80. FumaTech, Ion Exchange Membranes, 2016.
81. R. S. Reiter, W. White and S. Ardo, Communication—Electrochemical Characterization of Commercial Bipolar Membranes under Electrolyte

Conditions Relevant to Solar Fuels Technologies, *J. Electrochem. Soc.*, 2016, **163**(4), H3132–H3134.
82. X. Chen, Z. Zhang, L. Liu, R. Cheng, L. Shi and X. Zheng, RO applications in China: History, current status, and driving forces, *Desalination*, 2016, **397**(Supplement C), 185–193.
83. R. E. Moussaoui, G. Pourcelly, M. Maeck, H. D. Hurwitz and C. Gavach, Co-ion leakage through bipolar membranes Influence on I–V responses and water-splitting efficiency, *J. Membr. Sci.*, 1994, **90**(3), 283–292.
84. F. G. Wilhelm, N. F. A. van der Vegt, M. Wessling and H. Strathmann, Chronopotentiometry for the advanced current–voltage characterisation of bipolar membranes, *J. Electroanal. Chem.*, 2001, **502**(1), 152–166.
85. J. Pretz and E. Staude, Reverse electrodialysis (RED) with bipolar membranes, an energy storage system, *Ber. Bunsen-Ges. Phys. Chem.*, 1998, **102**(4), 676–685.
86. European Commision, Blue Acid/Base Battery: Storage and recovery of renewable electrical energy by reversible salt water dissociation. https://ec.europa.eu/inea/en/horizon-2020/projects/h2020-energy/storage/baobab (accessed 19-10-2017).
87. M. Schreier, F. Héroguel, L. Steier, S. Ahmad, J. S. Luterbacher, M. T. Mayer, J. Luo and M. Grätzel, Solar conversion of CO2 to CO using Earth-abundant electrocatalysts prepared by atomic layer modification of CuO, *Nat. Energy*, 2017, **2**, 228911.

CHAPTER 9

Modelling-derived Design Guidelines for Photo-electrochemical Devices

SOPHIA HAUSSENER,* YANNICK GAUDY AND SAURABH TEMBHURNE

Ecole Polytechnique Fédérale de Lausanne, Institute of Mechanical Engineering, Station 9, Lausanne 1015, Switzerland
*Email: sophia.haussener@epfl.ch

9.1 Introduction

The solar energy received on the earth's surface can meet humankind's current and future energy demand.[1,2] Direct conversion of solar energy and water (or other reactants such as CO_2 or N_2) into chemical energy *via* photo-driven electrochemical (PDEC) and photoelectrochemical (PEC) processes is one viable route for renewable fuel processing, for chemical commodities processing, and for energy storage.

The differentiation between PDEC and PEC can be made on the basis of two arguments: (i) the way of charge separation, which is at a solid–solid interface (PDEC case) or at a solid–liquid interface (PEC case), and (ii) the scale over which the transport of the separated holes and electrons to the catalytically active sites happens, which can be macroscopic and therefore allow for the incorporation of power electronics (PDEC case) or microscopic and therefore allow for heat transfer between the components (PEC case).

Energy and Environment Series No. 20
Advances in Photoelectrochemical Water Splitting: Theory, Experiment and Systems Analysis
Edited by S. David Tilley, Stephan Lany and Roel van de Krol

Published by the Royal Society of Chemistry, www.rsc.org

A clear separation between the two approaches is not always possible as photoelectrodes that are covered by appropriate protection layers can develop a buried junction, or photovoltaic devices can be closely integrated with the electrocatalytic components and therefore provide microscopic charge connections. Such hybrid cases are sometimes termed integrated PEC approaches and are equally interesting for investigation.

Complicated and coupled multi-physics processes, which happen on multiple time and spatial scales, govern the functionality of PEC and PDEC devices. It is obvious that device modelling can efficiently and actively support the choice of the most interesting—in terms of efficiency, cost, robustness, scalability, practicability, and sustainability—material choices, device designs, and operating approaches. Models can span a wide variety of scales and various physical details.[3] As will be shown, even relatively simple models can provide valuable guidance for feasible device designs and can help focusing the research priorities on the material research and development side.

9.2 Zero-dimensional Models

9.2.1 Governing Equations

Simple zero-dimensional models or lumped equivalent circuit models describe the relation between current density, i, and potential, V, within the semiconducting photoabsorber component as a diode-like component with imperfections (eqn (9.1)):

$$i = i_p \varepsilon_{d,1} - i_d \varepsilon_{d,2} \left(\exp\left(\frac{q(V + iR_s \varepsilon_3)}{zkT} \right) \right) - \frac{V + iR_s \varepsilon_{d,3}}{R_{sh} \varepsilon_{d,4}}, \tag{9.1}$$

where the photo current density, i_p, the dark saturation current density, i_d, the series resistance, R_s, the shunt resistance, R_{sh}, the ideality factor, z, and empirical degradation factors, $\varepsilon_{d,i}$, are used to characterise the material behaviour.

The relationship between current density and potential to drive the catalysis and overcome overpotentials and non-idealities within the solid or liquid electrolytic component can be described as eqn (9.2):

$$E = E_0 + \eta_a + |\eta_c| + \Delta E_\Omega + \eta_{deg} + \eta_{other}, \tag{9.2}$$

where the thermodynamic redox potential of the reaction under standard conditions, E_0, is added to various overpotentials such as the anodic and cathodic reaction overpotentials, η_a and η_c, ohmic losses in the conducting electrode and electrolytes, ΔE_Ω, degradation related increases in the potential which are lumped into one time-dependent overpotential, η_{deg}, and other overpotentials, η_{other}, accounting, for example, for concentration overpotentials or an empirical effect of the two-phase flow phenomena (bubbles) on the potential. The intersection between these two iV curves (eqn (9.1) and (9.2)) marks the operating point of the device, *i.e.* the operating current density, i_{op}, under which the potential produced in the photoabsorber at a given current density is able to drive the electrochemical reaction and all overpotentials in the electrocatalytically active part to produce the desired product. Based on this operating current density, a solar-to-fuel efficiency can be defined (for devices with no external bias) (eqn (9.3)):

$$\eta = \frac{i_{op}E_0}{I_{sol,in}}\eta_F\eta_{col} = \frac{nF\dot{m}''E_0}{MI_{sol,in}}\eta_F\eta_{col}, \tag{9.3}$$

where the solar irradiation over the complete spectrum, $I_{sol,in}$, is considered as input. Note that the surface area of the irradiation needs to be the same surface area which is used for the definition of the current density, otherwise the latter needs to be adjusted (in case when concentrated irradiation or current concentration is used). The spectrum of $I_{sol,in}$ can either be a normed AM1.5 spectrum, or the spectrum of the natural light at the current atmospheric conditions and solar position. The former would describe a more comparable benchmarking efficiency while the latter describes a more practical and *in-situ* efficiency. The Faradaic efficiency, η_F, and the collection efficiency, η_{col}, describe how well the charge is converted into the desired product and how purely it is collected at the outlet port of the system. Using Faraday's law of electrolysis, the efficiency can alternatively be defined as a function of the collected mass flux rate of product in $[\mathrm{kg\,s^{-1}\,m^{-2}}]$, $\dot{m}''$, the number of electrons involved in the electrochemical reaction, n, the Faradaic constant, F, and the molar mass of the product, M.

9.2.2 Example Results

Figure 9.1 shows two typical current density *vs.* voltage characteristics (in black and blue) for photoabsorbers (also termed power curve) and electrochemical (also termed load curve) components where the red circles indicate the operating current densities for the two cases. For example, the black curve can be interpreted as being the designed operating condition of a photoabsorber and electrocatalyst combination, while the blue curves show the same device design and components after several hours of operation affected by degradation.

Paramount for the understanding on how material choices, device design choices, operating conditions, or degradation can affect the device operation is the realisation that the photoabsorber curve as well as the electrocatalysis curves vary individually (and with time), affecting the performance of the device. As a consequence, different optimal choices of semiconductor materials for different electrolyser material choices, device designs and operational conditions are required.

Eqn (9.1) and (9.2) allow for a general understanding of how the various losses in a system can affect the device performance. Starting with the electrochemical part, it is apparent that more active co-catalysts can reduce

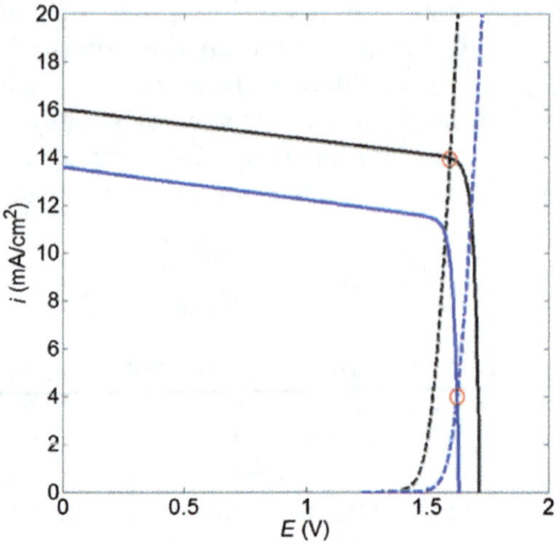

Figure 9.1 Current density's dependence on the potential for the photoabsorber (solid) and the electrolyser (dashed) for initial operation (black) and for operation after degradation or variation in operating conditions (blue). The operating points for the initial and degraded cases is indicated by a red circle.

the reaction overpotentials and hence the overall potential required. This explains the significant effort in the community on the development of effective co-catalysts that are additionally selective, stable, and earth-abundant. Note that the reaction overpotentials are usually non-linearly dependent on the current density. The ohmic losses can be reduced by minimising the electron or ion transport path lengths through the miniaturisation of the device (using mm to micron-sized components) or by increasing the electronic or ionic conductivity of the solid and liquid cell components, *e.g.* by using highly acidic or base electrolytes, or buffered solutions. The ohmic losses depend linearly on the current density. The degradation overpotentials can be reduced by utilising stabilisation layers or more moderate operating conditions (less corrosive electrolytes). The other losses such as the concentration overpotentials can be minimised by using electrolytes in which the protons or hydroxide ions are responsible for the ionic current, which requires highly acidic or base solutions, or buffered solutions in a device configuration where the operational conditions, *e.g.* convection, are able to counteract the build-up of concentration gradients.[4]

Similar considerations can be made based on the photoabsorber diode equation where a minimisation of the resistivity of the material and well-engineered contacts can affect the series and shunt resistances. The photo current density depends on the band gap of the material, possible surface or anti-reflection coatings, and also the carrier separation and collection efficiency. The dark current is affected by the quality of the junction as well as, for example, the temperature.

High efficiency requires that the intersection between the photodiode curve and the load curve is at a current density as high as possible. The only binary requirement is that the potential at this current is larger than the potential required for the reaction and the overpotentials. It should be noted that we do not strive for maximum power in a photo-electrochemical device (in contrast to PV devices), instead a maximisation of the current density is desired.

Finally, high efficiency is accompanied by η_F and η_{col} being close to 1. The former is achieved, for example, by using selective catalysts while the latter is influenced—for water splitting devices—by using hydrogen and oxygen non-permeable separators between the two chambers.

0D models have been used to study the efficiency limits of photo-electrochemical devices,[5] the effect of the choice of the semiconductor photoabsorber bandgap(s) of single or dual photoabsorbers on the performance,[6,7] the effect of the catalysis choice on the performance,[8,9] the effect of irradiation and current concentration on the performance,[10,11] or—more recently—to assess the effect of degradation on performance.[12] For the latter, a quasi-transient approach is required, where the degradation overpotential in eqn (9.2) and degradation factors in eqn (9.1) steadily increase and decrease, respectively, as a function of time.

0D PEC or PDEC device models incorporating degradation as a transient effect are simple and fast while still capturing the relevant physical phenomena in the device. They can therefore be nicely coupled to cost and sustainability analysis in order to provide a holistic view on the design of PEC and PDEC devices.[11] Such models incorporate and analyse the four pillars of an analysis of meaningful PEC and PDEC approaches of the future, namely that such a device needs to be efficient, cheap and scalable, sustainable, and robust and stable for a long operational time. Only a device approach respecting these four requirements simultaneously can represent a meaningful approach for the future and have an impact on our energy economy. The transient techno-economic-environmental study by Dumortier *et al.*[11] and its freely accessible web-implementation (http://specdo.epfl.ch), utilising such a 0D device model, predicted in a broader sense that devices made of rare materials (such as III–V photoabsorbers or Pt or Ru-based catalysts) can potentially still be competitive if irradiation concentration is used in the device (Figure 9.2(a)). However, the model predicted that lower performant devices with earth-abundant photo-absorbers and catalysts (Figure 9.2(b)) can be equally cost competitive utilising no irradiation concentration.

Based on this transient techno-economic-environmental modelling, two main avenues for interesting PEC and PDEC device design classes can be identified: (i) a highly engineered approach utilising concentrated irradiation and proper heat management, liquid reactant and dedicated auxiliary components, and (ii) a more low-tech, less complex solution with cheaper materials, reduced balance of system and less auxiliary components. The latter class can be pushed further to devices using a dilute vapour-phase

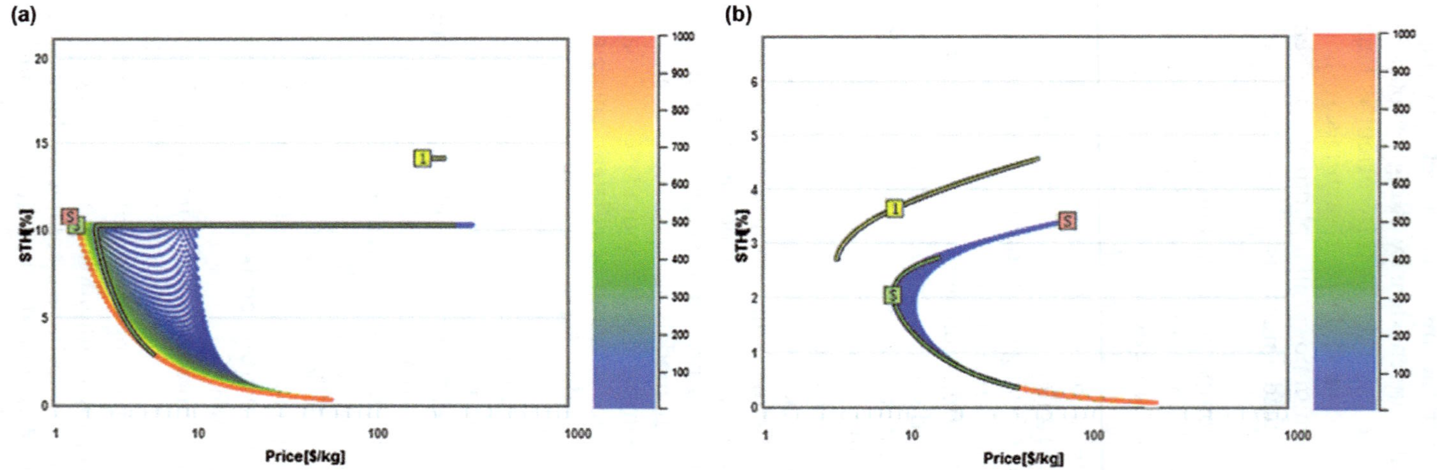

Figure 9.2 Operational time-averaged efficiency as a function of device exit hydrogen price and irradiation concentration (colour bar) for devices made of (a) rare III–V photoabsorbers and electrocatalysts, and (b) earth-abundant photoabsorber and electrocatalysts. The dots represent all possible solutions for a device utilising irradiation concentration and current dilution, the green line indicates a device utilising irradiation concentration but has an area-matched photoabsorber and electrocatalysts, and the yellow line represents a device utilising no irradiation concentration but current dilution. The yellow symbol indicates a device that does neither use irradiation concentration nor current dilution, the green and red symbols show the most cost competitive and most efficient, respectively, device if irradiation concentration and current dilution is used. The plots have been generated with the SPECDO tool.

reactant and natural convection as a transport mechanism to deliver the reactant to the surface.[13,14] The highly engineered device class could be envisaged to become a centralised fuel processing facility to provide transportation fuel or chemical commodities to a chemical processing plant, and the low-tech device class a decentralised plant potentially used in households as seasonal storage.

We finish the discussion on the usefulness of 0D modelling by returning to Figure 9.1. It becomes clear that we start with a well-matched combination of photoabsorbing and electrocatalytical components leading to a reasonably high efficiency (black curves). However, once the components are exposed to the challenging environmental conditions, they will degrade leading to a reduced performance of the components and, consequently, a significant drop in efficiency (blue curves). The photodiode and load curves drift away due to degradation and, even though the device had, in its initial condition, worked quite well and the components matched to achieve high efficiency, this is not true anymore after degradation starts to affect the components. Similarly, drifts of the performance can be observed as a result of changing operational conditions such as irradiation (during the year and during the day) or operating temperature.[15] Consequently, it becomes important to understand how conditions – different than the optimal ones used for designing the device – and degradation can affect the performance and, more importantly, simple engineering ways to counteract this. While there are a variety of ways to approach this problem, including controlling of operating and feeding conditions, here we use this observation as an argument to compare (integrated) PEC with PDEC devices for understanding whether the addition of power electronics (namely a maximum power tracker and a DC–DC converter) and therefore a macroscopic separation of the photoabsorber and the catalytically active sites can bring benefit. A DC–DC converter allows one to completely decouple the photoabsorber performance curve from the electrochemical component performance curve (at the expense of a loss of about 5% in the power). Power electronics provide—in this sense—the flexibility for a device to operate at a variety of conditions without severe impact on performance. We can use 0D modelling of PEC and PDEC devices to understand how the addition of power electronics impacts the performance. Figure 9.3(a) depicts the photoabsorber performance of two different types of absorber combinations under concentrated irradiation and one load curve. The operating point of a coupled device design (PEC or integrated PEC) is depicted as a green circle and the operating point one can achieve with an uncoupled PDEC device using power electronics is shown with a triangle. The solar-to-hydrogen efficiency of type 1 and type 2 devices run as (integrated) PEC or as PDEC with power electronics is shown in Figure 9.3(b). The components have been degraded with example rates typically found for individual PV ($\dot{\varepsilon}_{d,1} = 0.7\%$ per year) and electrolyser components ($\dot{\eta}_{deg} = 6\ \mu V\, h^{-1}$), which may or may not be realistic for future PEC devices. It becomes visible that, depending on the choice of the materials and the irradiation concentration used, the device

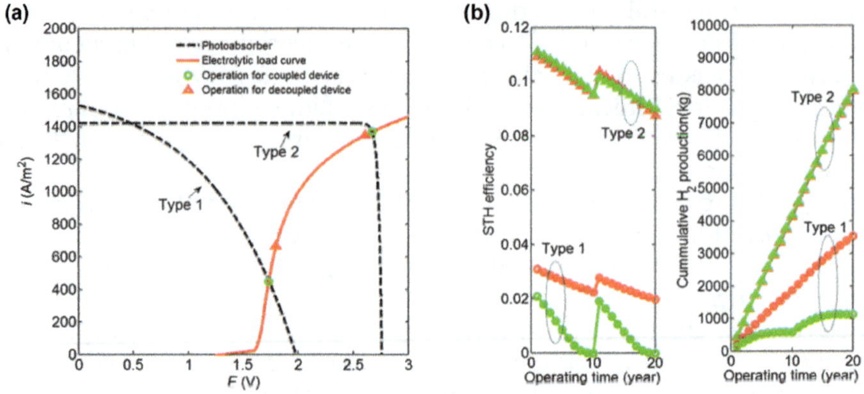

Figure 9.3 (a) Current density *versus* potential for two types of photoabsorbers (type 1: Si-based photoabsorber at irradiation concentration of 87; type 2: III–V based photoabsorber at irradiation concentration of 50), both interacting with a Pt-RuO_x electrocatalyst with a current dilution of 0.15. The operating points for a coupled device (PEC and integrated PEC) and for a PDEC with power electronics (95% efficiency) are indicated. (b) Operational time dependent solar-to-hydrogen efficiency and cumulative hydrogen production of device types 1 and 2 with coupled (green) and uncoupled design (red). For type 1, the photoabsorber and electrocatalysts are exchanged after 10 years, for type 2, the electrocatalyst is exchanged after 10 years.

can operate at a higher efficiency throughout its complete operation when operated decoupled (type 1), or work slightly better or similarly well when operated coupled (type 2).

It becomes apparent that it is not straightforward to conclude that either the coupled or decoupled approach is more beneficial for the device operation over the lifetime. This depends on the actual device design, operating condition, material choices, and actual degradation rates (which surely will be different for the coupled and decoupled cases). However, it becomes clear that the device performance has to be measured and predicted for a variety of different conditions and throughout the lifetime in order to get a complete picture of the feasibility of a design for a practical and robust system.

9.3 Multi-dimensional Models

Even though 0D models can provide useful insights into the photoelectrochemical process and provide general guidance on material choices, designs and interesting device groups, it is more difficult to use them to provide understanding of the detailed multi-physical coupling and multiscale effects. The *iV* curves are relatively insensitive to variations in conditions and the *iV* performance assessment is, in fact, an ill-defined inverse problem. Consequently, models with greater physical detail and multidimensions need to be considered to be more precise and specific in the device design and material guidance. Furthermore, only multi-dimensional models are able to resolve heterogeneities in the variable and vector fields (such as temperature and velocity), which may significantly affect the performance and degradation of devices. We will focus on two example cases: (i) a 1D model which resolves the semiconductor–electrolyte interface including the space charge layer and can provide guidance on semiconductor material targets, and (ii) a 2D model which resolves the energy conservation equation and can provide guidance on the thermal management in (integrated) PEC and PDEC devices.

9.3.1 One-dimensional Models

The semiconductor–electrolyte interface plays a key role in the performance of water-splitting photoelectrodes and therefore has been intensively studied over the last decades.[16–20] The first analytical model was based on a semiconductor-metal interface proposed by Gärtner[21] and was assumed to be valid for the semiconductor–electrolyte interface. This model assumed no majority carrier flux from the semiconductor to the electrolyte, no recombination in the space charge region, and a minority concentration at the interface of zero. Sah *et al.*[22] introduced a model accounting for recombination in the space charge layer for a *p–n* junction that could also be applied to semiconductor–electrolyte interface. Reichman[23] proposed a more precise model than the Gärtner model by adding a hole concentration at the semiconductor–electrolyte interface and the possibility for electrons to flow from the semiconductor to the electrolyte. He also included the model of Sah *et al.* to take into account the space charge region recombination. This model has been extensively used to describe semiconductor–electrolyte interfaces and to explain, for example, the photocurrent's dependency on the doping concentration of an *n*-GaN sample.[24] Wilson[25] has developed a model to describe the photocurrent–voltage curve of semiconductor that includes the effects of recombination at a discrete or continuous distribution of surface states in the band gap. Such 1D numerical models of the semiconductor–electrolyte interface can provide additional insights and predict local effects that cannot be captured by analytical models or experiments.

One of the first 1D models integrating light absorption, charge transport in the semiconductor and in the electrolyte, and charge transfer to the metallic catalyst has been used to quantify the dependency of the device performance on the choice of the light absorber.[26] Mills *et al.* have developed a 1D model that describes the current–voltage behaviour at various semiconductor–catalyst–electrolyte interfaces accounting for different types of catalysts, including metallic, adaptive, and molecular ones.[20] Cendula *et al.* have developed a 1D model of the charge transport in the semiconductor and have enhanced the understanding of the energy band dynamics of photoabsorbers in direct contact with an electrolyte.[27] More recently, we have contributed to these activities and have developed a 1D model that combines electromagnetic wave propagation and charge transport in the semiconductor considering surface recombination.[28]

We validated the model with experiments and used statistical methods to provide guidance on the significance of material characteristics and operating parameters on the performance of the device. Here, we show how such a model can be exploited to determine and quantify key factors (in terms of material characteristics) on the performance of water-splitting photoelectrodes.

9.3.1.1 Governing Equations

Photoabsorber—The generation rate of electron-hole pairs inside the semiconductor material is usually determined by the Beer–Lambert law in combination with ray tracing methods.[29] These methods are limited to planar films with a thickness larger than the wavelength of the incoming light. The solution of Maxwell's curl equations remains the only exact approach to solve for the absorbed radiation in cases where we deal with photoelectrodes made of morphologically complex structures in the range of nanometres or in thin films with multiple interferences between the forward and backward propagating waves:[30]

$$\nabla \times (\nabla \times \mathbf{E}(\mathbf{x},\nu)) - k_0^2\ \tilde{n}(\mathbf{x},\nu)^2 \mathbf{E}(\mathbf{x},\nu) = 0, \tag{9.4}$$

where $\mathbf{E}(\mathbf{x},\nu)$ is the spatial electric field depending on the frequency, ν, of the incoming light, k_0 is the free-space wavenumber, and $\tilde{n}(\mathbf{x},\nu) = n(\mathbf{x},\nu) - jk(\mathbf{x},\nu)$ is the complex refractive index. The relative permeability of the material is assumed $\mu_r = 1$ and the conductivity $\sigma = 0$ in eqn (9.4). The absorbed optical power is given by eqn (9.5):

$$P_{abs}(\mathbf{x},\nu) = -\pi\nu|\mathbf{E}(\mathbf{x},\nu)|^2\ \epsilon''(\nu), \tag{9.5}$$

where $\epsilon''(\mathbf{x},\nu)$ is the imaginary part of the complex permittivity, $\epsilon''(\mathbf{x},\nu) = 2n(\mathbf{x},\nu)k(\mathbf{x},\nu)$. The number of photons absorbed is given by the absorbed optical power divided by the energy of the photon, $g(\mathbf{x},\nu) = P_{abs}(\mathbf{x},\nu)/(h\nu)$. For non-monochromatic light sources, the generation rate is the integration of g over the entire spectrum (eqn (9.6)):

$$G(x) = \int_{\nu_{max}}^{\infty} g(\mathbf{x},\nu)\mathrm{d}\nu, \tag{9.6}$$

with the upper integration boundary $\nu_{max} \geq E_{gap}/h$.

Charge transport and conservation—The static behaviour of the charge carriers in the semiconductor is calculated by solving Poisson's equation (eqn (9.7)):[31]

$$\nabla \cdot (\epsilon_0 \epsilon_r \nabla \phi) = -\rho = q(n - p + N_A{}^- - N_D{}^+), \tag{9.7}$$

with the electron and hole density, n and p, and the ionised acceptor and donor concentrations, $N_A{}^-$ and $N_D{}^+$. The carrier density is given by integrating the product of the Fermi–Dirac distribution and the density of states over all energy states. The transport of the carriers is given by the drift-diffusion equation for electrons and holes inside the semiconductor and the steady-state charge conservation is given by eqn (9.8):

$$\frac{1}{q}\nabla \cdot i_{n/p} = U_{n/p}, \tag{9.8}$$

where $U_{n/p}$ is the net electron or hole recombination rate (eqn (9.9)),

$$U_{n/p} \equiv R_{n/p}^{\mathrm{SRH}} + R_{n/p}^{\mathrm{rad}} + R_{n/p}^{\mathrm{Au}} + R_{\mathrm{s},n/p} - G_{n/p}, \tag{9.9}$$

with $R_{n/p}^{\mathrm{SRH}}$ the bulk Shockley–Read–Hall recombination, $R_{n/p}^{\mathrm{rad}}$ the bulk radiative recombination, $R_{n/p}^{\mathrm{Au}}$ the Auger recombination, $R_{\mathrm{s},n/p}$ the surface recombination and $G_{n/p}$ the generation rate of electron–hole pairs. The single level surface recombination can be combined with the bulk Shockley–Read–Hall recombination using an effective lifetime, $\tau_{\mathrm{eff},n/p}$ (eqn (9.10)):[32,33]

$$\frac{1}{\tau_{\mathrm{eff},n/p}} = \frac{1}{\tau_{n/p}} + \frac{1}{\tau_{\mathrm{s},n/p}}, \tag{9.10}$$

where $\tau_{n/p}$ is the bulk electron or hole lifetime and $\tau_{\mathrm{s},n/p}$ is the electron or hole surface lifetime. The charge transfer *via* tunnelling from the semiconductor to the species in the electrolyte considering the redox level, the reorganisation energy and the concentration of the species can be implemented using the Gerischer theory.[34] Here, we present a simplified charge transport model using the Schottky contact mechanism.[28] It assumes no potential drop and no redox species limitation in the electrolyte. All current limitations are based on a single kinetic factor, termed the surface transfer kinetic velocity, ν_{s}. The current density at the semiconductor–electrolyte interface follows a Schottky contact mechanism (eqn (9.11)):

$$i_{\mathrm{sc}} = q(\nu_{\mathrm{s},p}(p_{\mathrm{s}} - p_{\mathrm{s,eq}}) - \nu_{\mathrm{s},n}(n_{\mathrm{s}} - n_{\mathrm{s,eq}})), \tag{9.11}$$

where $p_{\mathrm{s,eq}}$ and $n_{\mathrm{s,eq}}$ are the surface hole and electron concentrations at equilibrium, p_{s} and n_{s} are the surface hole and electron concentrations at non-equilibrium. The surface charge carriers at equilibrium are given by (eqn (9.12) and (9.13)):

$$p_{\mathrm{s,eq}} = N_{\mathrm{V}} e^{-(E_{\mathrm{gap}} - \varphi_{\mathrm{B}})/V_{\mathrm{th}}}, \tag{9.12}$$

$$n_{\mathrm{s,eq}} = N_{\mathrm{C}} e^{-\varphi_{\mathrm{B}}/V_{\mathrm{th}}}, \tag{9.13}$$

where the barrier potential height is $\varphi_{\mathrm{B}} = \varphi_{\mathrm{sc}} + E_{\mathrm{C}} - E_{\mathrm{F}}$ and $V_{\mathrm{th}} = k_{\mathrm{B}}T/q$. The space charge region potential, φ_{sc}, is given by the applied potential and the flatband potential by assuming no potential drop in the electrolyte, $\varphi_{\mathrm{sc}} = V_{\mathrm{a}} - V_{\mathrm{FB}}$ *vs.* RHE. The surface charge carriers at non-equilibirum are expressed by the quasi-Fermi levels, $E_{\mathrm{F}}^{\mathrm{S}}$, at the surface (eqn (9.14) and (9.15)):

$$p_{\mathrm{s}} = N_{\mathrm{V}} e^{-\left(E_{\mathrm{F},p}^{\mathrm{S}} - E_{\mathrm{V}}^{\mathrm{S}}\right)/V_{\mathrm{th}}}, \tag{9.14}$$

$$n_{\mathrm{s}} = N_{\mathrm{C}} e^{-\left(E_{\mathrm{C}}^{\mathrm{S}} - E_{\mathrm{F},n}^{\mathrm{S}}\right)/V_{\mathrm{th}}}, \tag{9.15}$$

where E_V^S and E_C^S are the valence and conduction band energy levels at the surface.

9.3.1.2 Example Results

Photogeneration—An interesting example of the electromagnetic wave propagation model is the case of photogeneration inside a thin film of $In_xGa_{1-x}N$. The band gap of $In_xGa_{1-x}N$ can be tuned from 0.7 eV ($x=1$) to 3.42 eV ($x=0$) by modifying the indium content, x.[35] The wide range of possible band gaps makes this material promising for high-performing photoelectrodes. A theoretical maximum efficiency of 22.5% can be predicted using Si as a bottom absorber and $In_xGa_{1-x}N$ as a top absorber with a band gap between 1.6 eV to 1.8 eV.[15] The generation rates inside the $In_xGa_{1-x}N$ layer for varying band gaps are depicted in Figure 9.4. Maxwell's curl equation (eqn (9.4)) is solved considering an incoming irradiation of AM1.5G ($\lambda \geq 400$ nm) travelling through a layer of water with a thickness of 1 cm, a 200 nm thin layer of $In_xGa_{1-x}N$ and a 350 μm thick Si layer. The model accounts for the reflection losses at the semiconductor–electrolyte interface and the absorption losses in the electrolyte. The optical properties of $In_xGa_{1-x}N$ for varying indium content were taken from Hazari *et al.*[36] and the Si optical properties from Edwards.[37] The generation rate is wavy due to wave interferences (Figure 9.4). $In_xGa_{1-x}N$ deposited on Si shows no

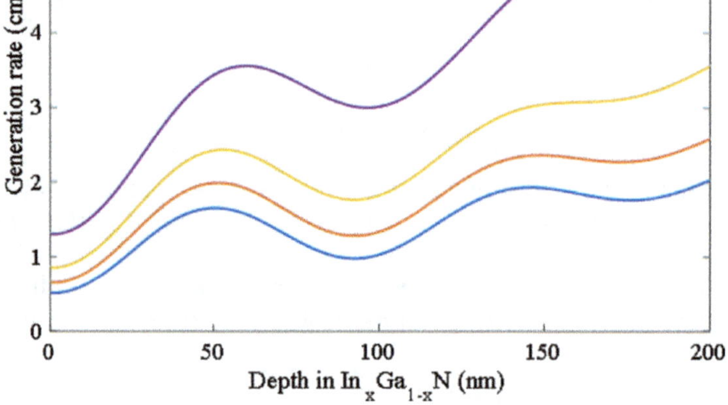

Figure 9.4 Carrier generation rate along the depth of the $In_xGa_{1-x}N$ layer for varying band gap.

resonant light trapping[38] since the Si layer is not reflective enough. This electromagnetic wave propagation model can predict complex generation rates due to light interferences where simpler model such as the Beer–Lambert law or ray-tracing approaches would fail.

Photocurrent—The presented numerical model combining electromagnetic wave propagation and charge transport and conservation was validated using a 1 μm thick non-intentionally doped *n*-GaN film immersed in 1 M H_2SO_4 exposed to a UV light emitting diode.[28] The validated numerical model was then used to predict the impact of key material characteristics (such as hole surface lifetimes) on the performance of the GaN photoanodes (Figure 9.5(a)). A larger hole lifetime does not only increase the photocurrent but it also benefits the onset potential and it impacts the photocurrent at larger potentials. This effect has been experimentally observed for titania and hematite photoelectrodes with phosphate ions deposited on the surfaces. Phosphate ions were found to have a beneficial impact on the photocurrent at low as well as high potentials.[39,40] For these two materials, it could be assumed that the phosphate ions increase the hole surface lifetime.

The impact of doping concentration on the photocurrent for GaN photoanodes is depicted in Figure 9.5(b). At a doping concentration of $1\times10^{14}\,\mathrm{cm}^{-3}$, the photocurrent–voltage behaviour follows a linear dependency. The semiconductor is fully depleted (Figure 9.6). There is no band bending at the semiconductor–electrolyte interface but instead a linear potential drop along the entire semiconductor. The photocurrent is mostly driven by the drift current in the developed space charge region at the semiconductor–electrolyte interface. These two effects lead to a linear dependency of the photocurrent on the potential. An optimal doping concentration of $1\times10^{16}\,\mathrm{cm}^{-3}$ was found above a potential of 0.2 V *vs.* RHE caused by different and opposing effects. On one hand, the recombination rate increases at higher doping concentrations due to higher electron concentrations and the onset potential is reduced (Figure 9.5). The charge mobility is also decreased at higher doping concentration.[24] On the other hand, the doping concentration affects the space charge layer width given by (eqn (9.16)):[34]

$$W = \sqrt{\frac{2\epsilon_0\epsilon_r}{qN_\mathrm{D}^+}(\phi_\mathrm{sc} - V_\mathrm{th})}, \tag{9.16}$$

and therefore, the electric field. A higher doping concentration will lead to a thinner space charge region but a larger electric field (Figure 9.6). The latter will enhance the charge extraction in the space charge region but will reduce it outside of the space charge region. Therefore, the optimal space charge region width is highly depending on the generation rate profile along the semiconductor and is obtained for a doping concentration of $1\times10^{16}\,\mathrm{cm}^{-3}$ in our case (Figure 9.6). These different effects must be carefully considered when optimising the doping concentration for enhanced photocurrent.

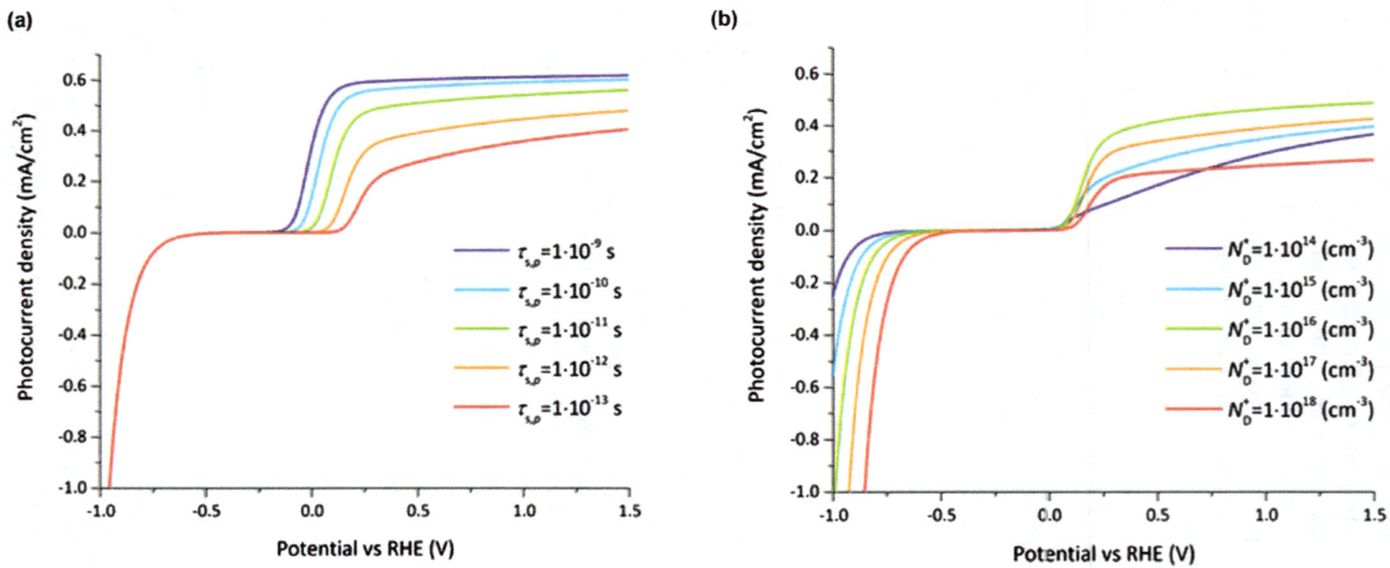

Figure 9.5 Numerically simulated linear sweep voltammograms for varying (a) hole surface lifetimes, and (b) doping concentrations, both based on non-intentionally doped *n*-GaN film of 1 μm thickness immersed in 1 M H_2SO_4. Reproduced from ref. 28 with permission from The Royal Society of Chemistry.

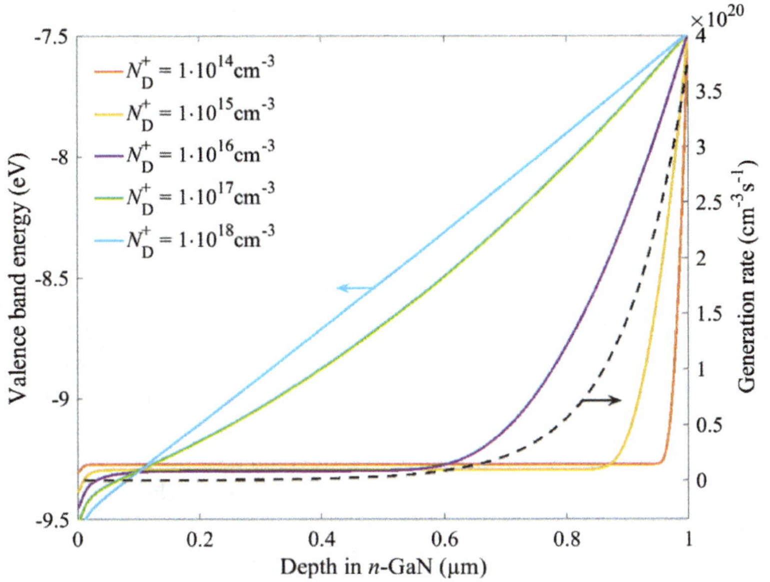

Figure 9.6 Valence band bending for different doping concentration and the generation rate along the *n*-GaN film thickness.

9.3.2 Two-dimensional Models

As discussed, the economic competitiveness of PEC and PDEC devices can be increased by considering concentration of irradiation.[11,41,42] Two examples of experimental investigation of decoupled PDEC devices using concentrated irradiation demonstrate that high efficiencies are achievable.[43,44] However, a variety of challenges are related to the actual implementation of such concentrated PEC or PDEC devices: (i) large current densities (approximately proportional to the concentration factor) which result in large overpotentials and mass transport limitations,[45] (ii) decreased performance of the photoactive components due to increased temperature, and (iii) dry-out of the (polymeric) solid electrolyte at temperatures above 120 °C.[15,46,47] However, higher temperatures are expected to enhance the reaction kinetics and ionic transport. This competing and coupled behaviour of the components requires a detailed understanding of the heat transfer, charge transport, fluid flow, and reaction kinetics in order to formulate performance optimisation strategies for concentrated PEC and PDEC devices. Furthermore, local hot spots in temperature or current can lead to overstraining of parts of the components (such as the catalyst or the photoabsorber) and might lead to faster degradation, hampering the performance of the overall device. There are few modelling investigations of PEC and PDEC devices which consider more than 1D. Haussener *et al.*[15,48] developed a 2D PEC device model focusing on the charge transport in the electrolyte and reaction kinetics incorporating the idealised Shockley-Queisser limit for the photoabsorber approximation. In Haussener *et al.*,[15] an isothermal model was used to provide predictions for the temperature-dependent performance behaviour of PEC devices. Carver *et al.*[49] incorporated detailed fluid dynamics in their model of a flow device in order to understand the conditions under which they were able to drive the device in such a way that the solution was never saturated and therefore bubble build-up and associated challenges were suppressed. Hankin *et al.*[50] used 2D and 3D models coupled to detailed interface kinetic models for hematite photoanodes to predict the performance when scaling these photoelectrodes.

We contributed to the multi-dimensional efforts by incorporating the detailed solution of the energy conservation equations in 2D PEC and PDEC device models allowing for the calculation of the spatial variations in temperature.[51,52] Understanding the details of the energy balance allows for a better—in fact alternative—understanding (compared to eqn (9.3)) on where the energy is lost and how more efficient devices can be designed and what materials they should be composed of. Furthermore, understanding how the temperature is varying spatially in the device allows for the development of smart thermal management strategies in which the energy is transported away from the photoabsorber—whose performance decreases with increase in temperature—towards the electrocatalysts—whose performance increases with temperature. Such strategies allow guiding the operation and design to maximise high-temperature advantages in kinetics and transport while minimising high-temperature disadvantages in charge transport and recombination in the photoabsorber.

Here, we discuss the advantages of a multi-dimensional solution of the energy equation in an integrated PDEC device model. Figure 9.7 depicts the considered device design. Concentrated solar irradiation irradiates the device from the *x*-direction where it is absorbed by a buried dual/triple-junction (*e.g.* $Ga_{0.51}In_{0.49}P$/GaAs or a-Si/μc-Si/μc-Si) PV cell. The generated electron–hole pairs are separated in the *p*–*n* junctions and transported across the various interfaces to the solid electrolyte and the catalysts. The PV and electrolyser can be arranged either in *p*–*n*-cathodic–anodic configuration or in *n*–*p*-anodic–cathodic configuration. The holes, which are delivered to the anode, lead to oxidation of water and production of oxygen and protons at the catalytic sites. The protons pass through the polymeric electrolyte to reach the cathode side where the electrons delivered from the PV's *n*-terminal cause their reduction leading to production of hydrogen at the catalytic sites. We have introduced a water channel between the PV cell's top and the concentrator's exit for cooling the PV cell and to preheat the water (reactant) before it enters the anodic electrolyser channel.

The governing equations are detailed in the corresponding section below and include the energy conservation equation. Additionally, the 2D solution of the coupled equations presented above for light propagation, absorption, and semiconductor physics were taken into account. Fluid flow and reaction kinetics in the solid and liquid electrodes and electrolytes, and at their interfaces were also considered.

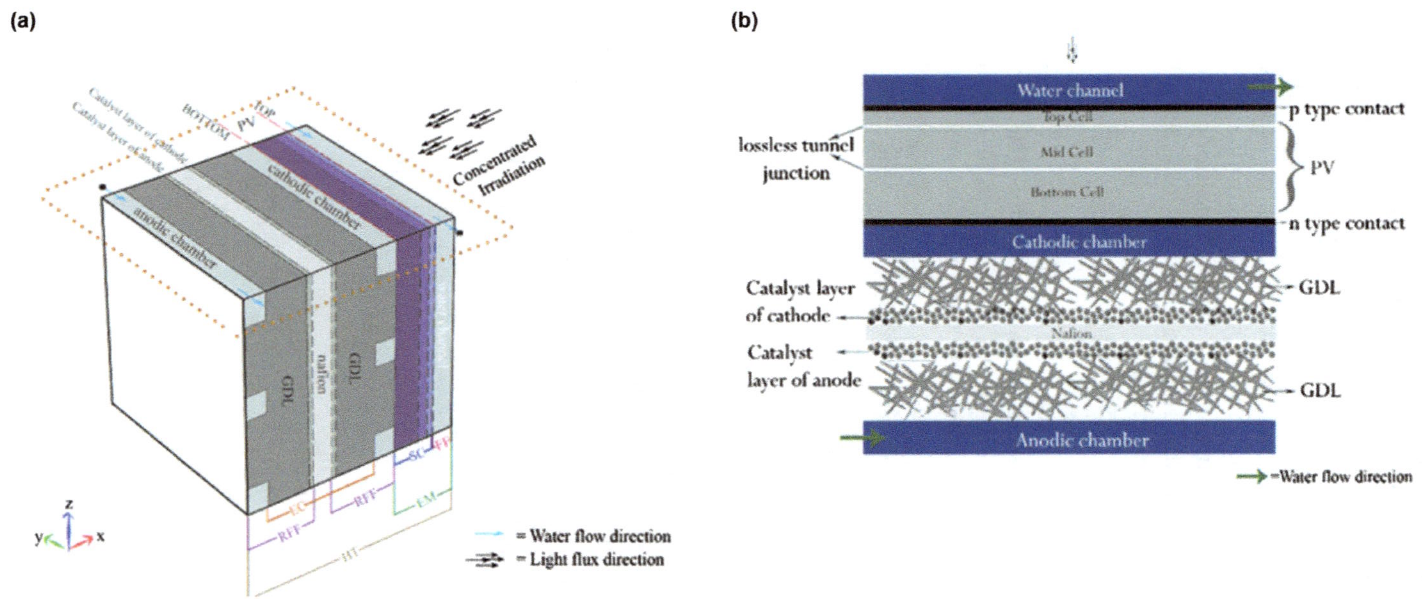

Figure 9.7 Schematic (not to scale) showing (a) 3D view, and (b) the top view of the integrated PEC device depicting incoming concentrated irradiation, cooling and preheating water channel, triple/dual-junction PV, and the integrated electrochemical system consisting of anodic and cathodic channels, gas diffusion layers (GDL), catalyst layers, and a polymeric electrolyte (Nafion®).

Such advanced multi-physics, multi-dimensional models require focusing on the accurate definition of the boundary conditions between the components and the consistency of the physical phenomena. The various conservation and transport equations must be solved with accurate interface conditions for component coupling. This coupling introduces additional complexity, as detailed component models accounting for a subset of physical phenomena rely on the solution of other subsets of equations. Consequently, such coupled modelling efforts require additional external iterative solution steps, which generally increase computational efforts and poses challenges related to model robustness.

9.3.2.1 Governing Equations

Here, we focus on the energy conservation equation only. The net heat source term coming from EM wave propagation is defined to be the sum of the individual source and sink terms (eqn (9.17)–(9.19)):

$$Q_{\mathrm{EM}_\lambda}^{PV} = Q_{TH}|^{\lambda < hc/E_g} + Q_R|^{\lambda > hc/E_g} + Q_M|^{\lambda > hc/E_g}, \tag{9.17}$$

$$Q_R + Q_M = -\nabla \cdot \mathbf{S}_{av_\lambda}|^{\lambda > hc/E_g}, \tag{9.18}$$

$$Q_{TH} = \left(1 - \frac{E_g}{h\nu}\right)(-\nabla \cdot \mathbf{S}_{av_\lambda})|^{\lambda < hc/E_g}. \tag{9.19}$$

Q_R are electrical (resistive) losses, Q_M are magnetic losses, and Q_{TH} are thermalisation losses. For simulation domains not involving semiconductors (such as water), the net heat source is simply $Q_{\mathrm{EM}_\lambda} = -\nabla \mathbf{S}_{\mathrm{av}_\lambda}$, where $\mathbf{S}_{\mathrm{av}_\lambda}$ is the time averaged Poynting vector. The overall Q_{EM} is calculated by summing the net heat source terms over the complete spectrum.

The internal heat source term in the semiconductor is given by eqn (9.20):

$$Q_{\mathrm{PV}} = \mathbf{J} \cdot \nabla V + (E_g + 3k_{\mathrm{B}}T)U, \tag{9.20}$$

where U is the net recombination rate. The first term ($\mathbf{J} \cdot \nabla V$) represents the ohmic losses, and the second term ($E_g + 3k_{\mathrm{B}}T$) gives the recombination losses in the semiconductor.

The heat source in the electrochemical system,[53] comprised of electrode ohmic losses, electrolyte ohmic losses and kinetic losses, is given by eqn (9.21):

$$Q_{\mathrm{EC}} = -\mathbf{J}_{\mathrm{s}} \cdot \nabla \phi_{\mathrm{s}} - \mathbf{J}_{\mathrm{l}} \cdot \nabla \phi_{\mathrm{l}} + \left(\phi_{\mathrm{s}} - \phi_{\mathrm{l}} - E_{\mathrm{eq}} + T\frac{\partial E_{\mathrm{eq}}}{\partial T}\right) i_{\mathrm{loc}}. \tag{9.21}$$

For heat transfer, the steady-state energy conservation equation[54] needs to be solved in order to calculate the temperature field in the PDEC device. $Q_{\mathrm{H}} = Q_{\mathrm{EM}} + Q_{\mathrm{PV}} + Q_{\mathrm{EC}}$ describes the total heat which includes the heat from electromagnetic heating, charge transport in the semiconductor, electrolyte and electrode, and electrochemical reactions.

Non-isothermal treatment needs to be considered in the entire modelling framework. Variations of temperature-dependent parameters, detailed simulation flow, and coupling between different physical nodes have been shown in Tembhurne *et al.*[51]

9.3.2.2 Example Results

A triple junction thin film a-Si/μc-Si/μc-Si cell having a *p–n*-cathodic–anodic configuration is considered under concentrated irradiation. The results predicted that the operational temperature of all components could be maintained below 80 °C, implying that it can be operated with liquid water and no drying out of the polymeric electrolyte membrane. Generally, the model predicts that operation of such a device with high irradiation concentration is feasible.

The water mass flow rate played a major role in the thermal management. Water with an initial temperature of 20 °C was fed through the inlet with varying velocities (range: 0.03 $m\,s^{-1}$ to 2 $m\,s^{-1}$) corresponding to varying mass flow rates per length (range: 6 $g\,s^{-1}\,m^{-1}$ to 400 $g\,s^{-1}\,m^{-1}$). The objective functions H_2 production rate and solar-to-hydrogen efficiency were studied for different water mass flow rates and different irradiation concentrations (1–1000). We noted two different regimes when changing the water mass flow rate. These two regimes resulted from the behavior of the *iV* curves of the electrolyser component, which intersected for varying mass flow rates: at low potentials, low mass flow rates resulted in smaller overpotentials while at high potential low mass flow rate resulted in larger overpotentials. At low irradiation concentrations, a reduction in water mass flow rate leads to increased performance as the temperatures in the electrolysing component increase and therefore reducing its ohmic losses and the reaction overpotentials. At high concentrations, increased water mass flow rates increased the performance due to the fact that we run into mass transport limitations. The formation of these two characteristic operational regimes leads to trend reversals in the objective functions of solar-to-hydrogen efficiency and H_2 production when increasing the irradiation concentration.

The solar-to-hydrogen efficiency and the mass production rate of hydrogen behave opposite to each other with respect to irradiation concentration: The hydrogen production rate is proportional to the operating current density, while the efficiency generally suffers with higher concentrations. This leads to the formation of a Pareto frontier when plotting solar-to-hydrogen efficiency as a function of hydrogen production rate (Figure 9.8). This front consists of device configurations with the largest mean flow velocity. The effective cooling provided by the water channel on the top of the PV, with a mass flow rate of 40 $g\,s^{-1}\,m^{-1}$ (equal to a velocity of 0.2 $m\,s^{-1}$), is found to be optimal at all irradiation concentrations. Large water mass flow rates help in alleviating mass transport limitation and have better cooling capability than smaller flow rates. In terms of irradiation concentration, the designer will have to choose according to what she or he values more: efficiency or

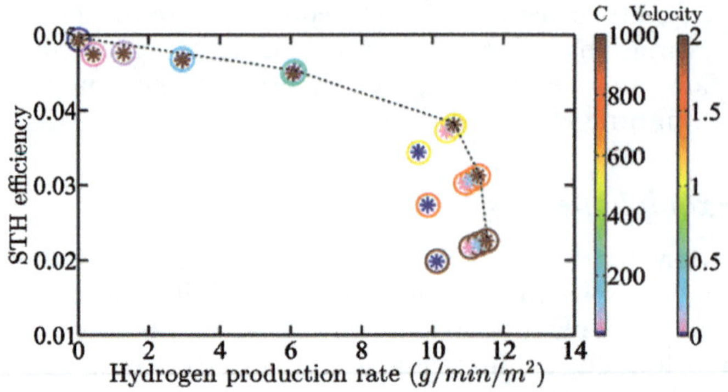

Figure 9.8 Operating STH efficiency for varying mean flow velocities and C as a function of hydrogen production rate. The Pareto front in (e) is shown by the dotted black line. The color of the circles represents concentration and the color of the stars represents the mean flow velocity [m s^{-1}]. Reproduced from S. Tembhurne and S. Haussener, Integrated Photo-Electrochemical Solar Fuel Generators under Concentrated Irradiation—Part II: Thermal Management a Crucial Design Consideration, *J. Electrochem. Soc.*, 2016, **163**, H999–H1007,[52] http://dx.doi.org/10.1149/2.0321610jes.

production rate. Based on this choice, a design point on the front can be chosen.

The formation of two distinctive regimes of operation of PDEC allowed water flow rate-based control of the operating point. Efficiency improvements as high as 12–17%[52] could be achieved just by water flow rate control. In fact, the water flow rate could act as an internal optimum power point tracker eliminating any need for external DC–DC based power point trackers. Flow based control of the PEC/PDEC device would ultimately help in getting increased and stabilised efficiency and productions rates in spite of daily and seasonal irradiance variations. It also helps in alleviating the degradation-related drop in efficiency occurring during the lifetime of the device and, hence, ultimately improves the device's operational time.

Utilising the model for parametric studies indicated that the hydrogen production and solar-to-fuel efficiency increased for more active catalysts, *i.e.* increased exchange current density and active specific surface area (ASSA), at a particular irradiation concentration. Hydrogen production was maximised at large irradiation concentration while solar-to-fuel efficiency was maximised at small irradiation concentration. Consequently, again a Pareto front appeared which consists of cases with large exchange current densities and large ASSA. Interestingly, at low and very high concentrations, the ASSA and exchange current density did not influence the performance, therefore the device performance was independent of any choice of catalyst material. The membrane thickness was important for the performance of the integrated device, as its variation led to significant changes in the operating points. The mass transport limitations were instigated earlier with increased membrane thickness, leading to reduced saturation currents. Significant changes in the heat sources within the photoabsorber and electrocatalysts were observed with varying membrane thicknesses due to large operating current density changes. However, the water mass flow rate ensured proper device cooling, resulting in small temperature variations (a few Kelvin only). The highest H_2 production occurred at the highest irradiation concentration and for the thinnest membranes, whereas the maximum solar-to-hydrogen efficiency was observed at lower concentrations (below 200), and was found to be independent of the membrane's thickness. The objective functions of H_2 production and solar-to-hydrogen efficiency formed also a Pareto front which consisted mainly of the cases with the thinnest membrane. The other dimensional properties, GDL and catalyst thicknesses, did not lead to significant changes in the operating point with parameter variations. The heat sources, and hence temperature, were found to be insensitive to these parameters. The Pareto front for these two cases consisted mainly of the thinnest GDL and thicker catalyst.

For increasing dimensional, operational, and material parameters, the H_2 production and solar-to-fuel efficiency exhibited a saturating trend, shown in Figure 9.9. A GDL thickness of 300 μm, a catalyst thickness of 200 nm, a mean flow velocity of 0.2 m s^{-1} (40 g s^{-1} m^{-1}), an exchange current density of 50 times the standard value, and an ASSA of 10^5 were found to be

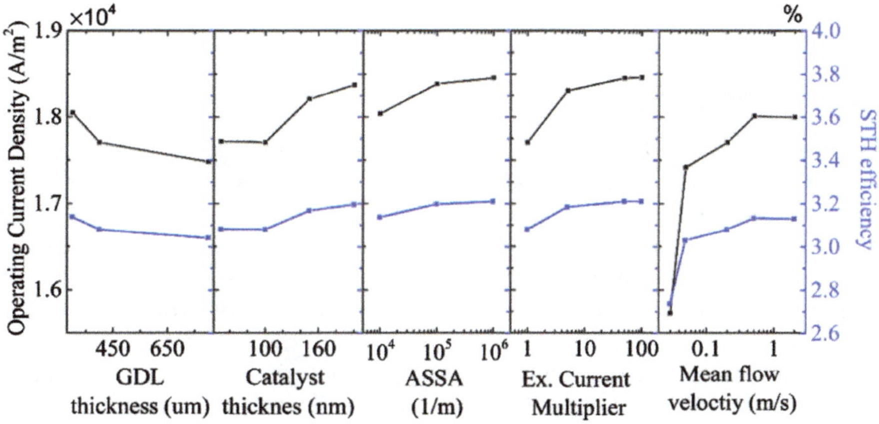

Figure 9.9 Operating current density (left *y*-axis) and solar-to-hydrogen (STH) efficiency (right *y*-axis) as a function of GDL thickness, catalyst thickness, active specific surface area, exchange current density multiplier, and mean flow velocity, for a-Si/μc-Si/μc-Si based device at $C = 707$.
From S. Tembhurne and S. Haussener, Integrated Photo-Electrochemical Solar Fuel Generators under Concentrated Irradiation—Part II: Thermal Management a Crucial Design Consideration, *J. Electrochem. Soc.*, 2016, **163**, H999–H1007,[52] http://dx.doi.org/10.1149/2.0321610jes.

the optimum values for enhanced performance of the PDEC device for an irradiation concentration of 707. Similar values were observed for other irradiation concentration factors.

Our 2D model allowed for the calculation of the spatially resolved temperature profiles. Such information helps in choosing a concentrated PDEC device configuration (*e.g.* *p–n*-cathodic–anodic or *n–p*-anodic–cathodic) which reduces the thermal hot spots under high irradiation concentration operation. More homogeneous temperature fields will also decrease the thermal and operational stress of the photoabsorber or catalyst. For any PEC/PDEC system, the developed 2D model shows that the optimised rather than the maximised dimensional, operational and material parameters should be chosen. Also, the operational parameters such as input water flow rates should be controlled or adapted—and not be kept constant—depending on the transient to which the system is exposed (transients in irradiation or time-dependent changes in material properties through degradation). For any PEC or PDEC system, external peripherals such as DC–DC based maximum power point trackers could be eliminated with smart controlling loops acting, for example, on the water flow rate. Employing concentrated irradiation increases the economic competitiveness of any PEC or PDEC system and smart thermal management and controlling strategies might provide a competitive advantage for cost effective fuel production. Ultimately, smart thermal management and controlling might become a rationale for integration in PEC and PDEC devices (compared to completely separated PV plus electrolysis systems).

9.4 Conclusion

We have discussed the merits of computational modelling of PEC and PDEC devices. Specifically, we have shown that different levels of complexity (in terms of dimensionality as well as incorporation of physical phenomena) can provide different insights and guidance for the development of efficient, cost competitive, sustainable, robust, and practical PEC and PDEC devices.

0D models are a computationally inexpensive way of studying PEC and PDEC devices. They can provide insights on the maximum achievable efficiencies, they can predict the impact of materials choices (for example photoabsorber, catalyst, or electrolyte) on the device performance, they can guide the choice of operational conditions (concentration of irradiation or current), and they can estimate how degradation can affect the device performance. Coupling these simple models to economical and life cycle assessments allow for providing holistic design guidelines for PEC and PDEC devices and their implementation. Holistic approaches consider efficiency but also cost, sustainability and lifetime. The shortcoming of 0D models is related to their inability to include more complex physical phenomena or to account for spatial variability of, for example, the temperature, charge or flow fields in the device. Consequently, in cases where design questions or detailed multi-dimensional understanding of the variable fields is required, multi-dimensional models are required.

1D models are particularly suited to look into the details of the semiconductor–electrolyte interface, the core of any PEC device. Such models can resolve the space charge layer and can provide detailed guidance on the material requirements and required surface treatments of photo-electrochemical components. Furthermore, they can provide guidance on the dimensions of the device. In devices where two leading physical phenomena have perpendicular directions (often chosen by design), it becomes imperative to go to 2D (or more complex) geometries for understanding such coupled behaviour. 2D or 3D models allow for the investigation of any complex design and geometry, and simulation in fine physical detail. We presented a 2D model of a PDEC device as an example, in which we specifically solved for the energy conservation equation. Our developed 2D model proved to be a valuable tool for the design of a PDEC device that was exposed to highly concentrated irradiation. The results not only provided guidance on design, component dimensions, and material choices, but it also predicted that smart thermal management can enhance the efficiency of such devices and provide an approach for adaptive controlling of the device performance. This observation was only possible due to the multi-dimensional nature of the PDEC device model.

Overall, the development of PEC and PDEC device models is extremely useful in exploring theoretical limits of such devices, in guiding the choice of practical device classes, in supporting the detailed design of devices, in conducting feasibility studies, and generally in guiding research and development on a materials' level as well as a device and technology level. The choice of the complexity of the model in terms of dimensionality and incorporation of physical phenomena depends on the research questions to be answered.

Acknowledgements

This material is based upon work performed with the financial support of the Nano-Tera.ch initiative as part of the SHINE project (Grant #145936), the Starting Grant of the Swiss National Science Foundation as part of the SCOUTS project (Grant #155876), and the Project Grant of the Swiss National Science Foundation (Grant #159547).

References

1. N. S. Lewis and D. G. Nocera, Powering the planet: chemical challenges in solar energy utilization, *Proc. Natl. Acad. Sci. U. S. A.*, 2006, **103**, 15729–15735.
2. N. S. Lewis, Research opportunities to advance solar energy utilization, *Science*, 2016, **351**, aad1920.

3. C. Xiang, *et al.*, Modeling, Simulation, and Implementation of Solar-Driven Water-Splitting Devices, *Angew. Chem. Int. Ed.*, 2016, **55**, 12974–12988.
4. M. A. Modestino, *et al.*, Robust production of purified H2 in a stable, self-regulating, and continuously operating solar fuel generator, *Energy Environ. Sci.*, 2014, **7**, 297–301.
5. K. T. Fountaine, H. J. Lewerenz and H. A. Atwater, Efficiency limits for photoelectrochemical water-splitting, *Nat. Commun.*, 2016, **7**, 13706.
6. R. E. Rocheleau and E. L. Miller, Photoelectrochemical production of hydrogen: engineering loss analysis, *Int. J. Hydrogen Energy*, 1997, **22**, 771–782.
7. S. Hu, C. Xiang, S. Haussener, A. D. Berger and N. S. Lewis, An analysis of the optimal band gaps of light absorbers in integrated tandem photoelectrochemical water-splitting systems, *Energy Environ. Sci.*, 2013, **6**, 2293–2984.
8. M. T. Winkler, C. R. Cox, D. G. Nocera and T. Buonassisi, Modeling integrated photovoltaic-electrochemical devices using steady-state equivalent circuits, *Proc. Natl. Acad. Sci.*, 2013, **110**, E1076–E1082.
9. Y. Surendranath, D. K. Bediako and D. G. Nocera, Interplay of oxygen-evolution kinetics and photovoltaic power curves on the construction of artificial leaves, *Proc. Natl. Acad. Sci. U. S. A.*, 2012, 1–5.
10. C. A. Rodriguez, M. A. Modestino, D. Psaltis and C. Moser, Design and cost considerations for practical solar-hydrogen generators, *Energy Environ. Sci.*, 2014, **7**, 3828–3835.
11. M. Dumortier, S. Tembhurne and S. Haussener, Holistic design guidelines for solar hydrogen production by photo-electrochemical routes, *Energy Environ. Sci.*, 2015, **8**, 3614–3628.
12. F. Nandjou and S. Haussener, Degradation in photoelectrochemical devices: review with an illustrative case study, *J. Phys. D. Appl. Phys.*, 2017, **50**, 124002.
13. J. Ronge, *et al.*, Solar Hydrogen Reaching Maturity, *Oil Gas Sci. Technol. – Rev. IFP Energies Nouv*, 2015, **70**, 863–876.
14. J. Ronge, *et al.*, Air-based photoelectrochemical cell capturing water molecules from ambient air for hydrogen production, *RSC Adv*, 2014, **4**, 29286–29290.
15. S. Haussener, S. Hu, C. Xiang, A. Z. Weber and N. Lewis, Simulations of the irradiation and temperature dependence of the efficiency of tandem photoelectrochemical water-splitting systems, *Energy Environ. Sci.*, 2013, **6**, 3605–3618.
16. D. Vanmaekelbergh, Direct and surface state mediated electron transfer at semiconductor/electrolyte junctions—I. A comparison of steady-state results, *Electrochim. Acta*, 1997, **42**, 1121–1134.
17. D. Vanmaekelbergh, Direct and surface state mediated electron transfer at semiconductor/electrolyte junctions—II. A comparison of the interfacial admittance, *Electrochim. Acta*, 1997, **42**, 1135–1141.

18. R. Memming, *Semiconductor Electrochemistry*, Wiley-VCH, 2001.
19. A. J. Nozik and R. Memming, Physical Chemistry of Semiconductor – Liquid Interfaces, *J. Phys. Chem.*, 1996, **100**, 13061–13078.
20. T. J. Mills, F. Lin and S. W. Boettcher, Theory and Simulations of Electrocatalyst-Coated Semiconductor Electrodes for Solar Water Splitting, *Phys. Rev. Lett.*, 2014, **112**, 148304.
21. W. W. Gärtner, Depletion-Layer Photoeffects in Semiconductors, *Phys. Rev.*, 1959, **116**, 84–87.
22. C. Sah, R. N. Noyce and W. Shockley, Carrier Generation and Recombination in P-N Junctions and P-N Junction Characteristics, *Proc. IRE*, 1956, **1**, 1228–1243.
23. J. Reichman, The current-voltage characteristics of semiconductor-electrolyte junction photovoltaic cells, *Appl. Phys. Lett.*, 1980, **36**, 574–577.
24. M. Ono, *et al.*, Photoelectrochemical reaction and H2 generation at zero bias optimized by carrier concentration of n-type GaN, *J. Chem. Phys.*, 2007, **126**, 54708.
25. R. H. Wilson, A model for the current-voltage curve of photoexcited semiconductor electrodes, *J. Appl. Phys.*, 1977, **48**, 4292–4297.
26. A. Berger and J. Newman, An Integrated 1-Dimensional Model of a Photoelectrochemical Cell for Water Splitting, *J. Electrochem. Soc.*, 2014, **161**, 3328–3340.
27. P. Cendula, *et al.*, Calculation of the Energy Band Diagram of a Photoelectrochemical Water Splitting Cell, *J. Phys. Chem. C*, 2014, **118**, 29599–29607.
28. Y. K. Gaudy and S. Haussener, Utilizing modeling, experiments, and statistics for the analysis of water-splitting photoelectrodes, *J. Mater. Chem. A*, 2016, **4**, 3100–3114.
29. M. Modest, *Radiative Heat Transfer*, Academic Press, 2003.
30. D. M. Pozar, *Microwave Engineering*, Wiley, 2012.
31. S. M. Sze and K. K. Ng, *Physics of Semiconductor Devices*, Wiley-Interscience, 2007.
32. K. Gullbinas, V. Grivickas, H. P. Mahabadi, M. Usman and A. Hallen, Surface Recombination Investigation in Thin 4H-SÍC Layers, *Mater. Sci.*, 2011, **17**, 119–124.
33. A. B. Sproul, Dimensionless solution of the equation describing the effect of surface recombination on carrier decay in semiconductors, *J. Appl. Phys.*, 1994, **76**, 2851–2854.
34. R. van de Krol and M. Grätzel, *Photoelectrochemical Hydrogen Production*, Springer, 2012, DOI: 10.1007/978-1-4614-1380-6.
35. P. Aseev, *et al.*, Uniform low-to-high in composition InGaN layers grown on Si, *Appl. Phys. Express*, 2013, **6**, 115503-1–115503-4.
36. A. Hazari, *et al.*, Optical constant of $In_xGa_{1-x}N$ $(0 \leq x \leq 0.73)$ in the visible and near-infrared wavelength regimes, *Opt. Lett.*, 2015, **40**, 3304–3307.

37. D. F. Edwards, Silicon (Si)*, in *Handbook of Optical Constants of Solids*, ed. E. D. Palik, Academic Press, 1985, pp. 547–569.
38. H. Dotan, *et al.*, Resonant light trapping in ultrathin films for water splitting, *Nat. Mater.*, 2013, **12**, 158–164.
39. J. Y. Kim, J.-W. Jang, D. H. Youn, G. Magesh and J. S. Lee, A Stable and Efficient Hematite Photoanode in a Neutral Electrolyte for Solar Water Splitting: Towards Stability Engineering, *Adv. Energy Mater.*, 2014, **4**, 1–7.
40. L. Jing, *et al.*, Dynamics of photogenerated charges in the phosphate modified TiO2 and the enhanced activity for photoelectrochemical water splitting, *Energy Environ. Sci.*, 2012, **5**, 6552–6558.
41. B. A. Pinaud, *et al.*, Technical and economic feasibility of centralized facilities for solar hydrogen production via photocatalysis and photoelectrochemistry, *Energy Environ. Sci.*, 2013, **6**, 1983–2002.
42. M. Dumortier and S. Haussener, Design guidelines for concentrated photo-electrochemical water splitting devices based on energy and greenhouse gas yield ratios, *Energy Environ. Sci.*, 2015, **8**, 3069–3082.
43. A. Nakamura, *et al.*, A 24.4% solar to hydrogen energy conversion efficiency by combining concentrator photovoltaic modules and electrochemical cells, *Appl. Phys. Express*, 2015, **8**, 107101.
44. O. Khaselev and J. A. Turner, A Monolithic Photovoltaic-Photoelectrochemical Device for Hydrogen Production via Water Splitting, *Science*, 1998, **280**, 425–427.
45. H. Nakajima, *Mass Transfer: Advanced Aspects*, InTech, 2011.
46. P. Singh and N. M. Ravindra, Temperature dependence of solar cell performance—an analysis, *Sol. Energy Mater. Sol. Cells*, 2012, **101**, 36–45.
47. Y. Sone, P. Ekdunge and D. Simonsson, Proton Conductivity of Nafion 117 as Measured by a Four-Electrode AC Impedance Method, *J. Electrochem. Soc.*, 1996, **143**, 1254.
48. S. Haussener, *et al.*, Modeling, simulation, and design criteria for photoelectrochemical water-splitting systems, *Energy Environ. Sci.*, 2012, **5**, 9922.
49. C. Carver, *et al.*, Modelling and development of photoelectrochemical reactor for H2 production, *Int. J. Hydrogen Energy*, 2012, **37**, 2911–2923.
50. A. Hankin, *et al.*, From millimetres to metres: the critical role of current density distributions in photo-electrochemical reactor design, *Energy Environ. Sci.*, 2017, **10**, 346–360.
51. S. Tembhurne and S. Haussener, Integrated Photo-Electrochemical Solar Fuel Generators under Concentrated Irradiation – Part I: 2-D Non-Isothermal Multiphysics Modelling, *J. Electrochem. Soc.*, 2016, **163**, H988–H998.
52. S. Tembhurne and S. Haussener, Integrated Photo-Electrochemical Solar Fuel Generators under Concentrated Irradiation – Part II: Thermal

Management a Crucial Design Consideration, *J. Electrochem. Soc.*, 2016, **163**, H999–H1007.

53. A. J. Bard and L. R. Faulkner, *Electrochemical Methods: Fundamentals and Applications*, Wiley, 2001.
54. F. P. Incropera, D. P. DeWitt, T. L. Bergman and A. S. Lavine, *Fundamentals of Heat and Mass Transfer*, John Wiley & Sons, 6th edn, 2007.

CHAPTER 10

Technoeconomic Analysis of PEC Water Splitting at Various Scales

ARTJOM MALJUSCH*[a] AND MICHAEL WULLENKORD[b]

[a] Evonik Creavis GmbH, Paul-Baumann-Str. 1, 45772 Marl, Germany;
[b] Institute of Solar Research, German Aerospace Center (DLR), Linder Hoehe, 51147 Koeln, Germany
*Email: artjom.maljusch@evonik.com

10.1 Introduction

Hydrogen production *via* photoelectrochemical (PEC) water splitting represents a promising route to store intermittent solar energy in a versatile chemical energy carrier and could significantly contribute to a future sustainable energy infrastructure. The transition of a technology from the lab to a real plant, *i.e.* running a plant instead of running an experiment, involves numerous technical and economic aspects which need to be taken into account in order to achieve the final intention: economically viable, ecologically beneficial, and industrially applicable processes.

This work refers to technoeconomic findings obtained within the EU project PECDEMO (2014–2017), which addressed hydrogen production in hybrid photoelectrochemical-photovoltaic (PEC-PV) tandem devices. Section 10.2 compiles some practical considerations and boundary conditions relevant for the realisation of extensive hydrogen production by PEC water splitting. Different hydrogen production and application scenarios, which cover a wide range of hydrogen production capacity, are considered.

Energy and Environment Series No. 20
Advances in Photoelectrochemical Water Splitting: Theory, Experiment and Systems Analysis
Edited by S. David Tilley, Stephan Lany and Roel van de Krol

Published by the Royal Society of Chemistry, www.rsc.org

Section 10.3 comprises information about plant design, main plant components, and their required sizes. The global warming potential and levelized hydrogen production costs represent important quantities for the assessment of hydrogen production technologies. Sections 10.4 and 10.5 provide brief introductions of life-cycle analysis (LCA) and cost estimation, respectively, as well as specific data for the considered scenarios, which are subsequently evaluated in the context of benchmarking hydrogen production processes (Section 10.6).

10.2 Basic Considerations and Definitions

Many different PEC water-splitting configurations have been introduced and discussed. We here exemplarily focus on water splitting in hybrid photoelectrochemical–photovoltaic (PEC–PV) tandem devices which feature at least one photoelectrode in contact with an electrolyte solution. The photoelectrode absorbs the shorter wavelength part of the incoming solar light. A fraction of the transmitted light finally reaches the PV-module behind the photoelectrode, where it is converted to electrical power with a certain efficiency. In this chapter, neither technical details of photoelectrodes and counter electrodes, nor any sophisticated cell designs reported in literature are analysed or discussed; for this, we refer the reader to the preceding chapters in this book. At this stage, the PEC–PV device is treated as a 'black box' system which uses solar irradiation to convert water (H_2O) to hydrogen (H_2) and oxygen (O_2) in separate cell compartments with given solar-to-hydrogen efficiency (STH) following the overall reaction (eqn (10.1)):

$$H_2O \rightarrow H_2 + 0.5\ O_2 \quad (10.1)$$

A schematic of the considered PEC–PV system, or production plant, with its principal mass and energy flows is shown in Figure 10.1. Hydrogen is the main product of the process, whereas oxygen represents a by-product which could, as a side effect, generate additional revenue. Electrical decoupling of the PEC and PV components allows the *in-situ* generation of excess electricity *via* active power management (APM) with a specific solar-to-electricity

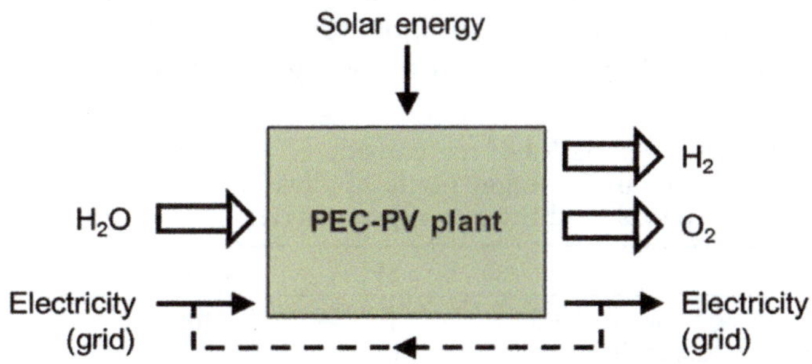

Figure 10.1 Principal mass and energy flows of a PEC–PV plant.

efficiency (STE) as presented in ref. 1. The generated electricity can primarily be used to power plant components and can secondarily be sold to the grid leading to further returns.

10.2.1 Hydrogen Production and Application Scenarios

Numerous scenarios for the generation and utilisation of hydrogen have been introduced.[2] In this chapter, the following three use cases for PEC-PV water splitting are considered: a single home application (SHA), a hydrogen refuelling station (HRS), and an industrial process (IP). The SHA represents a decentralised approach of hydrogen production to meet the energy demand of domestic buildings, *e.g.* by subsequent use in a fuel cell. While the HRS provides hydrogen at a high pressure to fuel diverse vehicles with a pressurised tank, *e.g.* cars and buses, the IP refers to hydrogen usage as feedstock for further processing. The main characteristics of the three scenarios are presented in Table 10.1.

10.2.2 Location

The location for a PEC–PV hydrogen generation plant has to be carefully chosen, since many technical and economic aspects are directly connected to it. Technical issues are *e.g.* the solar irradiance (global and direct), ambient temperature and other conditions such as pressure, wind, and dust, availability of land and cooling water, surface characteristics, and risk of abrasive storms. Additionally, also infrastructure (such as roads and shops) as well as utilities and required site preparation activities are relevant. Economic aspects include *e.g.* market situation, hydrogen demand and need of transport, costs of labour and land, taxes and fees, insurance costs, required contingencies, political stability, subsidies, and feed-in tariffs.

10.2.3 Specific Collector Size and Number of PEC-PV Units

Depending on the STH and the site location, or rather the locally available solar irradiation, the collector size required to achieve a specified hydrogen output can be determined. For example, for Seville (Spain) with an annual global horizontal irradiation of about 1900 $kWh\,m^{-2}\,a^{-1}$ (derived from ref. 3), an assumed 8% STH (based on the higher heating value of hydrogen

Table 10.1 Main characteristics of the considered hydrogen production and application scenarios: single home application (SHA), hydrogen refuelling station (HRS), and industrial process (IP).

Scenario	SHA	HRS	IP
Nominal production rate in $kg\,d^{-1}$	1	400	4000
Absolute outlet pressure in bar	6	810	20
Purity	>85%	99.995%	99.995%

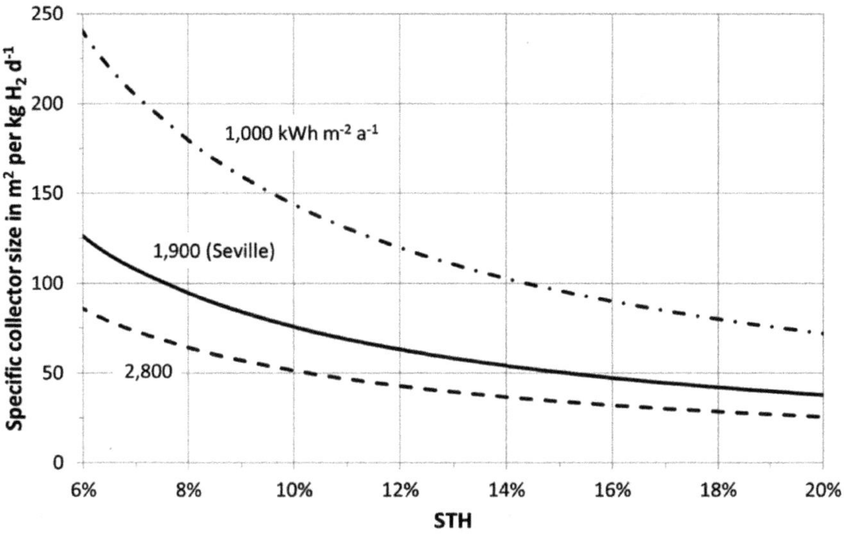

Figure 10.2 Specific collector size as a function of the STH based on the HHV of hydrogen and the annual global irradiation.

of $HHV_{H2} = 285.8\ kJ\,mol^{-1}$), and an average operation of 8 $h\,d^{-1}$, the required specific size of the collector field corresponding to the active PEC–PV area can be calculated to 94.6 m^2 per kg H_2 d^{-1}. As can be seen in Figure 10.2, the specific collector size decreases with rising STH and increasing global irradiation. Doubling the efficiency leads to a reduction of the collector field to half of the original size. An equal influence can be extracted for the global irradiation. Even at high STH and high global irradiation, tens of m^2 of active area are required for an average hydrogen production of 1 $kg\,d^{-1}$. It becomes clear that a massive PEC–PV (and electrolyte solution) inventory far beyond from what is common in the laboratories has to be installed in order to meet hydrogen production capacities appropriate for commercial application and that loss effects might become relevant which are negligible in a lab-scale set-up of the size of several cm^2. The SHA as defined above requires—when realised in Seville—a collector size of 94.6 m^2, which matches well with commonly available roof areas. With 37 825 m^2 and 378 250 m^2 significantly larger collector fields have to be built for the HRS and the IP, respectively, due to the higher specified hydrogen production rates.

In the following we consider a plant with 8% STH based on the HHV, which corresponds to about 6.8% STH based on the lower heating value of hydrogen (LHV_{H2}, 241.8 $kJ\,mol^{-1}$). By using the specific collector size and the active area of one PEC-PV unit the number of PEC–PV units per kg H_2 d^{-1} can be estimated. Respective results are shown in Figure 10.3. If PEC–PV devices with 50 cm^2 active area were used, more than 18 900 units would have to be installed per kg H_2 d^{-1} in Seville (Spain), which is a remarkable number unlikely to be realised at industrial scale. The required number

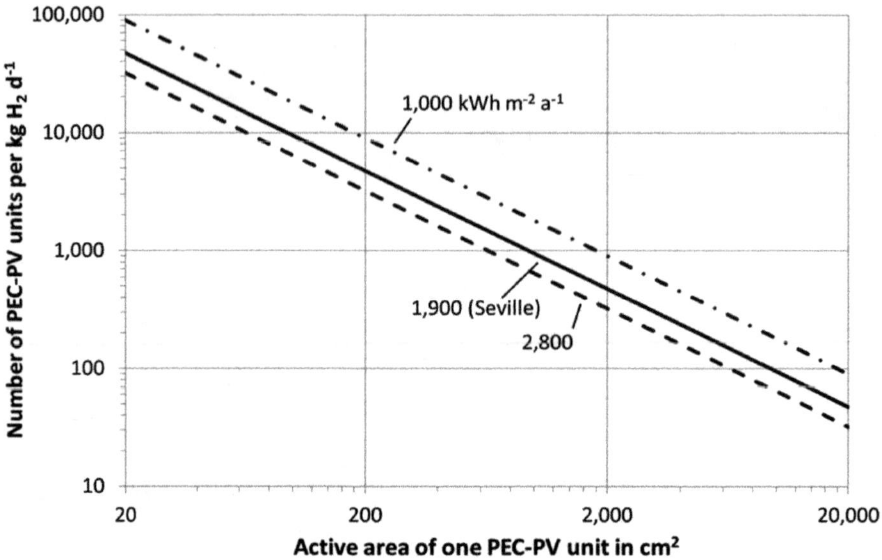

Figure 10.3 Number of PEC–PV units per $kg\,H_2\,d^{-1}$ operating with 8% STH based on the HHV of hydrogen as a function of the active area of one PEC–PV unit and the annual global irradiation.

drops significantly when larger active areas are provided by each single PEC-PV cell, leading to 1051 units with 900 cm^2 active area each (*e.g.* 30 cm×30 cm) and 95 units with 1 m^2 active area each (*e.g.* 1 m×1 m). The number of PEC–PV units and the global irradiation are inversely proportional.

10.2.4 Fundamental Economic Constraints under Non-concentrated Sunlight

An average productivity of 1 $kg\,H_2\,d^{-1}$ leads to a total production of approximately 3650 kg, 7300 kg, and 10 950 kg hydrogen within assumed lifetimes of a suitable solar hydrogen generation plant of 10, 20, and 30 years, respectively. Correspondingly, depending on the target hydrogen production costs, the allowable total costs of the solar plant can be calculated. These total costs comprise not only the capital expenses (CAPEX) for the PEC–PV devices but also the operating expenses (OPEX) as well as the respective costs of all other plant components, insurances and contingencies *etc.*—generally speaking the complete costs associated with the installation and successful operation of a solar plant during the entire lifetime. Figure 10.4 shows the dependency of the aforementioned allowable total plant costs on the target hydrogen costs and the lifetime of the plant built in Seville and operating with 8% STH, which results in a specific collector size of 94.6 m^2 per $kg\,H_2\,d^{-1}$ as shown above. Target hydrogen production costs of 5 € kg^{-1} lead to allowable total plant costs of not more than 193 €, 386 €,

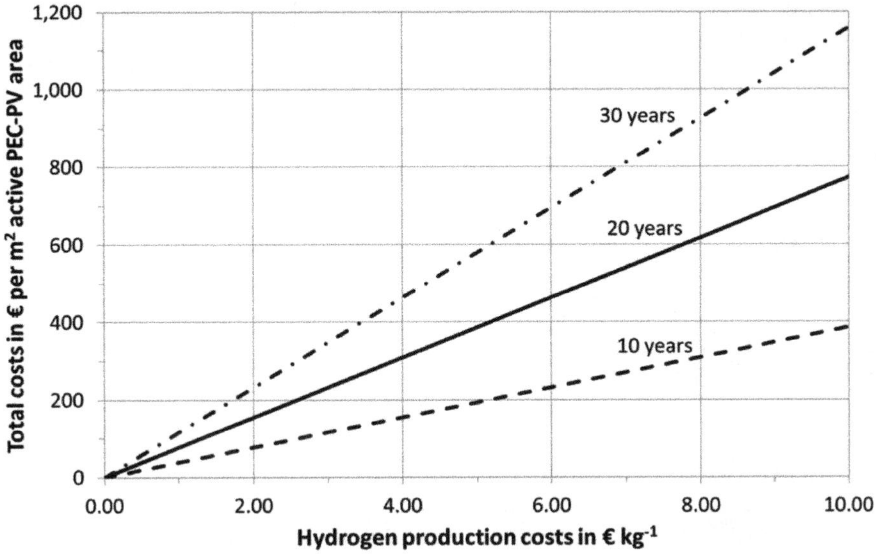

Figure 10.4 Allowable total costs associated with the installation and operation of a solar plant per m^2 active PEC–PV area as a function of the target hydrogen production costs and the plant lifetime based on a specific collector size of 94.6 m^2 per kg H_2 d^{-1}.

and 579 € per m^2 active PEC–PV area for lifetimes of 10, 20, and 30 years, respectively.

Figure 10.5 shows the allowable total plant costs per PEC–PV unit as a function of the active area of one PEC–PV device and the lifetime of the plant considering target hydrogen production costs of 5 € kg^{-1} and a specific collector size of 94.6 m^2 per kg H_2 d^{-1}. If 50 cm^2 devices were used, the total plant costs must be kept below 1 €, 2 €, and 3 € per PEC–PV unit for plant lifetimes of 10, 20, and 30 years, respectively, which suggests that such a configuration will never be commercially available. As the active area per PEC–PV unit increases the allowable total plant costs per device rise and reach about 35 € and 386 € for 900 cm^2 and 1 m^2 active area per PEC–PV unit, respectively, when the plant can be operated for 20 years.

10.2.5 Fundamental Economic Constraints under Concentrated Sunlight

Solar concentrators, composed of mirrors or lenses, shift low irradiance on a large aperture to high irradiance on a small-sized focus with a concentration ratio *C*, which corresponds to the ratio between these two irradiances (and to the ratio between the irradiated surfaces neglecting any losses associated with concentration). The utilisation of solar concentrators thus allows a more compact design of the PEC–PV system which then receives a higher irradiance compared to operation under non-concentrated sunlight. An

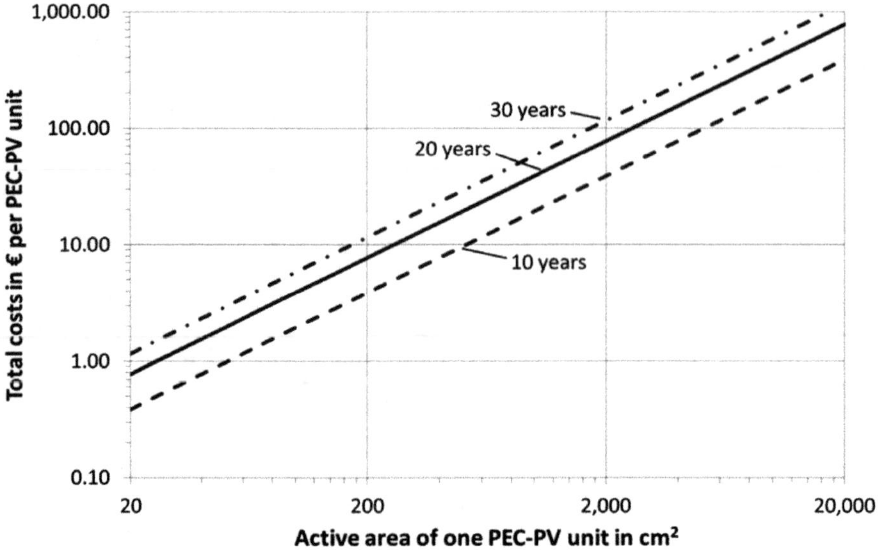

Figure 10.5 Allowable total costs associated with the installation and operation of a solar plant per PEC–PV unit as a function of the active area of one PEC–PV unit and the plant lifetime based on a specific collector size of 94.6 m^2 per kg H_2 d^{-1} and target hydrogen production costs of 5 € kg^{-1}.

added benefit is that the complexity and overall length of the piping system connected to the PEC–PV devices is correspondingly reduced under concentrated sunlight.

Well-known and proven technologies for solar concentration are the parabolic trough characterised by a collective of one-axis tracking parabolic reflectors and an oblong receiver in the focal line as well as the solar tower characterised by a collective of two-axis tracking heliostats pointing at an absorber on top of a tower. Costs for parabolic trough fields and heliostat fields can be found in ref. 4. Without receiver costs, which are not relevant here, CAPEX of $146 per m^2 aperture for a parabolic trough field and $143 per m^2 reflective surface for a heliostat field were estimated for 2015. Based on these numbers we assume the CAPEX of a solar concentrator applicable for PEC–PV technology to currently amount to about 146 € per m^2 aperture area. The OPEX for the concentrating system are neglected here, because the efforts for maintenance are low compared to other components of the plant, in particular the PEC–PV part. Figure 10.6 shows the allowable total plant costs (with and without costs for the concentrating system) per PEC–PV unit considering target hydrogen production costs of 5 € kg^{-1}, a specific collector size of 94.6 m^2 per kg H_2 d^{-1}, and 20 years lifetime of the solar plant. For this approximate illustration, we neglect that solar concentrators utilise only the direct fraction of the solar input, since diffuse light cannot be focused, as well as the optical losses of the concentrating system.

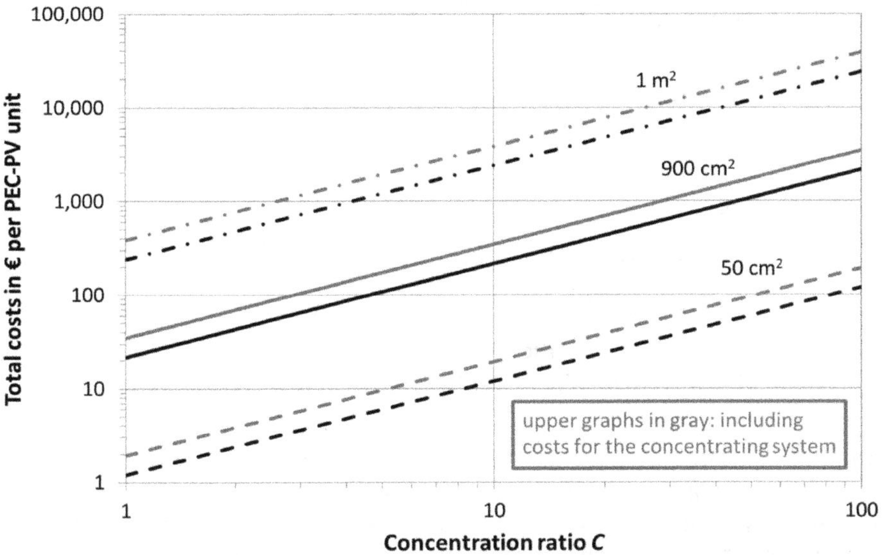

Figure 10.6 Allowable total costs associated with the installation and operation of a solar plant per PEC–PV unit as a function of the concentration ratio and the active area of one PEC–PV unit based on a plant lifetime of 20 years, a specific collector size of 94.6 m^2 per kg H_2 d^{-1}, and target hydrogen production costs of 5 € kg^{-1}.

We also assume that PEC–PV components are capable of receiving and converting high photon fluxes under concentrated sunlight with the same STH as under non-concentrated sunlight. Compared to operation under non-concentrated sunlight the total CAPEX and OPEX of the plant that can be spent per PEC–PV device to reach the target hydrogen costs of 5 € kg^{-1} clearly rises with increasing concentration ratio. The allowable total plant costs reach 36 €, 648 €, and 7200 € per PEC–PV device at $C = 30$ and 120 €, 2160 €, and 24 000 € per PEC–PV device at $C = 100$ for active areas of 50 cm^2, 900 cm^2, and 1 m^2 per PEC–PV unit, respectively, after subtracting the costs for the concentrator from the total budget.

For the two solar concentrator technologies considered here as a model for a solar concentrator suitable for PEC–PV water splitting the dominating materials with respect to mass and energy are steel, glass, and concrete. Based on ref. 5–10 the material demands of a solar concentrator in association with PEC-PV devices can be estimated to 22.9–25.5 kg steel, 10.1–10.2 kg glass, and 1.4–77.9 kg concrete per m^2 aperture area.

10.3 Plant Design and Components

The PEC–PV system has to be implemented in a plant that allows safe operation leading to a specified output of hydrogen under practical conditions. The hydrogen production part and the hydrogen conditioning part of the

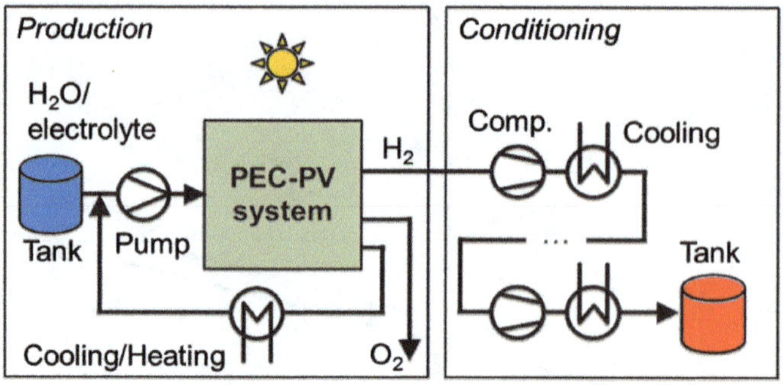

Figure 10.7 Simplified schematic of a solar hydrogen production plant with production and conditioning part (Comp.: compressor).

plant can be distinguished, the latter ensuring that the produced hydrogen has at the gate of the plant the desired characteristics for the considered application scenario. Figure 10.7 shows a simplified schematic of a solar hydrogen production plant, which features a common multi-stage compression with intercooling in the conditioning part. A pump transports the electrolyte solution, with water as the main fraction, from a tank through a suitable piping system to the PEC–PV system. There, a part of the water is converted to hydrogen and oxygen. The unconverted water passes a heat exchanger which allows temperature controlling before being introduced to the PEC–PV system again together with electrolyte solution from the tank, which compensates the water loss due to the water splitting reaction. The generated hydrogen enters the conditioning section, where it undergoes further processing before reaching a storage tank. The product gas can be extracted from the tank as demanded and employed according to the application scenario either on-site or at a distant location after transporting.

A solar hydrogen production plant based on PEC water splitting not only involves the mentioned tanks, pump, heat exchangers and compressors as well as the PEC–PV system itself featuring photoelectrodes, counter electrodes, PV-modules, reactor bodies, covers and if applied the concentrating or photon management system, but also numerous auxiliary parts and components. Piping for the electrolyte solution and for the product gases, appropriate adaptors, components associated with heating and cooling such as air blowers, additional pumps and heat sources, electrolyte solution processing to maintain the required quality and to avoid the formation of biofilms, instrumentation, safety equipment, and control systems as well as civil and structural work have to be factored in. The sizes of the particular components should be carefully assessed by design point calculations and subsequent dynamic simulations, the latter covering realistic conditions in the course of a year at a reasonably fine resolution. Safety issues and cooling/heating requirements are addressed in more detail below. Furthermore, an example of component sizing is conducted for a hypothetical hydrogen production plant in Seville.

10.3.1 Safety

Hydrogen represents the main product of PEC–PV water splitting. Special measures have to be taken in order to avoid explosions of gas mixtures involving hydrogen and oxygen, the latter being abundantly present in the ambience as well as a by-product of the water splitting reaction. Leakages in the hydrogen piping system have to be quickly detected with suitable sensors and eliminated by specialised staff. Fire protection equipment and procedures are required. Materials and components certified for hydrogen application have to be used. Safety regulation codes for hydrogen systems such as the NFPA Hydrogen Technologies Code[11] as pointed out in ref. 12 and others[13–15] have to be considered. Depending on the nature of the electrolyte solution, suitable precautions against hazards such as discharge to the environment or injury of personnel have to be taken.

10.3.2 Cooling and Heating

The components of the PEC–PV plant, in particular components of the PEC–PV devices such as photoelectrodes, membranes, PV-modules, *etc.*, require operation within specified limits. An important limit is given by the maximum operation temperature that can be tolerated with respect to stability and efficiency. It has to be guaranteed that actual temperatures in the PEC–PV field can be kept below the defined maximum operation temperature even under the severest operating conditions imaginable for the chosen location. The PEC–PV devices considered here work at 8% STH and 5% STE. Thus, neglecting optical losses, 87% of the solar input represents a heat load on the system. If passive cooling due to convective and radiative heat transfer from the PEC–PV devices (and piping) to the ambience does not suffice to reliably compensate the heat load and consequently to maintain a desired temperature, active cooling capacity has to be additionally installed.

Similarly, it has to be ensured that a defined minimum operation temperature of the PEC–PV system is respected. As a first approximation, the minimum acceptable operation temperature can be assumed to be 0 °C, the freezing temperature of water. This has to be taken into account, when erecting a plant at a location where ambient temperatures below 0 °C can be expected, especially during night-time without solar input. The heat capacity of the system would of course avoid instant freezing at low temperatures. However, the heating requirements have to be carefully assessed in the context of the defined minimum temperature.

10.3.3 Component Sizing

The operation of a PEC–PV plant is predominately influenced by the solar input and the ambient temperature. First insight into the required sizes of core components of a plant for full load operation as well as mean load operating conditions can be assessed after reasonably defining the severest and mean operating conditions with respect to the aforementioned parameters based on available local data. Required capacities of core plant components exemplarily calculated for the SHA, the HRS, and the IP defined above realised in Seville with and without the application of solar concentration at $C=30$ are presented in Table 10.2. Data for average operating conditions and further information relevant for the calculations can also be found in this table. For cooling of the electrolyte, we consider a solution based entirely on air-cooling due to the guaranteed availability of the cooling medium. It ensures that temperatures in the PEC–PV system do not exceed the 60 °C that we defined as the maximum operating temperature. An air blower in the production part—being an element of the air cooling system—dominates the electricity demand of the plant under severest operating conditions as well as under mean operating conditions in case of $C=30$. Otherwise the compressors in the conditioning part cause the major electricity demand. APM covers the complete electricity demand of the plant except for the HRS operated at $C=30$ and at severest conditions, for which at least about 94% are provided. The compiled data is used for the subsequent life-cycle analysis and cost estimation.

10.4 Life-cycle Analysis

The life-cycle analysis (LCA) quantifies the environmental impact associated with all the stages during the life of the product, *i.e.* from the raw material extraction, going to materials processing, manufacture, distribution, use, repair and maintenance, ending up with disposal or recycling (so-called 'cradle-to-gate' cycle). It is a figure of merit that has received increasing attention in the last years. The LCAs of the three hydrogen production scenarios described above were performed in accordance with ISO 14040[21] to estimate the environmental impact of hydrogen production using PEC–PV technology. We have selected the global warming potential (GWP) as the main performance metric for the LCA. The GWP is a measure of the amount of heat trapped by a certain mass of the gas in question to the amount of heat trapped by a similar mass of carbon dioxide expressed as kg CO_2 equivalent per kg of produced hydrogen. The calculations of the GWP were conducted using GaBi 7.0 software.[22]

Table 10.3 summarises the GWP impacts of the different H_2 generation scenarios for both small PEC cells with 50 cm^2 active area each and large PEC cells with 1 m^2 active area each, with and without implementation of a solar concentrator ($C=30$), with and without consumption of electricity from the local electricity grid. For comparison, the GWP of the 'best in class' H_2 production from wind electrolysis is around 1.0 kg CO_2 eq per kg H_2 and the 'worst in class' H_2 production from coal gasification is around 23.7 kg CO_2 eq per kg H_2).

If grid electricity from country specific mixed sources is used to meet the electricity demand of plant components, in all analysed scenarios the GWP impact of the PEC–PV technology is higher than the 'best in class' wind

Table 10.2 Maximum electricity demand of air blowers and electrolyte solution pumps as well as maximum heat flow associated with heat exchangers (HX) for electrolyte solution (E) and hydrogen cooling corresponding to severest operating conditions (values in brackets refer to mean operating conditions).

Scenario		SHA	HRS		IP	
Solar input		99.7 kW (61.5 kW)	39.9 MW (24.6 MW)		399 MW (246 MW)	
APM (5% STE)		5.0 kW (3.1 kW)	2.0 MW (1.2 MW)		19.9 MW (12.3 MW)	
H_2 production in g s^{-1}		0.0563 (0.0347)	22.5 (13.9)		225 (139)	
C		1	1	30	1	30
Production components (and their efficiency)	Blower (69%)	2.5 kW (79 W)	1.0 MW (31 kW)	1.6 MW (420 kW)	10.1 MW (314 kW)	16.4 MW (4.2 MW)
	Pump (75%)	107 W (5 W)	43 kW (2 kW)	69 kW (41 kW)	426 kW (22 kW)	693 kW (413 kW)
	HX_E (100%)	53 W (3 W)	21 kW (1 kW)	34 kW (20 kW)	211 kW (11 kW)	342 kW (204 kW)
Conditioning components (and their efficiency)	Comp. (75%)	216 W (134 W)	409 kW (252 kW)	409 kW (252 kW)	1.7 MW (1.0 MW)	1.7 MW (1.0 MW)
	HX_{H2} (100%)	357 W (151 W)	465 kW (259 kW)	465 kW (287 kW)	2.2 MW (1.1 MW)	2.2 MW (1.4 MW)
Specification of severest (and average) operating conditions		41.0 °C (24.9 °C) ambient temperature,[16,17] 1055 W m^{-2} (651 W m^{-2}) global irradiation,[3,18,19] 0 m s^{-1} (0 m s^{-1}) wind velocity				
Further specifications for calculation		PEC-PV system: planar, characteristic length (l_c) 1 m, 52–60 °C Electrolyte solution piping: outer diameter (OD) = 0.1 m, $l_c = 0.5\ \pi * OD$, 1 m per m^2 PEC–PV area Convective and radiative heat transfer from PEC–PV system and electrolyte solution piping to ambience according to ref. 20				

Table 10.3 Overview of the GWP impacts in kg CO_2 eq per kg H_2 of the analysed H_2 production scenarios considering PEC–PV modules with an active area of 50 cm^2 and 1 m^2. No APM means that all electricity needed for electrical hardware was taken from the local electricity grid, while with APM means that all needed electricity was produced by the PV part of the PEC–PV system.

Scenario	SHA (50 cm^2)	SHA (1 m^2)	HRS (50 cm^2)	HRS (1 m^2)	IP (50 cm^2)	IP (1 m^2)
$C=1$ (no APM)	3.23	2.24	4.42	3.43	3.33	2.34
$C=30$ (no APM)	—	—	5.67	5.63	4.58	4.55
$C=1$ (with APM)	2.63	1.64	2.47	1.48	2.40	1.41
$C=30$ (with APM)	—	—	0.50	0.47	0.44	0.40

electrolysis technology. However, if no external electricity is needed due to active power management the GWP impact of PEC–PV technology becomes slightly worse than that of the wind electrolysis technology in all three cases, assuming implementation of 1 m^2 PEC–PV cells. Moreover, in the case of implementation of the solar concentration ($C=30$) combined with implementation of APM, a new state of the art technology with the lowest GWP impact reported to date could be obtained.

Hydrogen production in the case of the HRS scenario would lead to highest GWP impact among all analysed scenarios. Detailed analysis showed that 45% (50 cm^2 PEC–PV cell) and 58% (1 m^2 PEC–PV cell) of the overall GWP impact of the PEC–PV system is due to the electrical energy from the local grid (non-renewable generation of electrical energy has high GWP impact) while only 12% and 20% of the overall GWP impact are due to hardware supporting hydrogen production (*i.e.* air blower, compressor, pumps and heat exchanger). The remaining 43% and 22% are caused by components of the PEC–PV cell itself. Implementation of solar concentration changes the contribution of single components and especially of the PEC–PV cell. In this case, the contribution of the electrical energy for the hardware to the overall GWP impact becomes higher (above 90%) due to the increased cooling demand, while the contribution of the PEC–PV cell is less than 1% due to the decreased number of required PEC–PV cells.

In order to be able to design the PEC–PV cell with small GWP impact the contributions of its single components were analysed. In the case of 50 cm^2 PEC–PV cells the highest GWP impact is caused by the cell frame, which is assumed to be made of PVC, followed by the contribution of the solar cells. An analysis of the GWP impact of alternative polymeric materials for the PEC–PV cell showed that PVC has smallest GWP impact of 4.2 kg CO_2 eq m^{-3} polymer followed by polystyrene (4.5 kg CO_2 eq m^{-3}), PMMA (9.8 kg CO_2 eq m^{-3}), polyamide (12.4 kg CO_2 eq m^{-3}), and PTFE (25.8 kg CO_2 eq m^{-3}). In case of 1 m^2 PEC–PV cells the solar cells contribute most, followed by the glass windows. A comparison of the GWP impact of perovskite and Si based solar cells showed a potential of saving of up to 0.4 kg CO_2 eq m^{-2} if Si based solar cells are used. Counter electrodes and ion-conducting membranes do not have a significant contribution. The large impact of the housing material (PVC) can be explained by the small design of the 50 cm^2 PEC–PV cells leading to a high contribution of the cell frames to the total amount of materials used. Therefore, implementation of PEC–PV cells with 1 m^2 active area would be beneficial from the environmental point of view.

10.5 Cost Estimation

A solar PEC–PV plant will only be realised if hydrogen can be produced at acceptable costs. First insight into economic figures were obtained using the H2A Hydrogen Production model (version 3.1) provided by the US Department of Energy (DoE) on its web page.[23] The H2A Hydrogen Production model is based on process design assumptions, which were verified by an international H2A team. Required input parameters to the H2A models include capital and operating costs, efficiencies of the processes used, and plant lifetime as well as financial parameters such as the type of financing, discounted cash flow rate, and desired internal rate of return. An economic analysis with the H2A Hydrogen Production model was performed for the three hydrogen production scenarios described above and the corresponding levelized costs of hydrogen production (LCHP) were estimated. For the calculation of LCHP many default input parameters provided by H2A model were used, while process efficiency was set to 8% and facility operation for 8 $h\,d^{-1}$ was assumed. Lifetime of production facilities was assumed to be 20 years, while the equipment with shorter lifetime will be replaced on demand.

For the worked-out designs of the hydrogen production facilities, the cost of the required hardware (air blowers, pumps, heat exchangers, PV panels, PEC cells, H_2 piping *etc.*) was estimated based on pricing information for large scale quantities provided by vendors of industrial grade equipment, as well as on information available in scientific literature and in engineering text books.[24]

Two sizes of PEC cells with active area of 50 cm^2 and 1 m^2 were considered for the cost analysis. According to fire protection standards of hydrogen technologies[11] stainless-steel hydrogen piping was considered (see also ref. 12). This leads to rather high hydrogen piping costs per single PEC cell—33 € per 50 cm^2 PEC–PV cell and 107 € per 1 m^2 PEC–PV cell. Hydrogen piping costs per small PEC–PV cell is almost three times lower than for a bigger PEC–PV cell, however, 200 of such cells are required for the production of the same amount of H_2 as can be produced by a single large PEC cell. This would lead to extremely high LCHP if small PEC cells were used in H_2 production facility.

The economic analysis of H_2 production using 1 m^2 PEC–PV cells provided LCHP values of 9 € kg^{-1} for the SHA scenario (1 kg H_2 per day), LCHP of 19–23 € kg^{-1} for the HRS scenario (400 kg H_2 per day) and LCHP of 16–20 € kg^{-1} for the IP scenario (4000 kg H_2 per day). In the case of the HRS and IP scenarios, the higher LCHP values correspond to the case when electricity from the grid is used and no solar concentration is applied, while lower LCHP values correspond to the case when no electricity from the grid is used due to implementation of APM and solar concentration with $C=30$ is used. The higher LCHP values for the HRS scenario are caused by high electrical energy consumption due to compression of H_2 to 810 bars, while the comparably low LCHP value for the SHA scenario is mainly caused by rather simple hardware required.

Further analysis was performed to identify main contributors to LCHP. Results reported below refer to the HRS scenario with large PEC cells and supply of electric hardware with electricity from the local grid. From this analysis, we conclude that:

- Lowering of costs for hardware required for active cooling by a factor of two would lower the LCHP by up to 2.8 € kg^{-1}, while complete removal of active cooling (*i.e.* operation of PEC–PV cell at elevated temperatures without the need of active cooling) would lower the LCHP by up to 5 € kg^{-1}
- Lowering the costs of the PV modules by a factor of two would lower the LCHP by up to 4.3 € kg^{-1}
- Implementation of solar concentrators with $C=30$ leads to cost reduction up to 2.1 € kg^{-1}, while lowering of costs of solar concentrators by a factor of two would further lower the LCHP by up to 3.6 € kg^{-1}
- Lowering the costs of H_2 piping by a factor of two would lower the LCHP by up to 2.7 € kg^{-1}
- Implementation of an APM system (*i.e.* supply of all electrical hardware with excess electricity produced by the solar cells of the PEC–PV device) leads to a LCHP reduction up to 1.7 € kg^{-1}

Due to the much smaller complexity of SHA scenario, analysis of the main contributors to the LCHP provided different results:

- Due to very high contribution of costs of PV module to overall PEC–PV cell costs (around 73%), lowering the PV costs by a factor of two would lower the LCHP by up to 2.4 € kg^{-1}
- Lowering of H_2 piping costs by a factor of two would lead to lowering of the LCHP by up to 1.2 € kg^{-1}
- Implementation of an APM system has no significant influence on the LCHP value due to the small electrical energy consumption in the SHA scenario; lowering of costs for hardware required for active cooling will not have a significant influence on the LCHP value due to rather simple hardware requirements

An increase of solar-to-hydrogen conversion efficiency (STH) from 8% to 12% or even 15% would have a significant influence on the LCHP value. Additionally, selling of excess electrical energy produced by the PV module would generate extra revenue and would lead to a further decrease of the LCHP value.

Table 10.4 Global warming potential and hydrogen production costs associated with diverse hydrogen production technologies and PEC–PV water splitting as considered in this study.

H_2 production technology	GWP[32] in kg CO_2 eq per kg H_2	H_2 production costs[25–31] in € kg^{-1}
Steam methane reforming	14.5	0.8–3.0
Coal gasification	23.7	0.9–2.1
Biomass gasification	8.0	1.0–4.3
PV electrolysis	2.5	5.3–22.4
Wind electrolysis	1.0	4.2–6.4
PEC–PV water splitting	0.4–5.7	9–23

10.6 Benchmarking

Hydrogen production *via* PEC water splitting has to be assessed in the context of other hydrogen production technologies. Steam reforming of methane, which uses as coal gasification fossil feedstocks and therefore inherently involves the generation of carbon dioxide, is the dominant hydrogen production technology today. Prominent technologies which use renewable feedstocks are biomass gasification and electrolysis powered by electricity produced by wind turbines or PV modules.[25–31] The global warming potentials as well as hydrogen production costs of these approaches represent essential assessment categories and are summarised in Table 10.4. The table also includes the numbers for the PEC–PV water splitting approach that were calculated in this work.

The GWP of PEC–PV water splitting reaches lower values compared to biomass gasification and significantly lower values compared to the benchmarking processes which rely on fossil fuels. APM and solar concentration allow reaching clearly lower GWP than PV electrolysis and even than wind electrolysis.

The hydrogen production costs for all three scenarios considered in this study are higher than the estimated costs for the benchmarking processes, except for PV electrolysis. The latter shows most similarity to PEC–PV water splitting since it uses the same feedstocks—sunlight and water. For PV electrolysis, a rather broad range of 5.3–22.4 € kg^{-1} of costs has been estimated, which shows a wide overlap with the cost ranges determined here for PEC–PV hydrogen production with 8% STH.

10.7 Conclusions and Outlook

Many aspects have to be considered when PEC–PV technology is transferred from the lab to a real plant. Large areas and complex plant designs are involved. The conducted life-cycle analysis showed that PEC–PV technology for hydrogen generation *via* water splitting has the potential to establish a new 'best in class' process for H_2 generation when taking the global warming potential as the main indicator for environmental impact. However, an economic analysis showed that PEC–PV generation of hydrogen as considered here leads to higher hydrogen production costs than wind electrolysis and biomass gasification, and to significantly higher costs than established H_2 generation technologies based on the utilisation of fossil fuels. At the same time, PEC–PV water splitting could compete with PV electrolysis. The specified conditions of hydrogen at the plant gate have a great influence on plant complexity and consequently on the environmental impact and the levelized hydrogen production cost.

Though solar concentration, active power management, and the utilisation of large cells with 1 m^2 size represent effective measures to reduce costs and should be pursued, achieving competitiveness in the context of alternative hydrogen production technologies is a demanding challenge. Pushing the solar-to-hydrogen efficiency of the PEC–PV system, maintaining high efficiency also at concentration ratios above 30 and in large area cell designs, reducing plant complexity and cooling requirements, increasing lifetime, and developing suitable solar concentrators and PV modules with lower costs represent important steps forward.

Acknowledgements

The authors would like to thank Europe's Fuel Cell and Hydrogen Joint Undertaking (FCH JU) for co-funding the PECDEMO project under Grant Agreement No. 621252. Michael Wullenkord would like to thank Dr Christian Jung and Anis Houaijia for their assistance.

References

1. A. Rothschild and H. Dotan, *ACS Energy Lett.*, 2017, **2**, 45.
2. L. Bertuccioli, A. Chan, D. Hart, F. Lehner, B. Madden and E. Standen, Development of Water Electrolysis in the European Union, final report, E4tech Sàrl with Element Energy Ltd for the Fuel Cells and Hydrogen Joint Undertaking, Lausanne, Cambridge, 2014.
3. I. Heras-Saizarbitoria, E. Cilleruelo and I. Zamanillo, *Renewable Sustainable Energy Rev.*, 2011, **15**(9), 4685.
4. S. Dieckmann, J. Dersch, S. Giuliano, M. Puppe, E. Lüpfert, K. Hennecke, R. Pitz-Paal, M. Taylor and P. Ralon, LCOE Reduction Potential of Parabolic Trough and Solar Tower CSP Technology until 2025 in *Proceedings of the 22nd SolarPACES International Conference*, Abu Dhabi, 2016.
5. E. Pihl, D. Kushnir, B. Sandén and F. Johnsson, *Energy*, 2012, **44**(1), 944.
6. K.-J. Riffelmann, A. Schweitzer and A. Wohlfahrt, *Ultimate Trough – Entwicklung eines Designs und Prototyps für eine neue kostengünstige Generation von Parabolrinnenkollektoren*, final report BMU0325232, DOI: 10.2314/GBV:777402327, 2012.

7. P. Dallyn, A. El-Hamalawi and A. Palmeri, Cost-effective parabolic trough foundations for concentrated solar power plants, in *Proceedings of the 3rd International Conference on Engineering, Project and Production Management*, Brighton, 2012.
8. http://www.flabeg-fe.com/en/engineering/ultimate-trough.html (last accessed Jan 2017).
9. G. J. Kolb, S. A. Jones, M. W. Donnelly, D. Gorman, R. Thomas, R. Davenport and R. Lumia, *Heliostat Cost Reduction Study*, Sandia National Laboratories, 2007.
10. http://www.titantracker.com/v_portal/inc/clicklink.asp?t=3&cod=2684&c=0&s=449996275 (last accessed Jan 2017).
11. The National Fire Protection Association, *Hydrogen Technologies Code*, 2016.
12. A. Landman, H. Dotan, G. E. Shter, M. Wullenkord, A. Houaijia, A. Maljusch, G. S. Grader and A. Rothschild, *Nat. Mater.*, 2017, **16**(6), 646.
13. International Code Council, *2015 International Fuel Gas Code (IFGC)*, 2014.
14. C. Rivkin, R. Burgess and W. Buttner, *Hydrogen Technologies Safety Guide NREL/TP-5400-60948*, National Renewable Energy Laboratory (NREL), 2015.
15. The American Society of Mechanical engineers (ASME), *Hydrogen Piping and Pipelines (B31.12-2014)*, 2014.
16. http://worldweather.wmo.int/en/home.html (last accessed May 2016).
17. http://www.aemet.es/es/-s:print/serviciosclimaticos/datosclimatologicos/efemerides_extremos?w=0&k=and&l=5783&datos=det (last accessed May 2016).
18. http://www.aemet.es/documentos/es/serviciosclimaticos/datosclimatologicos/atlas_radiacion_solar/atlas_de_radiacion_24042012.pdf (last accessed May 2016).
19. http://www.satel-light.com/ (last accessed May 2016).
20. (a) H. Klan, Fe. Wärmeübergang durch Mischkonvektion (überlagerte freie und erzwungene Konvektion) an umströmten Körpern, in *VDI-Wärmeatlas*, ed. Verein Deutscher Ingenieure VDI-Gesellschaft Verfahrenstechnik und Chemieingenieurwesen (GVC), Springer, Berlin, 10th edn, 2006; (b) S. Kabelac and D. Vortmeyer, Ka. Strahlung technischer Oberflächen, in *VDI-Wärmeatlas*, ed. Verein Deutscher Ingenieure VDI-Gesellschaft Verfahrenstechnik und Chemieingenieurwesen (GVC), Springer, Berlin, 10th edn, 2006.
21. ISO 14040:2006 Environmental management – Life cycle assessment – Principles and framework.
22. http://www.gabi-software.com (last accessed March 2017).
23. https://www.hydrogen.energy.gov/h2a_production.html (last accessed May 2017).
24. G. Towler and R. Sinnot, *Chemical Engineering Design. Principles, Practice and Economics of Plant and Process Design*, Butterworth–Heinemann, Oxford, UK, 2nd edn, 2013.

25. M. Ball, W. Weindorf and U. Bünger, in *The Hydrogen Economy: Opportunities and Challenges*, ed. M. Ball and M. Wietschel, Cambridge University Press, Cambridge, New York, 2009, pp. 277–308.
26. A. Körner *et al.*, *Technology Roadmap: Hydrogen and Fuel Cells*, The International Energy Agency (IEA), Paris, 2015.
27. P. Nikolaidis and A. Poullikkas, *Renewable Sustainable Energy Rev.*, 2017, **67**, 597.
28. C. Acar and I. Dincer, *Int. J. Hydrogen Energy*, 2014, **39**(1), 1.
29. I. Dincer and C. Acar, *Int. J. Hydrogen Energy*, 2015, **40**(34), 11094.
30. S. E. Hosseini and M. A. Wahid, *Renewable Sustainable Energy Rev.*, 2016, **57**, 850.
31. M. R. Shaner, H. A. Atwater, N. S. Lewis and E. W. McFarland, *Energy Environ. Sci.*, 2016, **9**(7), 2354.
32. Datasets provided by Thinkstep, Leinfelden-Echterdingen, Germany, 2017.

Subject Index